# TOTALITY

*The Great North American*
*Eclipse of 2024*

# TOTALITY

# Praise for *Totality*

"This is a book rich with wonders, revelations, and delights—visual as well as intellectual. There is just something *so* astounding about a total solar eclipse (I've been privileged to witness two), and Littmann and Espenak have captured that. They help us comprehend the how, the when, and the why of those few moments, as we gaze into the eyeball of majesty. Get ready for 2024."

David Quammen
Author of *Spillover* and *The Song of the Dodo*, among others

"The authors provide an essential run-up to the great total solar eclipse of April 8, 2024 with an interesting history of past solar eclipses together with predictions for where, and how, to best observe the coming event. There are also enough observer anecdotes and testimonies of past eclipses to convince even the most reluctant travelers to make an effort in 2024 to seek out Mother Nature's rare, but unparalleled, celestial performance."

Donald K. Yeomans
NASA/JPL Senior Research Scientist

"A total eclipse of the Sun is a fascinating and moving event of great scientific interest. This book is packed with information and is ideal for anyone hoping to witness the 2024 eclipse, whether an experienced eclipse chaser or a first timer."

Philippa Browning
Professor, Jodrell Bank Centre for Astrophysics, University of Manchester

"This is a richly informative guide for viewing the Total Solar Eclipse of 2024. It probably answers every question you might have had about this upcoming event as well as some that you probably have not thought of. For anyone planning to travel to see this eclipse the information on weather and viewing locations will be invaluable."

Joe Rao
Associate, Guest Lecturer, Hayden Planetarium, New York

"The authors serve up a splendid repast of eclipse adventure, history, science, travel, and story that captures the excitement and anticipation of one of nature's grandest spectacles. The evolution of the eclipse experience is reconstructed through the amulets of ancient China and the cuneiform tablets of Babylon, the geometric contrivances of Greek philosophers, and the discoveries of Renaissance,

Victorian, and 19th-century scientists, all interspersed with an emotional smorgasbord of personal eclipse stories. Written by two experienced 'eclipse chasers,' *Totality* offers suggestions on travel, site selection, photography, and, best of all, how to just watch and absorb the unfolding, too-short, celestial drama. An invaluable composition, to be read before the 2024 eclipse and those that follow."

Jay Anderson
Eclipse climatologist

"If there is a more complete modern book describing solar eclipses, I haven't found it. *Totality: The Great North American Eclipse of 2024* describes how, where, and why humans are privileged to see these events. This book offers much more than information on the 2024 total eclipse. Wrapped in history, it tells stories of observers' experiences during eclipses. I am impressed with how thorough these historical biographies and reports are. Successes, failures, surprises, coincidences . . . It has them all."

Stephen J. Edberg
Astronomer, NASA Jet Propulsion Laboratory (retired)

"Total eclipses of the sun are the most spellbinding sights in the heavens, and Littmann and Espenak's Totality is far and away the most complete and authoritative guide to why, where, and how to see them. Filled with useful observing tips and maps, it's a must-read for experiencing the great US solar eclipse of April, 2024, and an enduring reference for eclipse watching in the future."

Larry Marschall
Professor of Physics, Emeritus, Gettysburg College

Total solar eclipses have terrified and fascinated people for millennia. Littmann and Espenak describe the history and science of solar eclipses, with stories about how eclipse enthusiasts were affected by these spectacular events. If you haven't seen a total solar eclipse, this book will prepare you for the experience of a lifetime.

Ralph Chou
Professor Emeritus, School of Optometry & Vision Science, University of Waterloo

# TOTALITY

*The Great North American
Eclipse of 2024*

Mark Littmann and Fred Espenak

**OXFORD**
UNIVERSITY PRESS

# OXFORD
## UNIVERSITY PRESS

Great Clarendon Street, Oxford, OX2 6DP,
United Kingdom

Oxford University Press is a department of the University of Oxford.
It furthers the University's objective of excellence in research, scholarship,
and education by publishing worldwide. Oxford is a registered trade mark of
Oxford University Press in the UK and in certain other countries

Published in the United States of America by Oxford University Press
198 Madison Avenue, New York, NY 10016, United States of America

British Library Cataloguing in Publication Data
Data available

Library of Congress Control Number: 2023936239

ISBN 978–0–19–887908–4

DOI: 10.1093/oso/9780198879084.001.0001

Printed and bound in the UK by
Clays Ltd, Elcograf S.p.A.

To Peggy, to our children Beth and Owen, and to our grandchildren Liam, Adele, and Leah, with love.

*Mark Littmann*

To my wife Pat, and to our grandchildren Valerie and Maggie who saw their first total eclipse in 2017.

*Fred Espenak*

# Contents

◄○►

Acknowledgements     xi

1   **The Experience of Totality**     1

   *A Moment of Totality:* Reaction to Totality     7

2   **The Great Celestial Cover-Up**     9

   *A Moment of Totality:* A Lasting Impression     31

3   **Ancient Efforts to Understand**     33

   *A Moment of Totality*: The Value of Totality     43

4   **Eclipses in Mythology**     45

   *A Moment of Totality:* My Favorite Eclipse     57

5   **The Strange Behavior of Man and Beast—Long Ago**     59

   *A Moment of Totality:* Fashionably Late to an Eclipse     69

6   **The Sun at Work**     71

   *A Moment of Totality:* The Audio of the Video     83

7   **The First Eclipse Chasers**     85

   *A Moment of Totality:* Stumbling onto a Total Eclipse     107

8   **The Eclipse That Made Einstein Famous**     109

   *A Moment of Totality:* The Difference Between Partial and Total     123

9   **Observing a Total Eclipse**     125

   *A Moment of Totality:* The Accidental Eclipse Tourist     149

10   **Eye Safety During Solar Eclipses**     151

   *A Moment of Totality:* Unexpected Totality     161

11    The Strange Behavior of Man and Beast—Modern Times          163
      *A Moment of Totality:* Spoiled                              173

12    Eclipse Photography                                          175
      *A Moment of Totality:* The Sounds of Totality              199

13    Remembering the All-American Eclipse of 2017                 201
      *A Moment of Totality:* The Eclipse Trip from Hell          221

14    Coming Back to America—The Total Eclipse of 2024             223
      *A Moment of Totality:* Vacation Planning                   245

15    The Weather Outlook                                          247
      *A Moment of Totality:* Paris to Zambia for 38 Hours to See an Eclipse    261

16    When Is the Next One? Total Eclipses: 2025–2033              263
      *A Moment of Totality:* The Extinction of Total Solar Eclipses    295

17    Epilogue: Eclipses—Cosmic Perspective, Human Perspective     297

Appendix A:   Maps for Every Solar Eclipse 2024–2045              301
Appendix B:   Total, Annular, and Hybrid Eclipses: 2024–2070      319
Appendix C:   Total Eclipses in the United States: 1900–2100      325

Glossary                                                          329
Bibliography                                                      335
Index                                                             345

# Acknowledgements

———◄o►———

Enormous thanks to astronomers Charles Lindsey, Jay Pasachoff, and Larry Marschall for their generous advice and help through our earlier eclipse books. Thanks too to astronomers Joseph Hollweg and Alan Clark for sharing their expertise with us.

*Totality: The Great North American Eclipse of 2024* is a richer work because noted authorities contributed vignettes (sidebars) for the book. We are very grateful to them:

Lucian V. Del Priore, M.D.    Laurence A. Marschall
Stephen J. Edberg    Luca Quaglia
Patricia Totten Espenak    Ken Willcox
Carl Littmann

There are 220 illustrations in this book. The magnificent eclipse photographs were graciously contributed by:

Alex Barbovschi    Mark Kidger
Catalin Beldea    Dave Kodama
Donald G. Bruns    Steve Lang
Chris Cook    Thierry Legault
Ben Cooper    Jeanne Loring and David Barker
Margaret Delos-Reyes    Sarah Marwick
Miloslav Druckmüller    Andreas Möller
Alan Dyer    Jay M. Pasachoff
Patricia Totten Espenak    Dimitry Rotstein
Judy Flayderman    Thanskrit Santikunaporn
Stephan Heinsius    Mary Lea Shane
P. Horálek/ESO    Robert Slobins
Philippe Jacquot    Larry Stevens
Tunç Tezel    M. Zamani/ESO
Tommy Tat-Fung Tse    Michael Zeiler
Jaime Villinga    Evan Zucker
Alson Wong

The handsome and informative maps of the 2024 eclipse path are the work of cartographer Michael Zeiler. He created them for this book and we admire his achievement.

The excellent diagrams are primarily the work of solar astronomer Charles A. Lindsey, Will Fontanez and Tom Wallin of Cartographic Services at the University of Tennessee, and Fred Espenak. We are very grateful to them.

Special additional thanks to John R. Beattie, who provided such a ringing moment-by-moment description of a total eclipse that it became the nucleus of our first chapter.

A number of eclipse veterans—both professional and amateur astronomers—graciously offered their experiences and advice for our chapters "Observing a Total Eclipse" and "The Strange Behavior of Man and Beast—Modern Times." We thank them for their insight and eloquence:

| | |
|---|---|
| Jay Anderson | Joseph V. Hollweg |
| Paul & Julie (O'Neil) Andrews | Xavier Jubier |
| Dave Balch | Jean Marc Larivière |
| John Beattie | Dawn Levy |
| Richard Berry | Charles Lindsey |
| Satyendra Bhandari | George Lovi |
| P.M.M. (Ellen) Bruijns | David Makepeace |
| Joe Buchman | Larry Marschall |
| Kristian Buchman | Sarah Marwick |
| Dennis di Cicco | Frank Orrall |
| Stephen J. Edberg | Jay M. Pasachoff |
| Alan Fiala | Patrick Poitevin |
| George Fleenor | Luca Quaglia |
| Ruth S. Freitag | Joe Rao |
| Thomas Hockey | Leif J. Robinson |
| Michael Rogers | Roger W. Tuthill |
| Gary & Barbara (Schleck) Ropski | Ken Willcox |
| Walter Roth | Sheridan Williams |
| Glenn Schneider | Michael Zeiler |
| Mike Simmons | Jack B. Zirker |
| Gary Spears | Evan Zucker |

Other professional and amateur astronomers graciously agreed to be interviewed about their experiences with the 2017 eclipse. We just wish we could have included all of them and their remarkable stories:

| | |
|---|---|
| Alex Barbovschi | Mark Kidger |
| Eric (Rick) and Janice Brown | Michael Landier |
| Donald Bruns | Dawn Levy |
| Joe Buchman | Mike Lohse |
| Ralph Chou | Jeanne Loring and David Barker |
| Tony Crocker and Liz O'Mara | Joe Rao |
| Toby Dittrich | Dimitry Rotstein |
| Charles Fulco | Robert and Elisabeth Slobins |
| Michael Gill | Tunç Tezel |
| Tora Greve | |

There are a host of other folks who helped us greatly in so many ways—astronomy, geology, archeoastronomy, history, mythology, translating, critiquing—in the course of earlier eclipse books and *Totality: The Great North American Eclipse of 2024*. Thanks so much to:

| | |
|---|---|
| Julie Andsager | Kevin Dieke |
| Anthony F. Aveni | Patricia Totten Espenak |
| Eric Becklin | Andrew Fraknoi |
| Kenneth Brecher | Geoffrey K. Gay |
| John Carper | Gerry Grimm |
| B. Ralph Chou | Robert S. Harrington |
| Karen Harvey | Jon Miller |
| Don Hassler | Larry November |
| Anne Hensley | Bea & Tom Owens |
| Kevin Krisciunas | David Pankenier |
| Carl Littmann | Ann and Paul Rappoport |
| David & Esther Littmann | Michael Rogers |
| Jane Littmann | Gary Rottman |
| Owen and Machiko Littmann | Christopher Samoray |
| Peggy Littmann | Glenn Schneider |
| Eli Maor | Sean Simoneau |
| Richard E. McCarron | Sabatino Sofia |
| Bruce McClure | E. Myles Standish, Jr. |
| Beth and James McGinnis | John Steele |
| John McNair | |

We are grateful to Philippa Browning, Professor of Astrophysics, Jodrell Bank Centre for Astrophysics, University of Manchester; and to Patricia Reiff, Professor of Physics and Astronomy, Rice University; who read portions of this manuscript as Oxford University Press referees.

They provided us with very helpful advice and encouragement. Thanks to Catherine Luther, Director of the School of Journalism & Electronic Media, and to Joe Mazer, Dean of the College of Communication & Information, University of Tennessee, for encouraging this book and finding resources to help its development.

The endowment for the Julia G. & Alfred B. Hill Chair of Excellence Professorship in Science Writing at the University of Tennessee was a gift of Tom Hill and Mary Frances Hill Holton, other benefactors, and the State of Tennessee. Their generosity provided funds to help with expenses for this book, for which we are truly grateful.

Our deep gratitude to Dr. Sonke Adlung, Oxford University Press' Senior Editor, Physical Sciences, for his interest in and encouragement for *Totality: The Great North American Eclipse of 2024*. We greatly appreciate also Giulia Lipparini, Senior Project Editor for Science and Medicine at Oxford University Press, for graciously and effectively guiding this project.

Our thanks too to the Oxford University Press production group: Jocelyn Cordova, Director of Publicity and Trade Marketing; Amy Packard Ferro, Senior Publicist; Sarah Butcher, Marketing Manager; Janet Walker, Copy Editor; Holly Day, Indexer; Marlene Taylor, Special Copy Editor; and to Gangaa Radjacoumar, Project Manager, and his team at Integra Software Services in India; and to others whose names we do not know but who brought *Totality* into this world.

To everyone who helped with this book, our profound thanks and our hope that you will enjoy many total eclipses of the Sun.

# 1

—◄○►—

# The Experience of Totality

*In rating natural wonders, on a scale of 1 to 10,*
*a total eclipse of the Sun is a million.*

An observer who has seen 30 total eclipses[1]

First contact. A tiny nick appears on the western side of the Sun.[2] The eye detects no difference in the amount of sunlight. Nothing but that nick portends anything out of the ordinary. But as the nick becomes a gouge in the face of the Sun, a sense of anticipation begins. This will be no ordinary day.

Still, things proceed leisurely for the first half hour or so, until the Sun is more than half covered. Now, gradually at first, then faster and faster, extraordinary things begin to happen. The sky is still bright, but the blue is a little duller. On the ground around you the light is beginning to dim. Over the next 10 to 15 minutes, the landscape takes on a steely gray metallic cast.

As the minutes pass, the pace quickens. With about a quarter hour left until totality, the western sky is now darker than the east, regardless of where the Sun is in the sky. The shadow of the Moon is approaching. Even if you have never seen a total eclipse of the Sun before you know that something amazing is going to happen, is happening now—and that it is beyond normal human experience.

Less than 15 minutes until totality. The Sun, a narrowing crescent, is still fiercely bright, but the blueness of the sky has deepened into blue-gray or violet. The darkness of the sky begins to close in around the Sun. The Sun does not fill the heavens with brightness anymore.

Five minutes to totality. The darkness in the west is very noticeable and gathering strength, a dark amorphous form rising upward and spreading out along the western horizon. It builds like a massive storm but in utter silence, with no rumble of distant thunder. And now the darkness begins to float up above the horizon, revealing a yellow or orange twilight beneath. You are already seeing through the Moon's narrow shadow to the resurgent sunlight beyond.

The acceleration of events intensifies. The crescent Sun is now a blazing white sliver, like a welder's torch. The darkening sky continues to close in around the Sun, faster, engulfing it.

Diamond ring effect at the total solar eclipse of August 21, 2017 from Casper, Wyoming. [Celestron Nexstar Evolution 6, Canon EOS 5D Mark iv, ISO 200, f/10 at 1/1000 second. © 2017 Evan Zucker]

Minutes have become seconds. A ghostly round silhouette looms into view. It is the dark limb of the Moon, framed by a white opalescent glow that creates a halo around the darkened Sun. The corona, the most striking and unexpected of all the features of a total eclipse, is emerging. At one edge of the Moon the brilliant solar crescent remains. Together they appear as a celestial diamond ring.

Suddenly, the ends of the bare sliver of the Sun break into individual points of intense white light—Baily's beads—the last rays of sunlight passing through the deepest lunar valleys. The beads flicker, each lasting but an instant and vanishing as new ones form. And now there is just one left. It glows for a moment, then fades as if it were sucked into an abyss.

Totality.

Where the Sun once stood there is a black disk in the sky, outlined by the soft pearly white glow of the corona, about the brightness of a full moon. Small but vibrant reddish features stand at the eastern rim of the Moon's disk, contrasting vividly with the white of the corona and the black where the Sun is hidden. These are the prominences, giant clouds of hot gas in the Sun's lower atmosphere. They are always a surprise, each unique in shape and size, different yesterday and tomorrow from what they are at this special moment.

You are standing in the shadow of the Moon.

It is dark enough to see Venus and Mercury and whichever of the brightest planets and stars happen to be close to the Sun's position and above the horizon. But it is not the dark of night. Looking across the landscape at the horizon in all directions, you see beyond the shadow to where the eclipse is not total an eerie twilight of orange and yellow. From this light beyond the darkness which envelops you comes an inexorable sense that time is limited.

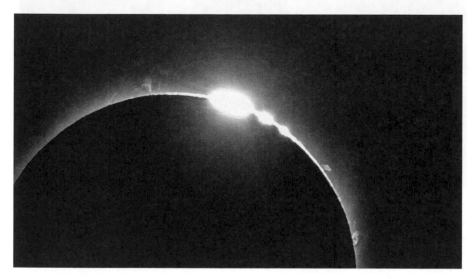

Baily's beads are seen amid a red prominences during the total solar eclipse of August 21, 2017 from Glendo, Wyoming. [Nikon D600, Borg 100ED refractor, f/6.5, 1/1000 second, ISO 100. © 2017 Dave Kodama]

Partial phases of the total solar eclipse of March 29, 2006 from Jalu, Libya. [Nikon D200 DSLR, Sigma 170–500 mm at 500 mm, f/11, 1/500 second, ISO 200, Thousand Oaks Type 3 solar filter. © 2006 Patricia Totten Espenak]

A spectacular HDR (High Dynamic Range) composite of the August 21, 2017 corona was captured by combining eight separate exposures of totality and enhancing the detail and contrast—Salem, Oregon [Nikon D810, Nikkor 800 mm f/5.6, 1/120 s to ½ second, ISO100, image by Jay Pasachoff, Vojtech Rušin, and the Williams College Solar Eclipse Expedition; computer image processing by Roman Hubčík]

Now, at the midpoint in totality, the corona stands out most clearly, its shape and extent never quite the same from one eclipse to another. And only the eye can do the corona justice, its special pattern of faint wisps and spikes on this day never seen before and never to be seen again.

Yet around you at the horizon is a warning that totality is drawing to an end. The west is brightening while in the east the darkness is deepening and descending toward the horizon. Above you, prominences appear at the western edge of the Moon. The edge brightens.

Suddenly totality is over. A point of sunlight appears. Quickly it is joined by several more jewels, which merge into a sliver of the crescent Sun once more. The dark shadow of the Moon silently slips past you and rushes off toward the east.

It is then you ask, "When is the next one?"[3]

## NOTES AND REFERENCES

1. Epigraph: Fred Espenak, October 24, 2014; updated June 26, 2022.
2. In sky observations, the western side of the Sun or Moon refers to the edge of the Sun or Moon closer to the western horizon. For observers in mid-northern latitudes, the Sun is usually to the south. When facing south, east is to the left and west is to the right. This south-looking orientation can briefly confuse readers who are used to maps that are oriented north, so that east is to the right and west to the left.
3. Special thanks to John Beattie of New York City, upon whose experience and description this chapter is based.

# A MOMENT OF TOTALITY

## Reaction to Totality

Eclipse veteran Sheridan Williams says that at the end of totality, the usual reaction of a first-time eclipse observer is "It was so short," "That was the most amazing thing I've ever seen," "I never realized it could be so beautiful," and "When is the next one?" Often they have tears in their eyes. There are not enough superlatives.

"After seeing a total eclipse," Sheridan says, "I have never, never heard anyone say, 'I don't see what all the fuss was about' or 'Why bother to see another one?'"*

---

* Sheridan Williams is a British rocket scientist (retired) and a Fellow of the Royal Astronomical Society.

# 2

## The Great Celestial Cover-Up

*If God had consulted me before embarking upon creation, I would have recommended something simpler.*

Alfonso X, King of Castile (1252)[1]

A total eclipse of the Sun is exciting and even profoundly moving.

But what causes a total solar eclipse? The Moon blocks the Sun from view. And that is all you absolutely need to know to enjoy a solar eclipse. So you can now skip to the next chapter.

If, however, you are reading this paragraph, you are right: there is more to tell—about dark shadows and oblong orbits and tilts and danger zones and amazing coincidences. Yet before you venture further, promise yourself one thing. If for any reason your eyes begin to glaze over, stop reading this chapter immediately and go right on to the next. You must not let celestial mechanics, or this explanation of it, stand in the way of your enjoyment of the wild, wacky, and wonderful things people have thought and done about solar eclipses.

### Moon Plucking

How big is the Moon in the sky? What is its angular size?

Extend your arm upward and as far from your body as possible. Using your index finger and thumb, imagine that you are trying to pluck the Moon out of the sky ever so carefully, squeezing down until you are just barely touching the top and bottom of the Moon, trapping it between your fingers. How big is it? The size of a grape? A plum? An orange?

It is the size of a pea. (You can win bets at cocktail parties with this question.) The Moon has an angular size of only half a degree.

Now, how large is the Sun in the sky? Your friends will almost all immediately guess that it is bigger. Before they damage their eyes by trying the Moon pinch on the Sun, just remind them that a total eclipse is caused by the Moon completely covering the Sun, so the Sun must appear no bigger

A total solar eclipse occurs when the Moon's umbra touches the Earth. A lunar eclipse occurs when the Moon passes into the Earth's shadow (umbra). The relative sizes and distances of the Sun, Moon, and Earth in this diagram are not to scale.

The solar corona during the total solar eclipse of March 29, 2006 from Sallum, Egypt. [Nikon D70, Borg 77 mm refractor, ISO 200, f/6.5, 31 exposures: 1/2000 to 1 second, images combined into HDR image in Photomatix and Photoshop. © 2006 Alson Wong]

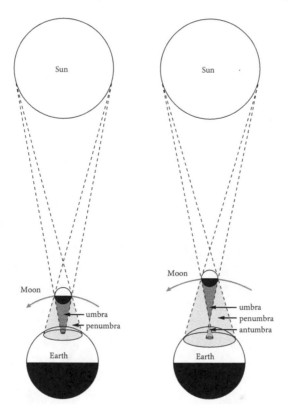

Configuration of a total solar eclipse (*left*) and an annular eclipse (*right*). Within the Moon's umbra (dark converging cone), the entire surface of the Sun is blocked from view. In the penumbra (lighter diverging cone), a fraction of the sunlight is blocked, resulting in a partial eclipse. When the Moon's umbra ends in space (*right*), a total eclipse does not occur. Projecting the cone through the tip of the umbra onto the Earth's surface defines the region in which an annular eclipse is seen.

than a pea in the sky as well. It is the brightness of the Moon and especially the Sun that deceives people into overestimating their angular size.

Now that you have collected on your bets and can lead a life of leisure, think about the remarkable coincidence that allows us to have total eclipses of the Sun. The Sun is 400 times the diameter of the Moon, yet it is about 400 times farther from the Earth, so the two appear almost exactly the same size in the sky. It is this geometry that provides us with the unique total eclipses seen on Earth when our Moon just barely covers the face of the Sun. If the Moon, 2,160 miles (3,476 km) in diameter, were 169 miles (273 km) smaller than it is, or if it were farther away so that it appeared smaller, people on Earth would never see a total eclipse.[2]

It is amazing that there are total eclipses of the Sun at all. As it is, total eclipses only just barely happen. The Sun is not always exactly the same angular size in the sky. The reason is that the Earth's orbit is not circular but elliptical, so the Earth's distance from the Sun varies. When the Earth is closest to the Sun (early January),[3] the Sun's disk is slightly larger in angular diameter, and it is harder for the Moon to cover the Sun to create a total eclipse.

An even more powerful factor is the Moon's elliptical orbit around the Earth. When the Moon is its average distance from the Earth or farther, its disk is too small to occult the Sun completely. In the midst of such an eclipse, a circle of brilliant sunlight surrounds the Moon, giving the event a ring-like appearance; hence the name *annular* eclipse (from the Latin *annulus*, meaning ring).

Because the angular diameter of the Moon is smaller than the angular diameter of the Sun on the average, annular eclipses are more frequent than total eclipses.

But the Moon does not just dangle motionless in front of the Sun. It is in orbit around the Earth. It catches up with and passes the Sun's position

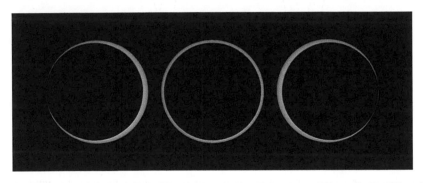

The beginning, middle, and end of annularity during the annular solar eclipse of October 3, 2005 from Spain. [Nikon D200 DSLR, Celestron 90 Maksutov, fl = 1000 mm, f/11, 1/125 second, ISO 200, Thousand Oaks Type II Filter, images combined in Photoshop. © 2005 Patricia Totten Espenak]

**Apparent Size of Moon and Sun in Sky**

|  | Diameter | Ratio—Moon to Sun |
|---|---|---|
| Moon | 2,160 miles (3,476 kms) | |
| Sun | 864,989 miles (1,392,000 km) | 1/400 |
|  | **Mean Distance from Earth** | |
| Moon | 238,870 miles (384,400 km) | |
| Sun | 92,960,200 miles (149,598,000 km) | 1/389 |

*Significance*

*The Sun has a diameter 400 times bigger than the Moon, but the Sun is about 400 times farther away than the Moon, so the Moon and Sun appear to be nearly the same size as seen from Earth.*

**The Shadow of the Moon**

|  | Maximum | Minimum | Mean |
|---|---|---|---|
| *Moon's distance from Earth* | | | |
| *(center to center)* | | | |
| Miles | 252,720 | 221,470 | 238,870 |
| Kilometers | 406,700 | 356,400 | 384,400 |
| *Length of Moon's shadow cone* | | | |
| *(umbra)* | | | |
| Miles | 236,050 | 228,200 | 232,120 |
| Kilometers | 379,870 | 367,230 | 373,540 |

*Significance*

Most of the time the Moon's shadow is too short to reach the Earth. Therefore, **total solar eclipses occur** *less* **often than annular solar eclipses.**

**Angular Size of the Sun and Moon** (as seen from Earth)

|  | Maximum | Minimum | Mean |
|---|---|---|---|
| Angular diameter of the Sun | 32'31.9" | 31'27.7" | 31'59.3" |
| Angular diameter of the Moon | 33'31.8" | 29'23.0" | 31'05.3" |

*Significance*

The Moon's angular diameter can *exceed* the Sun's angular diameter by as much as 6.6% (2.1 arc minutes), producing a *total* eclipse of the Sun.

The Sun's angular diameter can exceed the Moon's angular diameter by as much as 10.7% (3.1 arc minutes), producing an *annular* eclipse of the Sun.

Most of the time, the Moon's angular diameter is *smaller* than the Sun's angular diameter.

Therefore, **total solar eclipses occur** *less* **often than annular solar eclipses.**

in the sky about once a month. The word *month* comes from this circuit of the Moon.

The actual time for the Moon to complete this cycle is 29.53 days, and it is called a synodic month, after the Greek *synodos*, "meeting"—the meeting of the Sun and the Moon.

The Moon gives off no light of its own. It shines only by reflected sunlight. So half the Moon is always lit by the Sun. But as the Moon orbits the Earth, sometimes we see the Earth-facing side fully illuminated and sometimes we see only a thin crescent. As the days pass, the Moon changes phase—crescent, gibbous, full . . .

In 29.53 days, the Moon goes from new moon through full moon and back to new moon again. Solar eclipses can take place only at new moon (dark-of-the-moon) and lunar eclipses may occur only at full moon.

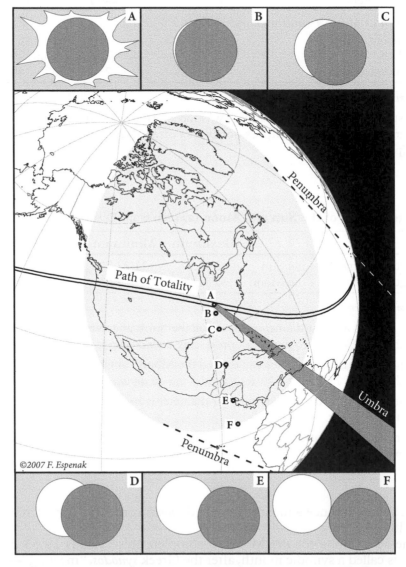

During a total eclipse of the Sun, the tip of the Moon's shadow touches the Earth and the Moon's orbital velocity carries the shadow rapidly eastward. Only along a narrow path is the eclipse total. Regions to the side of the path of totality experience varying degrees of partial eclipse. (This diagram illustrates the total eclipse of August 21, 2017.)

So why don't we have an eclipse of the Sun every 29.53 days—every time the Moon passes the Sun's position? The reason is that the Moon's orbit around the Earth is tilted to the Earth's orbit around the Sun by about 5°, so that the Moon usually passes above or below the Sun's position in the sky and cannot block the Sun from our view.

## "Danger Zones"

The Moon's tilted orbit crosses the plane of the Earth's orbit at two places. Those intersections are called *nodes*. Node is from the Latin word meaning knot, in the sense of weaving, where two threads are tied together. The point at which the Moon crosses the plane of the Earth's orbit going northward is the ascending node. Going south, the Moon crosses the plane of the Earth's orbit at the descending node.

A solar eclipse can occur only when the Sun is near one of the nodes as the Moon passes.

If the Sun stood motionless in a part of the sky away from the nodes, there would be no eclipses, and you would not be agonizing over this. But the Earth is moving around the Sun and, as it does so, the Sun appears to shift slowly eastward around the sky, through all the constellations of the zodiac, completing that journey in one year. In that yearly circuit, the Sun must cross the two nodes of the Moon. Think of it as a street intersection at which the Sun does not pause and runs the stop sign every time. It is an accident waiting to happen. When the Sun nears a node, there is the "danger" that the Moon will be coming and—crash!

No. The Moon is 400 times closer to the Earth than the Sun, so the worst—the best—that can happen is that the Moon will pass harmlessly but stunningly right in front of the Sun. The Sun's apparent pathway in the sky is called the *ecliptic* because it is only when the Moon is crossing

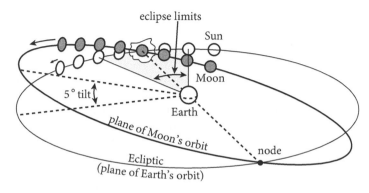

The paths of the Sun and Moon illustrate why eclipses occur only when the Sun is near the intersection (node) where the Moon crosses the ecliptic. The plane of the Moon's orbit is tilted approximately 5° to the ecliptic plane.

**Eclipse Limits ("Danger Zones")**

|  | Maximum* | Minimum* | Mean |
|---|---|---|---|
| A solar eclipse of some kind will occur at new moon if the Sun's angular distance from a node of the Moon is less than: | 18°31' | 15°21' | 16°56' |
| A central (total or annular) solar eclipse will occur at new moon if the Sun's angular distance from a node of the Moon is less than: | 11°50' | 9°55' | 10°52' |

Sun's apparent eastward movement in the star field each day (due to the Earth orbiting the Sun): 1°

Moon's synodic period (from new moon to new moon: the time the Moon takes to complete its eastward circuit of the star field and catch up with the Sun again): 29.53 days

*Significance*

The Sun cannot pass a node of the Moon without at least one solar eclipse occurring, and two are possible. If one occurs, it can be either a partial or a central eclipse (total or annular). If two occur, both will be partials, about one month apart.

The Sun can pass a node without a central solar eclipse (total or annular) occurring.

*\* Limits vary due to changes in the apparent angular size and speed of the Sun and Moon caused by the elliptical orbits of the Earth and Moon.*

the ecliptic that eclipses can happen. Thus twice a year, roughly, there is a "danger period," called an eclipse season, when the Sun is crossing the region of the nodes and an eclipse is possible.

The Sun comes tootling up to the node traveling about 1° a day.[4] The hot-rod Moon, however, is racing around the sky at about 13° a day. Now if the Sun and Moon were just dots in the heavens, they would have to meet precisely at a node for an eclipse to occur. But the disks of the Sun and Moon each take up about half a degree in the sky. And the Earth, almost 8,000 miles (12,800 km) across, provides an extended viewing platform. Therefore, the Sun needs only to be *near* a node for the Moon to sideswipe it, briefly "denting" the top or bottom of the Sun's face. That will happen whenever the Sun is within 15⅓° of a node.

An "eclipse alert" begins when the Sun enters the danger zone 15⅓° west of one of the Moon's nodes and does not end until the Sun escapes 15⅓° east of that node. The Sun must traverse 30⅔°. Traveling at 1° a day, the Sun will be in the danger zone for about 31 days. But the Moon completes its circuit, going through all its phases, and catches up to the

## Calendar of Solar Eclipses: 2022 – 2041

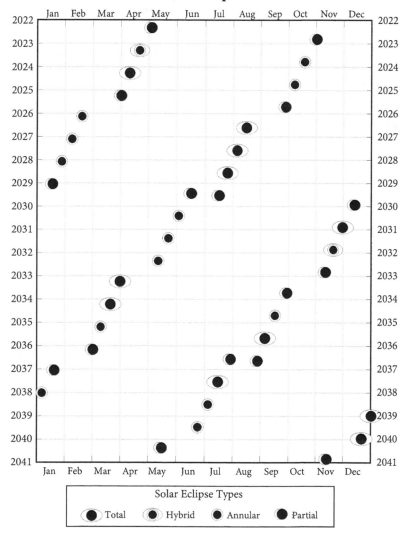

Solar eclipses 2022–2041 plotted to show eclipse seasons. Each calendar year, eclipses occur about 20 days earlier. Consequently, eclipse seasons shift to early months in the year. [Graph by Fred Espenak]

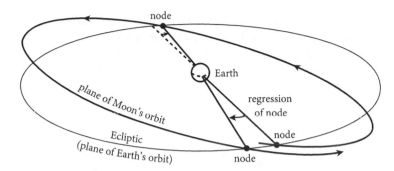

Each time the Moon completes an orbit around the Earth, it crosses the Earth's orbit at a point west of the previous node. Each year the nodes regress 19.4°, making a complete revolution in 18.61 years.

## Solar Eclipses Outnumber Lunar Eclipses

There are more solar than lunar eclipses. That surprises most people because they have seen an eclipse of the Moon but not an eclipse of the Sun. The reason is simple. When the Moon passes into the shadow of the Earth to create a lunar eclipse, the event is seen wherever the Moon is in view, which includes half the planet. Actually, a lunar eclipse is seen from more than half the planet because during the course of a lunar eclipse (up to 4 hours), the Earth rotates so that the Moon comes into view for additional areas.

In contrast, whenever the Moon passes in front of the Sun, the shadow it creates—a solar eclipse—touches only a small portion of the Earth. On average, your house will be visited by a *total* eclipse of the Sun only once in about 375 years.[a] To be touched by the dark shadow (umbra) of the Moon is quite rare.

But from either side of the path of a total eclipse, stretching northward and southward 2,000 miles (3,200 km) and sometimes more, an observer sees the Sun partially eclipsed. Even so, the zone of partial eclipse covers a much smaller fraction of the Earth's surface than a lunar eclipse. So more people have seen lunar eclipses than partial solar eclipses. Only a tiny fraction of people, about one in 50,000, have witnessed a total solar eclipse.

In his *Canon of Eclipses* (1887), Theodor von Oppolzer and his assistants computed with pen on paper all eclipses of the Sun and Moon from 1208 BCE to 2161 CE. He cataloged 8,000 solar eclipses and 5,200 lunar eclipses—about three solar eclipses for every two lunar eclipses.

This ratio can be misleading, however. Oppolzer counted all solar eclipses, whether they were total (the Moon's *umbra* touches the Earth) or partial (only the Moon's *penumbra* touches the Earth). But for lunar eclipses, Oppolzer counted only those in which the Moon was totally or partially immersed in the Earth's *umbra*. He did not count *penumbral* lunar eclipses because they are virtually unnoticeable. (If he had included penumbral lunar eclipses in his census, thereby counting *all* forms of solar and lunar eclipses, the ratio would have been close to even, but lunar eclipses would now outnumber solar eclipse by a small margin.)

Sun every 29.53 days. It is not possible for the Sun to crawl through the danger zone before the Moon arrives. A solar eclipse *must* occur each time the Sun approaches a node and enters one of these danger zones, about every half year.

In fact, if the Moon nips the Sun at the beginning of a danger zone (properly called an *eclipse limit*), the Sun may still have 30 days of travel left within the zone. But the Moon takes only 29.53 days to orbit the Earth and catch up with the Sun again. So it is possible for the Sun to be nipped by the Moon twice during a single node crossing; thereby creating two partial eclipses within a month of one another.

The closer the Sun is to the node when the Moon crosses, the more nearly the Moon will pass over the center of the Sun's face. In fact, if the Sun is within about 10° of the node at the time of the Moon's crossing, a

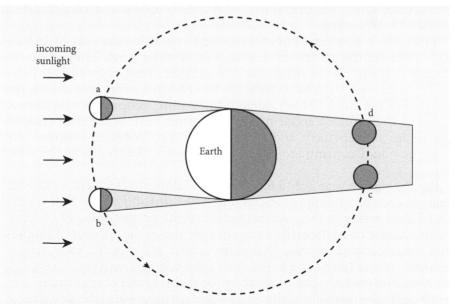

The reason why solar eclipses slightly outnumber lunar eclipses is most easily visualized if you imagine looking down on the Sun-Earth-Moon system and if you start by considering only total solar and total lunar eclipses.

The Moon will be totally eclipsed whenever it passes into the shadow of the Earth—between **c** and **d** on the diagram. At the Moon's average distance from Earth, that shadow is about 2.7 times the Moon's diameter. But there will be an eclipse of the Sun whenever the Moon passes between the Earth and Sun—between points **a** and **b**. The distance between **a** and **b** is greater than between **c** and **d**, so central solar eclipses (total, annular, and hybrid) must occur more often than total lunar eclipses.

[a] Jean Meeus: "The Frequency of Total and Annular Solar Eclipses at a Given Place," *Journal of the British Astronomical Association*, volume 92, April 1982, pages 124–126.

central eclipse will occur somewhere on Earth. Depending on the Moon's distance from the Earth and the Earth's distance from the Sun, this central eclipse will be total or annular.

## Nodes on the March

There should be a solar eclipse or two every six months, whenever the Sun crosses one of the Moon's nodes. Actually, the Sun crosses the ascending node of the Moon, then the descending node, and returns to the ascending node in only 346.62 days—the *eclipse year*. Within this eclipse year there are two *eclipse seasons*, intervals of 30 to 37 days as the Sun approaches, crosses, and departs from a node. All eclipses will fall within eclipse seasons. There can be no eclipses outside this period of time. Because the Sun crosses a node about every 173 days, the eclipse seasons are centered about 173 days apart.

The eclipse year does not correspond to the calendar year of 365.24 days because the nodes have a motion all their own. They are constantly shifting westward along the ecliptic. This regression of the nodes is caused by tidal effects on the Moon's orbit created by the Earth and the Sun. If the nodes did not shift, eclipses would always occur in the same calendar month year after year. If the Sun crossed the nodes in February and August one year to cause eclipses, the eclipses would continue to fall in February and August in succeeding years.

But the eclipse year is 346.62 days, 18.62 days shorter than a calendar year, so each ascending or descending node crossing by the Sun occurs 18.62 days earlier in the calendar year than the previous one of its kind.

This migration of the eclipse seasons determines the number of eclipses that may occur each year. One solar eclipse must occur each eclipse season, so there have to be at least two solar eclipses each year (although both may be partial). But because of the width of the eclipse limits—up to 19 days on either side of the Sun's node crossing—and the slowness of the Sun's apparent motion, there can be two solar eclipses at each node passage (both partials). Thus occasionally there will be four solar eclipses in one calendar year.

There can actually be five. Because the eclipse year lasts 346.62 days, almost 19 days less than a calendar year, if a solar eclipse occurs before or on January 18 (or January 19 in a leap year), that eclipse year could conceivably bring two solar eclipses in January and two more around July. That eclipse year would then end in mid December, and a new eclipse year would begin in time to provide one final solar eclipse before the end of December. At most, therefore, there can be five solar eclipses in a calendar year.

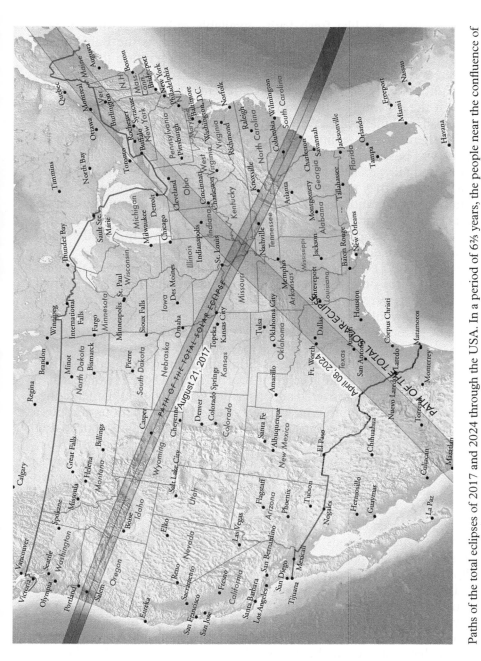

Paths of the total eclipses of 2017 and 2024 through the USA. In a period of 6⅔ years, the people near the confluence of the Ohio and Mississippi Rivers will see two total solar eclipses. [Map © 2016 Michael Zeiler]

## Heavenly Rhythm

Eclipses, then, are like fresh fruit—available only in season. Ancient peoples who kept written records of eclipses, such as the Babylonians did from 750 BCE on, noticed after decades of observation that eclipses happen only at certain intervals. These eclipse seasons are separated from one another by six or occasionally five new moons.[5]

From the expanse of their empire, the Babylonians[6] could see only about half the lunar eclipses and only a small fraction of the solar eclipses, so, for them at first, eclipse seasons were times of "danger" when an eclipse was possible.

But gradually, as they studied their eclipse records, the Babylonians realized that if they took the date of one eclipse and counted forward a certain interval in months, they would find the record of another eclipse. Eclipses have a long-term rhythm of their own. So they took the records and counted forward into the future a certain interval in months—and *predicted* that an eclipse would occur. And it did.

By about 600 BCE, the Babylonians could accurately predict the next eclipse—and the one after that and the one after that.[7] They realized that eclipse seasons were periods during which one to three eclipses (1 or 2 solar; 0 or 1 lunar) would necessarily occur. One of the two great celestial lights might be partially or totally darkened. They could tell which one would be eclipsed and, by 300 BCE, even to what extent. The Babylonians also realized that if they couldn't see an eclipse, it didn't mean that their prediction was wrong. It just meant that the eclipse occurred somewhere else on Earth.

The Babylonians discovered and inscribed on clay tablets in their cuneiform writing several different eclipse cycles. The most famous and most useful of these eclipse rhythms is the *saros*.[8] The Babylonians noticed that 6,585 days (18 years 11 days) after virtually every lunar eclipse, there was another very similar one. If the first was total, the next was almost always total. And these eclipses, separated by 18 years, occurred in the same part of the sky, as if they were related to one another.

In a sense, they were. Imagine that a total solar eclipse occurs one day at the Moon's descending node. After 6,585 days, the Moon has completed 223 lunations (synodic periods) of 29.53 days each and returned to new moon at that same node. In that same period of 6,585 days, the Sun has endured 19 eclipse years of 346.62 days each and has returned to the descending node, forcing another solar eclipse to occur. And because 6,585 days is very close to an even 18 calendar years, this solar eclipse occurs at the same season of the year as its predecessor, and with the Sun very close to the same position in the zodiac that it occupied at the eclipse 18 years earlier. Even though 18 years have intervened, these two eclipses certainly seem to be relatives.

All the more so because another lunar cycle crucial to eclipses has a multiple that also adds up to 6,585 days. That cycle is the *anomalistic month*—the time it takes the Moon in its elliptical orbit around the Earth to go from *perigee* (closest to Earth) to *apogee* (farthest) and back to *perigee*.[9] When the Moon is near perigee, its angular size in the sky is just slightly larger, creating solar eclipses that are total rather than annular. The anomalistic month is 27.55 days long and 239 of these cycles add up to 6,585 days. If the previous eclipse occurred with the Moon at perigee, the new eclipse will occur with the Moon near perigee, providing another total eclipse.

So, after a time interval of 6,585 days, the geometrical configuration of the Sun, Moon, and Earth is almost exactly repeated. Another solar eclipse occurs at the same node, at the same season, in the same part of the sky. If it was a total eclipse before, it will almost certainly be a total eclipse again.

---

**Frequency of Solar Eclipses**

**Solar eclipses by types**—for 5,000 years: 2000 BCE to 3000 CE

| | |
|---|---|
| Total: | 26.7% |
| Annular: | 33.2% |
| Hybrid (Annular/Total): | 4.8% |
| Partial: | 35.3% |

Solar eclipses outnumber lunar eclipses almost 3 to 2 (excluding penumbral lunar eclipses, which are seldom detectable visually)

Annular eclipses outnumber total eclipses about 5 to 4

Solar eclipses per century (average over 4,530 years): 238.9

Maximum number of solar eclipses per year: 5 (4 will be partial)

Minimum number of solar eclipses per year: 2 (both can be partial)

Maximum number of *total* solar eclipses per year: 2

Minimum number of *total* solar eclipses per year: 0

Maximum number of solar and lunar eclipses per year: 7 (4 solar and 3 lunar *or* 5 solar and 2 lunar)*

Minimum number of solar and lunar eclipses per year: 2 (2 solar and 0 lunar)*

Examples of years in which only 2 solar eclipses occur: 2017, 2020, 2021, 2022, 2023, 2024, 2025, 2026, 2027, 2028, 2030

Examples of years in which 4 solar eclipses occur: 2000, 2011, 2029, 2047

Examples of years in which 5 solar eclipses occur: 1805, 1935, 2206, 2709

Maximum diameter of the Moon's shadow cone (umbra) as it intercepts the Earth to cause a total eclipse: 170 miles (273 km)

Maximum diameter of the Moon's "anti-umbra" as it intercepts the Earth to cause an annular eclipse: 232 miles (374 km)

---

**Source:** Fred Espenak and Jean Meeus: *Five Millennium Canon of Solar Eclipses: –1999 to +3000* (2000 BCE to 3000 CE) (2nd Edition, Astropixels Publishing, Portal, AZ, 2021).

* Counts total and partial lunar eclipses but not penumbral lunar eclipses.

### Total Eclipses—Duration of Totality

| | |
|---|---|
| Longest duration (theoretical): | 7 minutes 32 seconds |
| Longest duration in 10,000 years:[a] | 7m 29s—July 16, 2186 CE |
| Longest duration by millennium: | |
| 3000 BC to 2001 BC: | 7m 21s—May 16, 2231 BCE |
| 2000 BC to 1001 BC: | 7m 05s—July 3, 1443 BCE |
| 1000 BC to 1 BC: | 7m 28s—June 15, 744 BCE |
| 1 AD to 1000 AD: | 7m 24s—June 27, 363 CE |
| 1001 AD to 2000 AD: | 7m 20s—June 9, 1062 CE |
| 2001 AD to 3000 AD: | 7m 29s—July 16, 2186 CE |
| 3001 AD to 4000 AD: | 7m 18s—July 24, 3991 CE |
| 4001 AD to 5000 AD: | 7m 12s—August 4, 4009 CE |
| Longest duration of the 20th century:[b] | 6m 29s—May 18, 1901 |
| (6 minutes or longer) | 6m 20s—September 9 1904 |
| | 6m 51s—May 29 1919 |
| | 6m 04s—June 8 1937 |
| | 7m 08s—June 20 1955 |
| | 7m 04s—June 30 1973 |
| | 6m 53s—July 11 1991 |
| Longest duration of the 21st century:[c] | 6m 39s—July 22, 2009 |
| (6 minutes or longer) | 6m 23s—August 2, 2027 |
| | 6m 06s—August 12, 2045 |
| | 6m 06s—May 22, 2096 |
| Number of eclipses with 7 minutes or more of totality in 21st century: | 0 |
| Number of eclipses with 7 minutes or more of totality from July 1, 1098 CE to June 8, 1937 CE (839 years): | 0 |

**Note:** All solar eclipses with long durations of totality have dates centered around July 4, the mean date for the Earth at aphelion (farthest from the Sun), when the Sun appears slightly smaller in size and is more easily covered by the Moon.

[a] Based on the 10,000-year period from 3000 BC to 7000 AD

[b] All the long total eclipses of the 20th and 21st centuries are members of saros series 136 except for September 9, 1904 and May 22, 2096.

[c] All the long total eclipses of the 20th and 21st centuries are members of saros series 136 except for September 9, 1904 and May 22, 2096.

## Do Your Own Eclipse Prediction

Take the date of any solar or lunar eclipse and add 6,585.32 days to it and you will accurately predict a subsequent eclipse of the same kind that will closely resemble the one 18 years earlier. Take the date of *every* solar and lunar eclipse that occurs and keep on adding 6,585.32 days to it and you will have, with rare exceptions, a quite reliable list of future eclipses.

## Flaws in the Rhythm

But these eclipses 6,585 days apart are not identical twins. The match between unrelated periods—the Moon's cycle of phases, the Sun's eclipse year cycle, and the Moon's cycle from perigee to perigee—is not perfect.

| | |
|---|---|
| 223 synodic periods of the Moon (lunations) at 29.5306 days each | = 6,585.32 days |
| 19 eclipse years of the Sun at 346.6201 days each | = 6,585.78 days |
| 239 anomalistic months of the Moon (revolution from perigee to perigee) at 27.55455 days each | = 6,585.54 days |

These synchronizing cycles are out of step with one another by fractions of a day. And those fractions of a day have their consequences.

Consider first that 223 synodic periods of the Moon amount to 6,585.32 days or, in calendar years, 18 years 11⅓ days (18 years 10⅓ days if five leap years intervene). Because the saros period is 18 years 11⅓ days, each subsequent eclipse occurs about one-third of the way around the world westward from the one before it. For a lunar eclipse, visible to half the planet, this westward shift usually does not push the eclipse out of view. For a solar eclipse, however, visible over a narrow swath of the Earth's surface, the subsequent eclipse in that saros series would almost never be visible from the initial site. The eclipse would be happening in an entirely different part of the world.

After three saros cycles—54 years 34 days—the eclipse would be back to its original longitude, but it would have shifted, on the average, about 600 miles (1,000 km) northward or southward, taking totality and even major partiality out of view for the original observer.[10]

A saros period truly does predict with accuracy that a solar eclipse will happen, but it would have been extremely hard for an ancient astronomer to confirm that the predicted eclipse had taken place and thus difficult for the astronomer or those he served to retain confidence in his solar eclipse predictions.

## The Evolving Saros

Now consider that 223 lunations (synodic periods) amount to 6,585.32 days, while 19 eclipse years of the Sun consume 6,585.78 days. The difference after 18 years 11⅓ days is 0.46 day. The result is that the Sun is not exactly where it was before in relation to the Moon's node. It is 0.477° farther west. If the Moon just barely grazed the western part of the Sun before, it will clip a little more the next time. At each return of the saros, the eclipses become larger and larger partials until the Moon is passing across the center of the Sun's disk, yielding total or annular eclipses. Then, with the passing generations, as the Sun is farther west within the eclipse limit, the eclipses return to partials. Finally, after about 1,300 years, the Sun is no longer within the eclipse limits when the Moon, after 223 lunations, arrives. After about 1,300 years of adding 6,585.32 days to the date of an eclipse to predict the next, the prediction fails. No eclipse occurs. That saros has died.[11]

Finally, consider that 223 lunations amount to 6,585.32 days while 239 anomalistic months of the Moon (perigee to perigee) total 6,585.54 days. If the Moon was at perigee and hence largest in angular size during one central eclipse, the eclipse must be total. But with each succeeding eclipse in that saros series, the Moon is a little farther from perigee, until finally the Moon's angular size is too small to completely cover the Sun, and eclipses become annular.

## The Saros Family

For people long ago, the discovery of the saros and other eclipse cycles probably brought some comfort, knowing that eclipses, however dire their interpretation, were part of nature's rhythms. Using these rhythms, it was possible to predict future eclipses without knowing anything about the mechanism that produced them. Even after people understood the causes of eclipses, it was far easier to predict their occurrence by the saros (or other cycles) than to calculate all the factors surrounding the varying motions and apparent sizes of the Sun, Moon, and Earth.

Today, with modern electronic computers to crunch orbits and tilts and wobbles into extremely accurate eclipse predictions, the saros would seem to be an anachronism. Yet it remains interesting to view any single total eclipse as a member of an evolving family that had its origin in the distant past, has gradually risen from insignificance to great prominence, and will inevitably decline into oblivion. Chapters 14 and 16 explore the genealogy of the 2024 and 2027 eclipses.

## Creating the Ultimate Eclipse

What conditions would provide the longest total eclipse of the Sun?

First, the Moon should be near maximum angular size, which means it should be near perigee—the point in its orbit when it is closest to Earth. That happens once every 27.55 days (an anomalistic month).

Second, the Sun should be near minimum angular size, which means the Earth should be near aphelion—the point in its orbit when it is farthest from the Sun. That happens once every 365.26 days (an anomalistic year). At present, aphelion occurs in early July.

Third, to prolong the eclipse as much as possible, the Moon's eclipse shadow must be forced to travel as slowly as possible. Here the observer enters the formula. At the time of a solar eclipse, the Moon's shadow is moving about 2,100 miles an hour (3,380 km an hour) with respect to the center of the Earth. But the Earth is rotating from west to east, the same direction that the eclipse shadow travels. At a latitude of 40° north or south of the equator, the surface of the Earth turns at about 790 miles an hour (1,270 km an hour), slowing the shadow's eastward rush by that amount. At the equator, the Earth's surface rotates at 1,040 miles an hour (1,670 km an hour), slowing the shadow's speed to only 1,060 miles an hour (1,710 km an hour), thereby prolonging the duration of totality, Baily's beads, and all phases of the eclipse.

Finally, to prolong the eclipse just a few seconds more, the Moon should be directly overhead, so that we are using the full radius of the Earth to place ourselves about 4,000 miles (6,400 km) closer to the Moon than the limb of the Earth, thus maximizing the Moon's angular size and therefore its eclipsing power. The latitude where the Moon stands overhead varies with the seasons as the Sun appears to oscillate 23½° north and south of the equator. Thus the maximum duration of totality always occurs in the tropics but rarely occurs exactly on the equator.

Eclipse expert Jean Meeus calculates that the maximum possible duration of totality in a solar eclipse is currently 7 minutes 32 seconds. The value slowly changes due to variations in the eccentricity of the Moon's orbit.[*]

[*] Jean Meeus: "The Maximum Possible Duration of a Total Solar Eclipse," Journal of the British Astronomical Association, volume 113, number 6, 2003; pages 343–348.

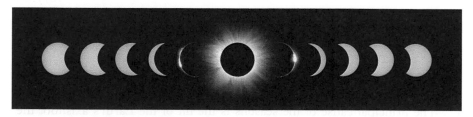

Phases of the 2001 total solar eclipse. A solar filter was used for the 8 partial phases; no filter for totality. Nikon N70, Vixen 90mm Refractor f/9, Chisamba, Zambia, © 2001 Fred Espenak

**Basic Sun and Moon Data Important to Eclipses**

Diameter

| | |
|---|---|
| Sun | 864,989 miles (1,392,000 km) |
| Moon | 2,160 miles (3,476 km) |
| Earth | 7,927 miles (12,756 km) |

Mean distance from Earth

| | |
|---|---|
| Sun | 92,960,200 miles (149,598,000 km) |
| Moon | 238,870 miles (384,400 km) |

Ratio of Sun's diameter to Moon's diameter:    400.5

Ratio of Sun–Earth mean distance to Moon–
Earth mean distance:    389.1

Orbital speed of Moon (mean):    2,290 miles per hour
(3,680 km per hour)

Speed through space of Moon's shadow during
a solar eclipse (not the same as the Moon's
orbital speed because of the orbital motion
of the Earth):    2,100 miles per hour
(3,380 km per hour)

Orbital eccentricity of Moon:    0.05490

Inclination of Moon's orbit to plane of Earth's
orbit (ecliptic) (varies due to tidal effects of
Earth and Sun):    5°08′ (mean)
5°18′ (maximum)
4°59′ (minimum)

Regression (westward drift) of Moon's
nodes:    19.4° per year

Period for Moon's nodes to regress all
the way around its orbit:    18.61 years

## NOTES AND REFERENCES

1.  Epigraph: Alfonso X, King of Castile as cited by Arthur Koestler: *The Sleepwalkers* (New York: Grosset & Dunlap, 1963), page 69.
2.  Alan D. Fiala, US Naval Observatory, personal communication, April 1990.
3.  The principal cause of the seasons is the tilt of the Earth's axis, not the Earth's rather modest change in distance from the Sun.
4.  The Sun takes 365¼ days to appear to go through the constellations of the zodiac once around the sky, so some ancient peoples measured it or rounded

it off to 360 days. Each day, the Sun goes one step around its great sky circle, which is why the Babylonians considered that a circle has 360 degrees. The Chinese circle had 365¼ degrees.

5.  Clemency Montelle: *Chasing Shadows: Mathematics, Astronomy, and the Early History of Eclipse Reckoning* (Baltimore: Johns Hopkins University Press, 2011), pages 71–74.

6.  *Babylonian* here refers to Mesopotamian culture. At various times during this period, Babylon was ruled by Assyrians and Chaldeans.

7.  Clemency Montelle: *Chasing Shadows: Mathematics, Astronomy, and the Early History of Eclipse Reckoning* (Baltimore: Johns Hopkins University Press, 2011), page 78. The earliest prediction of an eclipse in Babylonian records in 651 BCE.

8.  The Babylonians called this 223-lunar-month eclipse cycle "18 years." The use of the word *saros* to mean this 223-lunar-month eclipse cycle was erroneously introduced in 1691 by Edmond Halley when he applied it to the Babylonian eclipse cycle on the basis of a corrupt manuscript by the Roman naturalist Pliny. The Babylonian sign SAR has meaning as both a word and a number. As a word, it means (among other things) *universe*. As a number, it means 3,600, signifying a large number. But there is no evidence that the Babylonians ever applied the word *saros* to this 18-year eclipse cycle.

9.  This period is called anomalistic because it derives from the anomaly—or irregularity—of lunar motion due to the Moon's elliptical orbit.

10. The reason for the change in latitude after 54 years 34 days is the 34-day difference from a calendar year, which changes the eclipse's position within the season, so the altitude of the Sun is significantly different and the shadow is cast farther north or south. The triple saros cycle of 54 years 34 days was known to the Babylonians, who passed it along to the Greeks, who called it the *exeligmos*.

11. In a sense, it has not vanished altogether. With each period of 6,585.32 days, the Sun's position with respect to the node continues to slip westward without experiencing or causing any eclipses until, after about 5,500 years, it encounters the eclipse limit of the opposite node, and that saros may be said to be reborn. George van den Bergh: *Periodicity and Variation of Solar (and Lunar) Eclipses* (Haarlem: H. D. Tjeenk Willink, 1955).

# A MOMENT OF TOTALITY

## A Lasting Impression

Glenn Schneider saw his first total eclipse of the Sun in 1970 at age 14. He traveled with his amateur astronomy club from New York City to Greenville, North Carolina. They observed from the football field of East Carolina University. Glenn was loaded with still cameras, a movie camera, binoculars, and a telescope. He had carefully rehearsed over and over again every action he would take to capture each precious moment of the 2 minutes 54 seconds of totality.

As totality neared, he stationed himself high in the stands to see the horizon and watch the approach of the Moon's shadow. Then he raced down the steps to his telescope and cameras, checked the sky as the Sun blinked out—and froze. He stood and stared at the hole in the sky that hid the Sun, at the luminescent prominences briefly visible at the Sun's rim, at the delicate white tracery of the corona. He was breathless. He couldn't budge to activate his cameras. For 2 minutes 54 seconds, he stared at the sight he had heard about and studied—and for which all the reading, all the pictures, and all the stories were utterly inadequate.

He was profoundly moved—and in that moment he promised himself he would see every total eclipse of the Sun for the rest of his life. Glenn is now Dr. Glenn Schneider, University of Arizona astronomer and principal scientist for many Hubble Space Telescope projects. He continues to pursue solar eclipses. 2024 will be his 39th total.*

* Glenn Schneider, interview, May 5, 2015.

# 3

◄○►

# Ancient Efforts to Understand

> *I look up. Incredible! It is the eye of God. A*
> *perfectly black disk, ringed with bright spiky*
> *streamers that stretch out in all directions.*
>
> Jack B. Zirker, solar astrophysicist (1984)[1]

Ancient peoples around the world have left monuments and artifacts demonstrating their reverence for the sky and their efforts to record celestial motions. More than 2,500 years ago, the Babylonians could predict solar and lunar eclipses. And the ability to anticipate eclipses may go back further still.

## Stonehenge

The earliest and most famous of monuments that testify to a people's high level of astronomical knowledge is Stonehenge, in southern England. The builders of Stonehenge began work about 5,000 years ago.

The initial Stonehenge consisted of a circular embankment 360 feet (110 meters) in diameter, some postholes, and four markers set in a rectangle, later replaced by large standing stones.[2] To dig holes in the ground, the Stonehenge people used the antlers of deer. They had no metal tools, no wheeled vehicles, no draft animals—and no writing. Instead, they left their celestial knowledge set in stone.

About 2600 BCE, when the first pyramid was built in Egypt, the Stonehenge builders brought the 35-ton Heel Stone to the site, the first of the great boulders to be erected.[3] It stands outside the circular embankment to the northeast, 250 feet (77 meters) from the center of Stonehenge. But the Heel Stone was not alone. It had a twin that stood beside it, just to the north.[4]

As seen from the center of Stonehenge, the Sun at the beginning of summer—the longest day of the year—rose between the Heel Stone and its now-vanished companion. With this dramatic sunrise, the people of Stonehenge could celebrate the summer solstice. Using other alignments, they could also mark the beginning of summer and winter, creating an accurate solar calendar of great benefit to farmers and herders.

Aerial view of Stonehenge. [English Heritage]

For someone standing at the center of Stonehenge, the embankment around the monument served to level the horizon of rolling hills. Within the embankment, four stones—the Station Stones—outlined a rectangle within the circle. The sides of this rectangle offered interesting lines of sight. The short side of the rectangle pointed toward the same spot on the horizon that the two Heel Stones framed, the position where the Sun rose farthest north of east, marking the commencement of summer. Facing in the opposite direction along the short side of the rectangle, an observer would see the place where the Sun set farthest south of west, signaling the beginning of winter.

In contrast, the long sides of the rectangle provided alignments for crucial rising and setting positions of the Moon. Looking southeast along the length of the rectangle, an observer was facing the point on the horizon where the summer full moon would rise farthest south. In the opposite direction, looking northwest, this early astronomer's gaze was led to the spot on the horizon where the winter full moon would set farthest north. These positions marked the north and south limits of the Moon's motion.

The structure of Stonehenge offers additional testimony to its builders' efforts to understand the motion of the Moon. Evidence of small holes near the remaining Heel Stone strongly suggests that the users of Stonehenge observed and marked the excursion of the Moon as much as 5 degrees north and south of the Sun's limit.[5] This motion above and below the Sun's position is caused by the tilt of the Moon's orbit to the Earth's path around the Sun. Because of this tilt, the Moon does not pass

Viewed from the center of Stonehenge, the Sun rises just to the left of the remaining Heel Stone at the beginning of summer, the longest day of the year [English Heritage]

directly in front of the Sun (a solar eclipse) or directly into the Earth's shadow (a lunar eclipse) each month.

In the years that followed the installation of the Heel Stones and Station Stones, new generations of ancient Britons added two sets of giant archways. Those arches, completed by 2450 BCE, provide the familiar silhouette of Stonehenge we know today.

The outer set—the Sarsen Circle[6]—formed a ring of 30 linked archways, approximating the days in a lunar month. One of the 30 uprights was only half the diameter of the others, as if to suggest 29½ days, a more accurate record of the time it takes the Moon to complete a cycle of phases.

Inside this Sarsen Circle was a horseshoe of five even larger freestanding archways, the Trilithons, with uprights that weigh up to 50 tons—one pillar weighing as much as 25 cars. The Trilithon archways may have framed extreme setting positions of the Sun and Moon.

The massive, shaped boulders of the Sarsen Circle and Trilithons were dragged from a quarry 20 miles (32 km) away, carefully positioned, and set upright to codify in stone the discoveries made by the people before them.

In the last phase of building at Stonehenge, about 1800 BCE, two concentric circles of holes were dug just outside the Sarsen Circle—one with 30 holes and the other with 29. These circles reinforce the evidence that astronomers at Stonehenge were counting off the 29½-day cycle of lunar phases, from new moon to full moon and back to new moon again. Eclipses of the Sun can only take place at new moon; lunar eclipses can only occur at full moon.

Because the builders of Stonehenge had discovered and accurately recorded the range in rising and setting positions of the Sun and Moon and had built a monument whose alignments marked these positions quite well, they may have been able to recognize when the Moon was on course to intercept the position of the Sun, to cause a solar eclipse. Perhaps they could tell when the Moon was headed for a position directly opposite

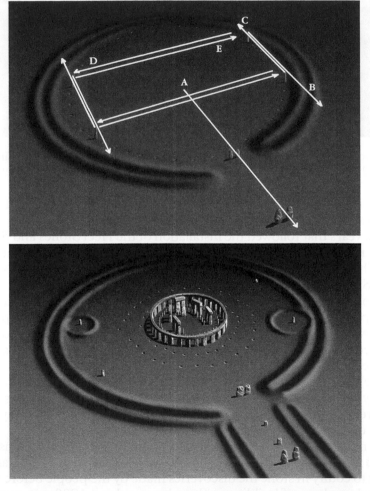

Stonehenge. *Top*: First phase of construction showing alignments A and B, northernmost sunrise: first day of summer; C, southernmost sunset: first day of winter; D, southernmost moonrise; E, northernmost moonset. *Bottom*: Final phase of construction. [© 1983 Hansen Planetarium]

the Sun, which would carry it into the shadow of the Earth for a lunar eclipse.[7] They almost certainly could not predict where or what kind of solar eclipse would be seen, but they might have been able to warn that on a particular day or night, an eclipse of the Sun or Moon was *possible*.

The builders of Stonehenge left no written records of their objectives or results, so we must judge from the monument and its alignments what they knew. Whatever that was, they thought it so worth celebrating that for 1,500 years the rulers and common people were willing to devote vast amounts of time, physical effort, and ingenuity to raising a lasting monument of awesome size, amazing precision, and haunting beauty.

Certainly Stonehenge was more than an astronomical observatory 4,500 years ago. For these Stone Age people, watching the Sun and Moon rise and set over the alignments of Stonehenge must have been a thrilling theatrical spectacle and a profoundly inspiring experience—to be so in touch with the rhythms of the universe.

## China

A frequently recounted Chinese story says that Hsi and Ho, the court astronomers, got drunk and neglected their duties so that they failed to predict (or react to) an eclipse of the Sun. For this, the emperor had them executed. So much for negligent astronomers.

If this story were an account of an actual event, the dynasty mentioned would place the eclipse somewhere between 2159 and 1948 BCE, making it by far the oldest solar eclipse recorded in history. But all serious attempts to identify one particular eclipse as the source of this story have been abandoned as scholars have recognized that the episode is mythological.

In ancient Chinese literature, Hsi-Ho is not two persons but a single mythological being who is sometimes the mother of the Sun and at other times the chariot driver for the Sun. Later, in the *Shu Ching* (Historical Classic), parts of which may date from as early as the 7th or 6th century BCE, this single character is split, not into two, but into six. In the *Shu Ching* story, the legendary Chinese emperor Yao commissions the eldest of the Hsi and Ho brothers "to calculate and delineate the sun, moon, the stars, and the zodiacal markers; and so to deliver respectfully the seasons to the people."[8] In further orders, he sends a younger Hsi brother to the east and another to the south; he orders a younger Ho brother to the west and another to the north. Each is responsible for a portion of the rhythms of the days and seasons, to turn the Sun back at the solstices and to keep it moving at the equinoxes.

These mythological magicians are also charged with the prevention of eclipses, hence the story that appears later in the *Shu Ching* about the emperor's anger with his servants for failing to *prevent* an eclipse, not just predict or respond ceremonially to it.[9]

The Hsi and Ho brothers receive their orders from Emperor Yao to organize the calendar.

The earliest Chinese word for eclipse, *shih*, means "to eat" and refers to the gradual disappearance of the Sun or Moon as if it were eaten by a celestial dragon.[10] It was a bad omen. The Chinese recorded more than a thousand sightings of solar and lunar eclipses—including more solar eclipses than any other civilization. Their accounts, mostly inscribed on animal bones, seldom contain the observational detail provided by the Babylonians. Perhaps there was more information in the original reports, but nearly all were lost long ago, leaving us with summaries in dynastic histories.

The earliest reliable accounts of Chinese eclipses come from *Spring and Autumn Annals (Ch'un-ch'iu)*, recording eclipses from 772 to 481 BCE, including a total solar eclipse in 709 BCE.[11] As they recorded more and more eclipses, the Chinese began to notice that over time eclipses occur in patterns. So, like the Babylonians, without yet understanding what caused an eclipse, the Chinese discovered by the late 1st century BCE that they could predict eclipses. They could take the date of an eclipse, count forward a certain number of months, and reliably predict when another eclipse would occur, even if it was not visible from their city.[12]

By the early 3rd century CE, the Chinese understood how the motions of the Sun and Moon cause eclipses and were able to use this knowledge to predict eclipses more accurately.[13]

## The Maya

In the New World, there were ancient people who, like the Babylonians and the Chinese, used writing to record eclipses and from these records detected a rhythm by which they could predict them or at least warn of their likelihood. Those people were the Maya and we know of their achievement through one of their books—one of only four that survived the Spanish conquest and its zealous destruction of the religious beliefs of the native peoples.

All that we know of Maya accomplishments in recognizing the patterns of eclipses comes from a manuscript called the Dresden Codex. Written in hieroglyphs and illustrated in color, the book was painted on processed tree bark with pages that open and shut in accordion folds. The Dresden Codex dates from the 11th century CE and is probably a copy of an older work.

We can only wonder what was lost when the conquering Spaniards destroyed thousands of books of the Maya and other Mesoamerican peoples. What remains is impressive enough. The Maya realized that discernible eclipses occur at intervals of 5 or 6 lunar months. Five or 6 full moons after a lunar eclipse, there was the *possibility* of another lunar eclipse. Five or 6 new moons after a solar eclipse, another solar eclipse was *possible*.

The Maya had discovered in practical, observable terms the approximate length of the eclipse year, 346.62 days, and the eclipse half year of 173.31 days. The interval for one complete set of lunar phases is 29.53 days. Six lunations amount to approximately 177.18 days, close enough to the eclipse half year (173.31 days) so that there is the "danger" of an eclipse at every sixth new or full moon, but not a certainty. After another 6 lunar months, the passing days have amounted to 354.36, nearly 8 days *too long* to coincide with the Sun's passage by the Moon's node. An eclipse is less likely. As the error mounts, the need increases to substitute a 5-lunar-month cycle into the prediction system rather than the standard 6-lunar-month count.

Some great genius must have noticed after recording a sizeable number of eclipses that major eclipses were occurring only at intervals of 177 days (6 lunar months) or 148 days (5 lunar months). Using the date of an observed solar or lunar eclipse, it would then have been possible to predict the likelihood of another eclipse, even though in some cases an eclipse would not occur at all and in others it would occur but not be visible from Mesoamerica.

In the Dresden Codex there are eight pages with a variety of pictures representing an eclipse. Each depiction is different, but most show the glyph for the Sun against a background half white and half black. In two of the pictures, the Sun and background are being swallowed by a serpent. Leading up to each picture is a sequence of numbers: a series of 177s ending with a 148. Each sequence adds up to the number of days in well-known 3- to 5-year eclipse cycles. At the end of each burst of numbers stands the giant, haunting symbol of an eclipse.

Astronomer-anthropologist Anthony Aveni notes that "[t]he reduction of a complex cosmic cycle to a pair of numbers was a feat equivalent to those of Newton or Einstein and for its time must have represented a great triumph over the forces of nature."[14]

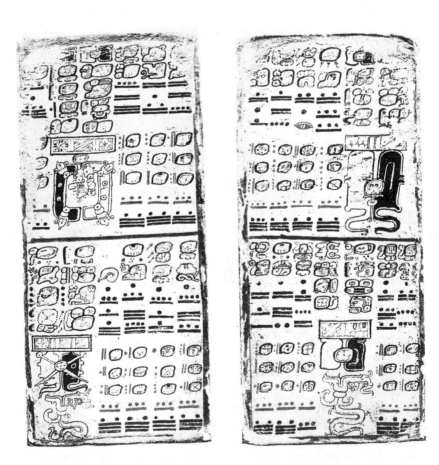

Portion of the Maya solar eclipse prediction tables from the Dresden Codex: at the bottom are the day counts that lead up to a solar eclipse, indicated, bottom right, by a serpent swallowing a symbol for the Sun. [American Philosophical Society]

From the Maya, we have the numbers that demonstrate one of the greatest of their many discoveries about the rhythms of the sky, but we have no account of the emotion the common folk felt when they observed an eclipse. Perhaps the closest we can come is a passage in the Florentine Codex of the Aztecs, who inherited and used the Mesoamerican calendar but apparently knew little of the astronomy discovered by the Maya a thousand years earlier.

> When the people see this, they then raise a tumult. And a great fear taketh them, and then the women weep aloud. And the men cry out, [at the same time] striking their mouths with [the palms of] their hands. And everywhere great shouts and cries and howls were raised. . . . And they said: "If the sun becometh completely eclipsed, nevermore will he give light; eternal darkness will fall, and the demons will come down. They will come to eat us!"[15]

## NOTES AND REFERENCES

1.  Epigraph: Jack B. Zirker: *Total Eclipses of the Sun* (New York: Van Nostrand Reinhold, (1984), page vi.
2.  There were also 56 chalk-filled Aubrey Holes just inside the embankment, but their function is still not generally agreed upon.
3.  Stonehenge construction dates come from Clive Ruggles: *Astronomy in Prehistoric Britain and Ireland* (New Haven, Connecticut: Yale University Press, 1999), pages 35–41; and Anthony Aveni: *Stairways to the Stars: Skywatching in Three Great Ancient Cultures* (New York: John Wiley & Sons, 1997), pages 57–91.
4.  The hole for this stone was discovered in 1979 under the shoulder of the road that passes close to the Heel Stone. The original stone is gone. Michael W. Pitts: "Stones, Pits and Stonehenge," *Nature*, volume 290, March 5, 1981, pages 46–47.
5.  The limits of the Moon's motion north and south of the ecliptic can differ by as much as 10 minutes of arc from the mean inclination of 5 degrees 8 minutes because of gravitational perturbations on the Moon caused by the Sun and the equatorial bulge of the Earth.
6.  In medieval times, people living near Stonehenge called these stones "sarsen," short for "Saracen," meaning Muslim—hence foreign, strange, pagan.
7.  A lunar eclipse prediction is much easier to verify because lunar eclipses are seen over half the Earth.
8.  James Legge, editor and translator: *The Chinese Classics*, volume 3, *The Shoo King* [Shu Ching] (Hong Kong: Hong Kong University Press, 1960), part 1, chapter 2, paragraphs 3–8 (pages 18–22). Legge romanizes Hsi's name as He. In some retellings, Hsi's name appears as Hi. The *Shu Ching* is one of five books in the *Wu Ching* (Five Classics), the sourcebooks of the Confucian tradition.

9. The Hsi and Ho story appears in a chapter that is an exhortation by the Prince of Yin, commander of the armies, to government officials to fulfill their duties to the administration, thereby making the emperor "entirely intelligent." If anyone neglects this requirement, "the country has regular punishments for you."

> Now here are Hsi and Ho. They have entirely subverted their virtue and are sunk and lost in wine. They have violated the duties of their office and left their posts. They have been the first to allow the regulations of heaven to get into disorder, putting far from them their proper business. On the first day of the last month of autumn, the sun and moon did not meet harmoniously in Fang. The blind musicians beat their drums; the inferior officers and common people bustled and ran about. Hsi and Ho, however, as if they were mere personators of the dead in their offices, heard nothing and knew nothing—so stupidly went they astray from their duty in the matter of the heavenly appearances and rendering themselves liable to the death appointed by the former kings. The statutes of government say, "When they anticipate the time, let them be put to death without mercy; when they are behind the time, let them be put to death without mercy."

We never hear whether Hsi and Ho were ever tracked down and executed.

10. Joseph Needham and Wang Ling: *Science and Civilisation in China*, volume 3, *Mathematics and the Sciences of the Heavens and the Earth* (Cambridge: Cambridge University Press, 1959), page 409. Most of this information is also available in Colin A. Ronan's abridgment of Needham's work, *The Shorter Science and Civilisation in China*, volume 2 (Cambridge: Cambridge University Press, 1981).

11. F. Richard Stephenson: *Historical Eclipses and Earth's Rotation* (Cambridge: Cambridge University Press, 1997), pages 58–59, 213, 220–221. Chinese eclipse recordkeeping continued through 1500 CE.

12. The Chinese used a 135-month (11-year) eclipse period rather than the 223-month (18-year) saros period favored by the Babylonians.

13. Joseph Needham and Wang Ling: *Science and Civilisation in China*, volume 3, *Mathematics and the Sciences of the Heavens and the Earth* (Cambridge: Cambridge University Press, 1959), page 421.

14. Anthony F. Aveni: *Skywatchers of Ancient Mexico* (Austin: University of Texas Press, 1980), page 181.

15. Bernardino de Sahagún: *Florentine Codex; General History of the Things of New Spain*, book 7, *The Sun, Moon, and Stars, and the Binding of the Years*, translated from the Aztec by Arthur J. O. Anderson and Charles E. Dibble (Santa Fe, New Mexico: School of American Research; Salt Lake City: University of Utah, 1953), pages 36, 38.

# A MOMENT OF TOTALITY

## The Value of Totality

In 2003, Xavier Jubier, a computer scientist who lives in Paris and created the interactive eclipse maps on Google, traveled with a group of 80 to a total eclipse in Antarctica. They made arrangements to fly on a Russian cargo plane from Cape Town, South Africa to a Russian research station in Antarctica—the equivalent of a coast-to-coast flight across the United States. In Antarctica, on eclipse day, they traveled 9 miles (15 km) further inland by snowcat. For the eclipse, the sky was perfectly clear and the temperature was −13 °F (−25 °C).

One of the people who accompanied Jubier into Antarctica was a 70-year-old American woman, recently widowed, who was living in Botswana. To raise money to travel to this eclipse she sold her house. After the eclipse, she told Jubier she loved it and had absolutely no regrets.[*]

[*] Xavier Jubier, interview, July 31, 2014.

# A MOMENT OF TOTALITY

## The Value of Totality

# 4

—◄○►—

# Eclipses in Mythology

*There was at the same time something in its singular
and wonderful appearance that was appalling: and
I can readily imagine that uncivilised nations may
occasionally have become alarmed and terrified at such
an object . . .*[1]

Francis Baily, on seeing his first total eclipse (1842)

Two great lights brighten the heavens. Life depends on them. Eclipses take
that light away. Through the ages, most cultures responded to eclipses of
the Sun and Moon with stories to explain the eerie events.

The mythology of solar eclipses might be divided into several themes,
and these themes are found scattered throughout the world:

- A celestial being (usually a monster) attempts to destroy the Sun.

- The Sun fights with its lover the Moon.

- The Sun and Moon make love and discreetly hide themselves in
  darkness.

- The Sun-god grows angry, sad, sick, or neglectful.[2]

Within these myths is a great truth. The harmony and well-being of Earth
rely on the Sun and Moon. Abstract science cannot convey this profound
realization as powerfully as a myth in which the celestial bodies come to
life.

## The Sun for Lunch

Most often in mythology a solar eclipse is considered to be a battle between
the Sun and the spirits of darkness. The fate of Earth and its inhabitants
hangs in the balance. With so much at stake, the people enduring an
eclipse are anxious to help the Sun in this struggle if they can.

In the mythology of the Norse tribes, Loki, an evil enchanter, is put in
chains by the gods. In revenge, he creates giants in the shape of wolves.

The mightiest is Hati, who causes an eclipse by swallowing the Sun. Sköll, the other wolf-like giant, follows the Moon, always seeking a chance to devour it.[3] Old French and German chants echo this belief: "God protect the Moon from wolves."

In the mythology of India, the demons and their younger brothers, the gods, fight over the possession of amrita, the nectar of immortality. Rahu, a demon, disguises himself and attends a gathering of the gods to steal the amrita. But the Sun and Moon recognize him and alert chief god Vishnu. Just as Rahu grabs the amrita and begins to drink it, Vishnu hurls a discus that slices through Rahu's neck. Not a drop of amrita gets past Rahu's throat, so his body dies. But Rahu has the nectar of immortality in his mouth, so his head doesn't die. Rahu's head flies off into the heavens to take revenge on the Sun and Moon. Whenever he catches them, he swallows them—and that's why we have eclipses. But Rahu has only a head and no body, so soon after he swallows them, the Sun and Moon reappear from his neck. Light returns to Earth, and the Sun and Moon continue their way across the sky.

In some versions of the myth, the headless body of Rahu has a name: Ketu. One retelling says that Ketu falls to Earth with a colossal thud, causing quakes. Another says that Ketu remains in the heavens as the constellations.[4]

The Indian myth of Rahu spread into China, Mongolia, and southern Siberia. Rahu took on different names but his severed head went on chasing and chewing on the Sun and Moon.

When the Sun and Moon scream for help, the people on the ground respond by hollering and by throwing stones and shooting arrows into the sky to scare away the monster.

In southern Siberia, a Buryat myth says the eclipse-maker is Arakho (Rahu), a beast who formerly lived on Earth. In those days long ago, human beings were quite hairy. Arakho roamed the Earth munching the hair off their bodies until people became the nearly hairless creatures we are today. This annoyed the gods, who chopped Arakho in two. So Arakho no longer grazes on human hair. Instead, his head, which is still alive, chomps on the Sun and Moon, causing eclipses.

Rahu appears again in Indonesia and Polynesia as Kala Rau, all head and no body, who eats the Sun, burns his tongue, and spits the Sun out.

## Other Sun-Eaters and Sun-Beaters

From around the world come stories of many different monsters intent on devouring the Sun and Moon. In China, it was a heavenly dog who causes eclipses by eating the Sun.[5] In South America, the Mataguaya

Egyptian emblem of the winged Sun at the top of the Gateway of Ptolemy at Karnak. [William Tyler Olcott: Sun Lore of All Ages]

Indians of the pampas saw eclipses as a great bird with wings outspread, assailing the Sun or Moon.[6] In Armenian mythology, eclipses were the work of dragons who sought to swallow the Sun and Moon. In contrast, another Armenian myth says that a sorcerer can stop the Sun or Moon in their courses, deprive them of light, and even force them down from the skies. Despite the Moon's size, once it has been brought down to Earth, the sorcerer can milk it like a cow.[7]

In an Egyptian myth, Ra, the Sun god, travels across the sky each day in a boat. His eternal enemy Apep is always watching for his opportunity to attack. Apep, god of chaos, hater of sunlight, is a whale-size serpent who lives in the depths of the Nile, At just the right moment, he leaps from the water, opens his jaws, and swallows the Sun—boat and all. But traveling with Ra are his defenders. They are always able—so far—to rip the Sun from Apep and cast the vicious serpent back into the depths. Ra, the Sun, continues his daily boat ride westward across the heavens. Apep hungers for another try.[8]

In North Africa, a Berber myth explains a solar eclipse like this: a huge, winged ifrit (evil jinni) zooms skyward from its underground lair and swallows the Sun. But the meal gives him a cosmic case of heartburn. He vomits up the Sun and it resumes shining.[9]

The Pomo Indians of northern California gave a solar eclipse a name that was also an explanation: Sun-got-bit-by-bear. This bear was walking in the sky when it bumped into the Sun. Stand aside, said the Sun. Get out of my way, said the bear. So, initiating a road rage tradition that lives on in America, they fought about it, with the bear chewing on the Sun, creating an eclipse. Then the bear and the Sun continued their journeys—until the bear bumped into the Moon. Which led to another argument. Which led to a Moon-got-bit-by-bear. Then the bear and the Moon continued on their way . . .[10]

On rare occasions in mythology, the Sun and Moon are not totally innocent victims. In a variant Hindu myth, the Sun and Moon once borrowed money from a member of the savage Dom tribe and failed to pay it back. In retribution, a Dom occasionally devours the two heavenly bodies.[11]

Drawing of the corona by Samuel P. Langley from the top of Pike's Peak, July 29, 1878. The elongated corona resembles the Egyptian emblem of the winged Sun. [Mabel Loomis Todd: Total Eclipses of the Sun]

## The Original Black Holes

In two instances, the hungry monster that swallows the Sun or Moon becomes quite scientifically sophisticated in character. A western Armenian myth, said to be borrowed from the Persians, tells of two dark bodies, the children of a primeval ox. These dark bodies orbit the Earth closer than the Sun and Moon. Occasionally they pass in front of the Sun or Moon and thereby cause an eclipse.[12]

Still more remarkable is a Hindu myth which speaks of the Navagrahas, the Nine Seizers. These nine "planets" that wander through the star field include the usual seven bodies familiar to the Greeks—Sun, Moon, Mercury, Venus, Mars, Jupiter, and Saturn—plus Rahu and Ketu. Rahu (eater of the Sun and Moon) and Ketu (Rahu's lower half) are the ascending and descending nodes of the Moon, the shifting points in the sky where the Moon crosses the apparent path of the Sun.[13] Thus, quite correctly, the Sun would be at risk of an eclipse whenever it passes by Rahu or Ketu.

Buddhism carried Rahu and Ketu from India to China in the 1st century CE, where they became Lo-Hou and Chi-Tu. They were imagined as two invisible planets positioned at the nodes in the Moon's path: Lo-Hou

The sudden appearance of the corona during totality would have been a terrifying sight to ancient civilizations. The corona was photographed from Australia during the total eclipse of November 12, 2012. [Nikon D700, Tamron 400 mm, f/6.3, multiple exposure composite: 1/2000 to 1 second, ISO 200. © 2016 Robert Slobins]

at the ascending node and Chi-Tu at the descending node. These "dark stars" were numbered among the planets and were considered to be the cause of eclipses.[14]

## Love, Marriage, and Domestic Violence

A Germanic myth explains eclipses differently. The male Moon married the female Sun. But the cold Moon could not satisfy the passion of his fiery bride. He wanted to go to sleep instead. The Sun and Moon made a bet: whoever awoke first would rule the day. The Moon promptly fell asleep, but the Sun, still irritated, awoke at 2 a.m. and lit up the world. The day was hers; the Moon received the night. The Sun swore she would never spend the night with the Moon again, but she was soon sorry. And the Moon was irresistibly drawn to his bride. When the two come together, there is a solar eclipse, but only briefly. The Sun and Moon begin to reproach one another and fall to quarreling. Soon they go their separate ways, the Sun blood-red with anger.

The native Awawak and Gê people of South America also have myths in which the Sun and Moon fight each other, creating an eclipse.[15] It's hard to know the antiquity of such stories, but if they are ancient, they may indicate a realization that the Sun and Moon *together* cause eclipses, which was beyond the understanding of most ancient people.

It is not always a fight between the Sun and Moon that causes an eclipse. Sometimes it is love, and modesty. The Tlingit Indians of the Pacific coast in northern Canada explained a solar eclipse as the Moon-wife's visit to her husband. Across the continent, in southeastern Canada, the Algonquin Indians also envisioned the Sun and Moon as loving husband and wife. If the Sun is eclipsed, it is because he has taken his child into his arms.[16]

The Fon people of Benin in western Africa say the male Sun rules the day and female Moon rules the night. They love one another but they are very busy, always in motion. They meet whenever they can, which is not often. But when they do, they modestly turn off the light.[17]

For the Tahitians, the Sun and Moon were lovers whose union creates an eclipse. In that darkness, they lose their way and create the stars in order to light their return.

## An Angry Sun

Sometimes, as in a folktale from eastern Transylvania, it is the perversion of mankind that brings on an eclipse. The Sun shudders, turns away in disgust, and covers herself with darkness. Stinking fogs gather. Ghosts appear. Dogs bark strangely and owls scream. Poisonous dews fall from the skies, a danger to man and beast. Neither humankind nor animals should consume water or eat fresh fruit or vegetables. Such beliefs persisted into the 19th century. This poisonous dew that supposedly accompanied eclipses could be the source of an outbreak of the plague or other epidemics. If people had to leave their homes, they wrapped a towel around their mouths and noses to strain out the noxious vapors. Clothes caught drying outdoors during a solar eclipse were considered to be infected.

Europeans were not alone in their belief that a solar eclipse brought a dangerous form of precipitation. Eskimos in southwestern Alaska believed that an unclean essence descended to Earth during an eclipse. If it settled on utensils, it would produce sickness. Therefore, when an eclipse began, every Eskimo woman turned all her pots, buckets, and dishes upside down.[18]

Yet it was not always an angry Sun that brought darkness to the Earth. When US Coast Survey scientist George Davidson observed the eclipse of August 7, 1869 from Kohklux, Alaska, he found that the Native Americans

When the weather gives you lemons, make lemonade! The thin cloud cover offers a more interesting photo than one in a clear sky. A large sunspot group on the Sun is just re-emerging from behind the lunar disk. The partial eclipse of the Sun, October 23, 2014, from Jasper, Alberta. [Canon 60Da, 66 mm apo refractor, f/6, 1/15 second, © 2014 Alan Dyer]

there attributed the eclipse to an illness of the Sun. He had alerted them to the impending eclipse, but they doubted it. Halfway through the partial phase, the Indians and their chief quit work and hid in their houses: "They looked upon me as the cause of the Sun's being 'very sick and going to bed.' They were thoroughly alarmed, and overwhelmed with an indefinable dread."[19]

Sometimes in mythology, an eclipse is not a monster devouring the Sun, not a sickness of the Sun, not a fight between the Sun and Moon, not even the result of the always abundant sins of mankind. Sometimes an eclipse is what in sports would be called an unforced error. The Nuxalk (Bella Coola) Indians of the Pacific coast in Canada have a myth that begins with a remarkable observational description of the Sun's apparent annual pathway though the sky. The trail of the Sun, they say, is a bridge whose width is the distance between the summer and winter solstices, the northernmost and southernmost positions of the Sun. In the summer, the Sun walks on the right side of this bridge; in the winter, he walks on the left. The solstices are where the Sun sits down. Accompanying the Sun on his journey are three guardians who dance about him. Sometimes the Sun simply drops his torch, and thus an eclipse occurs.[20]

## Warding Off the Evil

Corruption and death are a frequent theme of eclipse myths. Evil spirits descend to Earth or emerge from underground during eclipses.

On June 16, 1406, says an enlightened and bemused French chronicler, "between 6 and 7 a.m., there was a truly wonderful eclipse of the Sun which lasted nearly half an hour. It was a great shame to see the people withdrawing to the churches and believing that the world was bound to end. However the event took place, and afterward the astronomers gathered and announced that the occurrence was very strange and portended great evil."[21] (Half an hour is too long for totality, but the right length for conspicuously reduced light surrounding totality.)

For Hindus, the place to be during an eclipse was in the water, especially in the purifying current of the Ganges.[22] This Hindu practice of immersing oneself in water was known to the French philosopher and popularizer of science Bernard Le Bovier de Fontenelle, as recorded in his *Entriens sur la pluralité des mondes* (Conversations on the Plurality of Worlds) in 1686.

> All over India, they believe that when the sun and moon are eclipsed, the cause is a certain dragon with very black claws which tries to seize those two bodies, wherefore at such times the rivers are seen covered with human heads, the people immersing themselves up to the neck, which they regard as a most devout position, and implore the sun and moon to defend themselves well against the dragon.[23]

## A Participatory Event

From around the world come reports of people trying to assist the Sun and Moon against the peril of eclipse. Screaming, crying, and shouting are supposed to encourage the Sun and Moon to escape the clutches of the evil spirit. Historians recorded that Germans watching a lunar eclipse in the Middle Ages chanted in unison, "Win, Moon."

The Sun and Moon were the supreme gods of the Indians of Colombia. When these gods were threatened by an eclipse, the people seized their weapons and made warlike sounds on their musical instruments. They also shouted to the gods, promising to mend their ways and work hard. To prove it, they watered their corn and worked furiously with their tools during the eclipse.[24]

People frequently augmented their voices with the clanging of metal pots, pans, and knives. The Chippewa Indians in the northeastern United States and southeastern Canada went even further. Seeing the Sun's light being extinguished, they shot flaming arrows into the sky, hoping to rekindle the Sun.[25] The Sencis of eastern Peru also shot fire arrows toward the Sun, but not to rekindle it. They were trying to scare off a savage beast that was attacking the Sun.

Ethnographers descended on the Kalina tribe in Suriname to collect their folklore and watch their behavior as the total eclipse of June 30, 1973 neared. In Kalina mythology, the Sun and Moon are brothers. They usually get along well together, but occasionally they have sudden and ferocious quarrels that endanger mankind. At such times, it is important to separate the combatants by making a maximum of noise: banging on tools, hollow objects, and instruments. The fading of the Sun or Moon means that one has been knocked unconscious. When this happens, the tribesmen yell, "Wake up, Papa!" Papa, here, is a term of respect, not an indication that the Kalina consider themselves descended from the Sun or Moon. After the 1973 eclipse, invisible because of clouds but noticeable because of darkening, the old women (the pot makers) rounded up the children and used branches to spread white clay all over them, somewhat too vigorously for the children's enjoyment. Then they smeared the women and finally the men from head to foot with white clay. The white clay was the blood of the injured Moon that had dripped onto the ground. It was necessary to wash oneself with the Moon's blood to restore purity in man and whiteness to the Moon. After an hour, the tribesmen washed themselves off in the river.[26]

In ancient Mexico and Central America, the most important god was represented as a plumed serpent. For the Maya, he was Kukulcán; for the Aztecs, Quetzalcóatl. At an eclipse of the Moon and most especially during an eclipse of the Sun, a special snake was killed and eaten.[27]

In prehistoric times, screams, cries, banging noises, and prayer may not have been deemed adequate to ward off eclipses or their effects. In many places around the world, human sacrifice was performed at the appearance of unexpected and confusing sights, such as an eclipse or a comet. Yet few eclipse myths refer to human sacrifice, suggesting that this practice had largely been abandoned before most eclipse myths were preserved. An exception was in Mexico and Central America, where the Spanish invaders saw the Aztecs and their neighbors carry out human sacrifice in the early 16th century. For the Aztecs, almost any natural or political event was commemorated with a sacrifice. On the occasion of an eclipse, the Sun was in need of help from people, just as he had constant help from the dog Xolotl *(Sho-LOT-uhl)*. Xolotl was the god of human monstrosities (including twins), so it was humpbacks and dwarfs who were sacrificed to the Sun to help him prevail.[28]

Solar eclipses boded ill for everyone, it seems, except prospectors. In Bohemia, people believed that a solar eclipse would help them find gold.

## NOTES AND REFERENCES

1. Epigraph: Francis Baily: "Some Remarks on the Total Eclipse of the Sun, on July 8th, 1842," *Memoirs of the Royal Astronomical Society*, volume 15, 1846, page 6.

2. Viktor Stegemann: "Finsternisse" in Hanns Bächtold-Stäubli, editor: *Handwörterbuch des Deutschen Aberglaubens*, Bände 2 (Berlin: W. de Gruyter, 1930), columns 1509–1526. Germanic eclipse lore described in this chapter comes from this article unless otherwise noted.

3. In some versions of the myth, Sköll eats the Sun and Hati eats the Moon. See E. C. Krupp: *Beyond the Blue Horizon: Myths and Legends of the Sun, Moon, Stars, and Planets* (New York: HarperCollins, 1991), page 162.

4. Jan Knappert: *Indian Mythology: An Encyclopedia of Myth and Legend* (London: HarperCollins, 1991), page 203. Arthur Berriedale Keith: *Indian Mythology*, volume 6 of *The Mythology of All Races* (Boston: Marshall Jones, 1917), page 151.

5. John C. Ferguson: *Chinese Mythology*, volume 8 of *The Mythology of All Races* (Boston: Marshall Jones, 1928), page 84.

6. Hartley Burr Alexander: *Latin-American Mythology*, volume 11 of *The Mythology of All Races* (Boston: Marshall Jones, 1920), page 319.

7. Mardiros H. Ananikian: *Armenian Mythology*, volume 7 of *The Mythology of All Races* (Boston: Marshall Jones, 1925), page 48.

8. Harold Scheub: *A Dictionary of African Mythology: The Mythmaker as Storyteller* (Oxford: Oxford University Press, 2000), page 216.

9. Harold Scheub: *A Dictionary of African Mythology: The Mythmaker as Storyteller* (Oxford: Oxford University Press, 2000), page 74.

10. E. C. Krupp: *Beyond the Blue Horizon: Myths and Legends of the Sun, Moon, Stars, and Planets* (New York: HarperCollins, 1991), page 162.

11. *Arthur Berriedale Keith: Indian Mythology*, volume 6 of *The Mythology of All Races* (Boston: Marshall Jones, 1917), pages 232–233.

12. Mardiros H. Ananikian: *Armenian Mythology*, volume 7 of *The Mythology of All Races* (Boston: Marshall Jones, 1925), page 48.

13. Arthur Berriedale Keith: *Indian Mythology*, volume 6 of *The Mythology of All Races* (Boston: Marshall Jones, 1917), page 233.

14. Joseph Needham and Wang Ling: *Science and Civilisation in China*, volume 3, *Mathematics and the Sciences of the Heavens and the Earth* (Cambridge: Cambridge University Press, 1959), page 228.

15. E. C. Krupp: *Beyond the Blue Horizon: Myths and Legends of the Sun, Moon, Stars, and Planets* (New York: HarperCollins, 1991), page 162.

16. Hartley Burr Alexander: *North American Mythology*, volume 10 of *The Mythology of All Races* (Boston: Marshall Jones, 1916), pages 25, 277.

17. Harold Scheub: *A Dictionary of African Mythology: The Mythmaker as Storyteller* (Oxford: Oxford University Press, 2000), page 168.

18. James George Frazer: *Balder the Beautiful*, volume 1; *The Golden Bough*, volume 10 (London: Macmillan, 1930), page 162.

19. Mabel Loomis Todd: *Total Eclipses of the Sun*, revised edition (Boston: Little, Brown, 1900), page 131.

20. Hartley Burr Alexander: *North American Mythology*, volume 10 of *The Mythology of All Races* (Boston: Marshall Jones, 1916), page 255.

21. Paul Yves Sébillot: *Le folk-lore de France*, tome 1, *Le ciel et la terre* (Paris: Librairie orientale & américaine, 1904), page 52. The chronicler was Jean Juvénal des Ursins.

22. Arthur Berriedale Keith: *Indian Mythology*, volume 6 of *The Mythology of All Races* (Boston: Marshall Jones, 1917), page 234.

23. A free translation from the "Second Soir." Compare Bernard Le Bovier de Fontenelle: *A Plurality of Worlds*, translated by John Glanvill (England: Nonesuch Press, 1929), with the excerpt in François Arago: *Popular Astronomy*, volume 2, translated by W. H. Smyth and Robert Grant (London: Longman, Brown, Green, Longmans, and Roberts, 1858), page 349.

24. Hartley Burr Alexander: *Latin-American Mythology*, volume 11 of *The Mythology of All Races* (Boston: Marshall Jones, 1920), pages 277–278.

25. James George Frazer: *The Magic Art*, volume 1; *The Golden Bough*, volume 1 (London: Macmillan, 1926), page 311.

26. Patrick Menget: "30 juin 1973: station de Surinam," *Soleil est mort; l'éclipse totale de soleil du 30 jin 1973* (Nanterre, France: Laboratoire d'ethnologie et de sociologie comparative, 1979), pages 119–142.

27. Hartley Burr Alexander: *Latin-American Mythology*, volume 11 of *The Mythology of All Races* (Boston: Marshall Jones, 1920), page 135.

28. Hartley Burr Alexander: *Latin-American Mythology*, volume 11 of *The Mythology of All Races* (Boston: Marshall Jones, 1920), page 82.

# A MOMENT OF TOTALITY

## My Favorite Eclipse

*by Ken Willcox*

Every total eclipse is different – and wonderful – but if you see more than one, you probably have a favorite. Mine was the eclipse of November 3, 1994 on the Altiplano in Bolivia.

The Altiplano is a huge desert high in the Andes Mountains. It was there we had come, 110 of us from all over the United States and Europe, to a small plot of land 12,516 feet (3,815 meters) above sea level. I selected the Altiplano because it provided the best chance of clear skies along the eclipse path. But the price of clear skies was a major logistical problem: transporting 110 people safely and comfortably to a remote high-altitude viewing site in a foreign land.

On Wednesday, November 2, we boarded our own private train in La Paz, with a dining car and also a baggage car to accommodate all our equipment to record three precious minutes of time. At 4:50 p.m., our chartered narrow-gauge train began its spiraling climb out of the trench in which La Paz lies, then turned south toward our destination 200 miles (320 km) away, a spot on the central line, 7½ miles (12 km) south of Sevaruyo. In the middle of the night, using the global positioning system (GPS), Jim Zimbleman of the Smithsonian Institution and I navigated the train to a precise spot along the tracks selected two years earlier on a site inspection trip.

As we approached our site about 4 a.m., we were met by 70 gun-toting soldiers provided for our protection by the Bolivian army. Under floodlights, before dawn, in the middle of nowhere high in the Andes, amateur and professional astronomers and Bolivian soldiers dragged equipment from the train and prepared to record an extraordinary celestial event.

One hour after sunrise, the clouds began breaking up. Only high, scattered cirrus remained at first contact at 7:19 a.m., and those too were vanishing.

A family of Aymara Indians, descendants of the Incas, had been invited to join us. We gave them solar filters to view the partial phases. This Bolivian family huddled around their father as totality approached. We kept glancing at them to see what their response to the disappearing Sun and the onset of totality would be. A few minutes before totality, the father made his three boys look down at the ground until the eclipse was over. The father feared the souls of the boys would be unalterably affected.

"Here it comes!" someone shouted. "Where?" "Over there," pointing to the northwest. "Oh yes! I see it!" "It's getting dark now . . . it's getting real dark now . . . it's really getting . . . OH MY GOD!" Cheers and gasps accompanied the beginning of totality. It took everything I had to keep my mind focused on the task at hand, and even that didn't work when the lady behind me broke down crying. Her husband wrapped his arms around her.

As totality ended, I shot a last few exposures blinded by my own tears, gave up, and turned around and photographed the couple behind me.

The Aymara Indians were right. All our souls were unalterably affected by that eclipse. But it was—and is—a sublime experience, one that should be sought, not feared.*

---

* Ken Willcox was a chemist and leader of five solar eclipse expeditions. He died in 1999. Adapted from Mark Littmann, Fred Espenak, and Ken Willcox: *Totality: Eclipses of the Sun* (New York: Oxford University Press, 2008; updated 2009), pages 142–143.

# 5

---◀○▶---

# The Strange Behavior
# of Man and Beast—Long Ago

*The Sun . . .*
*In dim eclipse disastrous twilight sheds*
*On half the nations, and with fear of change*
*Perplexes monarchs.*

John Milton (1667)[1]

## The Human Reaction

One of the most dramatic responses in history to a total solar eclipse is presented by Herodotus, the first Greek historian, writing around 430 BCE.

> War broke out between the Lydians and the Medes [major powers in Asia Minor], and continued for five years . . . The Medes gained many victories over the Lydians, and the Lydians also gained many victories over the Medes. . . . As, however, the balance had not inclined in favour of either nation, another combat took place in the sixth year, in the course of which, just as the battle was growing warm, day was suddenly changed into night. This event had been foretold by Thales, the Milesian, who forewarned the Ionians of it, fixing for it the very year in which it actually took place. The Medes and Lydians, when they observed the change, ceased fighting, and were alike anxious to have terms of peace agreed upon.

A treaty was quickly made and sealed by the marriage of the daughter of the Lydian king to the son of the Median king.[2] This story indicates the awe that ancient people felt when confronted with a total eclipse of the Sun.[3]

Modern astronomers, armed with the dates of the kings described in the account and a knowledge of the dates and paths of ancient eclipses, have generally settled upon May 28, 585 BCE as the eclipse to which the story refers, if Herodotus, given to fanciful embellishment, can be trusted about an event that occurred a century before he was born.[4]

In his account, Herodotus credits the Greek philosopher Thales with predicting this eclipse. If so, Thales would have been the first person *known* to have calculated a future solar eclipse. But the Babylonians, not the Greeks, were the leaders in eclipse prediction. Their cuneiform writing on clay tablets from the period of the 585 BCE eclipse shows recognition of an 18-year-11-day rhythm in eclipses—the saros.[5] Might Thales have borrowed this eclipse rhythm from the Babylonians as it was being developed?

However, such a rhythm predicts not just the year but the month and precise day of the eclipse. Yet Herodotus seems amazed that Thales could be accurate to "the very year in which it actually took place." Was Herodotus so surprised that Thales could predict an eclipse accurate to the day that he simply could not believe that degree of precision and used the more conservative "year" instead? That would be out of character for the flamboyant Herodotus. Yet predicting a solar eclipse accurate to a year is not much of a trick since there is a minimum of two solar eclipses a year. The problem is to predict a total eclipse for a particular location on Earth. Could Thales have accomplished this? It is doubtful.

The saros period is actually 18 years 11⅓ days, so the Earth has spun through an extra 8 hours. Thus each subsequent eclipse falls about one-third of the way around the world westward from the one before it. The result is that successive eclipses in a saros series are almost never visible from the same site.

The eclipse in the same saros series that preceded 585 BCE occurred on May 18, 603 BCE, with an early morning path from the northern portion of the Red Sea to the northern tip of the Persian Gulf, about 600 miles (1,000 kilometers) distant from the end of the path of the May 28, 585 BCE eclipse. Thales could have heard reports of the 603 BCE eclipse and used it to calculate the date for the 585 BCE eclipse. But the saros projection would not have told him where the eclipse would be visible. Thales, then, first of the great Greek philosophers, could have warned of the *possibility* of a solar eclipse, but he could not predict from the saros period that it would be visible in Asia Minor. And there is no evidence that he had the celestial knowledge or the mathematics to calculate it from orbital considerations.

Of course the key to appreciating the story of the solar eclipse that stopped a war is the realization that people long ago were stunned by a total eclipse of the Sun and incredulous that someone could predict such an event. Quite often in ancient history, eclipses are reported to have played a decisive role in the turn of events.

Herodotus tells of another turning point in world history that he says hinged on a solar eclipse. Xerxes and his Persian army were about to march from Sardis to Abydos on their advance toward Greece.

During a battle between the Lydians and the Medes on May 28, 585 BCE, a total eclipse of the Sun occurred. It scared the soldiers so badly that they stopped fighting and signed a treaty. [Mabel Loomis Todd: *Total Eclipses of the Sun*]

At the moment of departure, the sun suddenly quitted his seat in the heavens, and disappeared, though there were no clouds in sight, but the sky was clear and serene. Day was thus turned into night; whereupon Xerxes, who saw and remarked the prodigy, was seized with alarm, and sending at once for the Magians, inquired of them the meaning of the portent. They replied—"God is foreshadowing to the Greeks the destruction of their cities; for the sun foretells for them, and the moon for us." So Xerxes, thus instructed, proceeded on his way with great gladness of heart.[6]

To disaster! He reached and burned Athens, but his navy was destroyed by the Greeks and his forces had to withdraw. Twice more Xerxes invaded Greece, but each time his armies were crushed. After his last defeat, his nobles assassinated him.

Xerxes' first march against Greece actually occurred in 480 BCE, but the only major eclipse visible in the region near that date was the total eclipse of February 17, 478 BCE. Thus the story tells us less about observational astronomy in that era than about the power exercised by eclipses over the minds of men and the effectiveness of their use to heighten the drama of a story.

A final story from Greece illustrates an advance from superstitious dread of eclipses to an understanding of what causes them. On August 3, 430 BCE, Pericles and his fleet of 150 warships were about to sail for a raid upon their enemies.

> But at the very moment when the ships were fully manned and Pericles had gone onboard his own trireme, an eclipse of the sun took place, darkness descended and everyone was seized with panic, since they regarded this as a tremendous portent. When Pericles saw that his helmsman was frightened and quite at a loss what to do, he held up his cloak in front of the man's eyes and asked him whether he found this alarming or thought it a terrible omen. When he replied that he did not, Pericles asked, "What is the difference, then, between this and the eclipse, except that the eclipse has been caused by something bigger than my cloak?" This is the story, at any rate, which is told in the schools of philosophy.[7]

The eclipse was a large partial at Athens and annular about 600 miles (1,000 km) to the northeast. This eclipse had also been recorded by Thucydides, without the didactic story, but exhibiting an increased awareness of the cause of eclipses: "The same summer, at the beginning of a new lunar month, the only time by the way at which it appears possible, the sun was eclipsed after noon. After it had assumed the form of a crescent and some stars had come out, it returned to its natural shape."[8]

Almost 2,000 years later, however, superstition still clouded the public's view of eclipses. On June 3, 1239, Thomas, Archdeacon of Split, Croatia, chronicled "the wonderful and terrible eclipse of the Sun." "Such great fear overtook everyone," he wrote, "that just like madmen they ran about to and fro shrieking, thinking that the end of the world had come."

On October 6, 1241, only 2½ years later, Split experienced a second total solar eclipse—a rare event—and Thomas once again recorded "great terror among everyone."[9]

Four centuries later, not much had changed. Paris was a center of education, yet on August 12, 1654, "at the mere announcement of a total eclipse, a multitude of the inhabitants of Paris hid themselves in deep cellars."[10]

No wonder then that the sight of a total eclipse on July 29, 1878 had a powerful effect on Native Americans near Fort Sill, Indian Territory (now Oklahoma). A non-Indian described it this way:

> It was the grandest sight I ever beheld, but it frightened the Indians badly. Some of them threw themselves upon their knees and invoked the Divine blessing; others flung themselves flat on the ground, face downward; others cried and yelled in frantic excitement and terror. Finally one old fellow stepped from the door of his lodge, pistol in hand, and, fixing his eyes on the darkened Sun, mumbled a few unintelligible words and raising his arm took direct aim at the luminary, fired off his pistol, and after throwing his arms about his head in a series of extraordinary gesticulations retreated to his own quarters. As it happened, that very instant was the conclusion of totality. The Indians beheld the glorious orb of day once more peep forth, and it was unanimously voted that the timely discharge of that pistol was the only thing that drove away the shadow and saved them from . . . the entire extinction of the Sun.[11]

## The Animal Response

Noting their own primal response to the daytime darkening of the Sun, people through the ages have been fascinated by the reaction of animals to a total eclipse. Reports go back more than 750 years. In describing the eclipse of June 3, 1239, Ristoro d'Arezzo wrote: "We saw the whole body of the Sun covered step by step . . . and it became night . . . and all the animals and birds were terrified; and the wild beasts could easily be caught . . . because they were bewildered."[12]

As a university student in Portugal, astronomer Christoph Clavius saw the total eclipse of August 21, 1560: "Stars appeared in the sky, and (miraculous to behold) the birds fell down from the sky to the ground in terror of such horrid darkness."[13]

In 1706 at Montpellier in southern France, observers reported that "bats flitted about as at the beginning of night. Fowls and pigeons ran precipitately to their roosts." In 1715, the French astronomer Jacques Eugène d'Allonville, Chevalier de Louville, traveled to London for the eclipse and observed that at totality "horses that were laboring or employed on the high roads lay down. They refused to advance."[14]

By 1842, some people were even conducting behavioral experiments on their pets. "An inhabitant of Perpignan [France] purposely kept his dog without food from the evening of the 7th of July. The next morning, at the instant when the total eclipse was going to take place, he threw a piece of bread to the poor animal, which had begun to devour it, when the sun's last rays disappeared. Instantly the dog let the bread fall; nor did he take it up again for two minutes, that is, until the total obscuration had ceased; and then he ate it with great avidity."[15]

William J. S. Lockyer, son of the pioneering solar spectroscopist, traveled to Tonga for an eclipse in 1911. The weather conditions were miserable and the insects numerous and very hungry. He and his colleagues caught only a brief view of the corona through thin clouds and the scientific results were meager. The only members of his team with good results were those studying animal behavior. The horses did not seem to notice the darkening, but fowl ran home to roost and pigs lay down. Flowers closed. But most memorable of all were the insects, which had been completely silent until the moment of totality and then sang as if it were night. "The noise," recalled Lockyer, "was most impressive, and will remain in my memory as a marked feature of that occasion."[16]

## A Cure for Eclipses

In 1834, the Indians of the Kiowa Tribe gathered together at the End of the Mountains, west of the Wichita Mountains in Oklahoma, for their annual Sun Dance ceremony. On November 30, right at noon, when people were about to eat dinner, the sky began to darken, but there were no clouds blocking the Sun. Darker and darker it got. "The Sun is dying," the people yelled. "A snake has come up from the Underworld," screamed others. "It is swallowing the Sun." "What can we do?" they cried.

A great medicine man came forward, his face painted as a bobcat—the strongest power there is. The medicine man began to shake his rattle and dance and sing, screeching like a bobcat—as if the Sun was sick and he was treating a sick person.

Soon, where the Sun had vanished, there was a sliver of light, like the edge of a fingernail. The medicine man danced and sang even harder. The light grew and grew. Birds flew from their nests and began to sing. The world was alive again. Young men and women laughed. Mothers went back to serving dinner. The Sun Dance ceremony continued.

People showered the medicine man with gifts. They didn't know what made the Sun go away, but they did know who brought the Sun back.[17]

To many early people, the eerie twilight sky of totality was interpreted as the end of the world. A wide-angle view to totality was shot from Dundlod, India on October 24, 1995. [Nikon FE SLR, Nikkor 50 mm, Ektachrome 100, f/5.6, 1/4 second. © 1995 Fred Espenak]

## The Shawnee Prophet Uses an Eclipse

Tenskwatawa, the Shawnee Prophet (1775?–1837?), was an important Indian religious leader in Ohio and Indiana in the early 19th century. He saw great danger for his people as they increasingly adopted the customs of the European settlers, especially alcohol. He urged them to return to traditional Indian ways and to unite into a single Indian nation under the leadership of his brother Tecumseh to resist the encroachment of white men with their fraudulent treaties.

General William Henry Harrison, later president of the United States, was at that time the governor of Indiana Territory, where the Shawnee Prophet was successfully recruiting converts to his Indian religious revival. Seeking to undermine the credibility of the Shawnee Prophet as a shaman, Harrison urged Indians to demand proof from the Prophet that he could perform miracles. Thinking in biblical terms, Harrison asked if Tenskwatawa could "cause the Sun to stand still, the Moon to alter its course, the rivers to cease to flow, or the dead to rise from their graves."

The followers of the Shawnee Prophet did not need such displays, but Tenskwatawa was a canny politician. He proclaimed that on June 16, 1806, he would blot the Sun from the sky as a sign of his divine powers. Whether he knew of this total eclipse from a British agent or from an almanac is uncertain, but a great many Indians gathered at the Shawnee Prophet's camp as the appointed day dawned clear.

At the proper moment, the Prophet, in full ceremonial regalia, pointed his finger at the Sun, and the eclipse began. When the Prophet called out to the Good Father of the Universe to remove his hand from the face of the Sun, the light gradually returned to the Earth. Response to the Prophet's performance was overwhelming and his fame spread rapidly and widely. Harrison's condescension had backfired, to his embarrassment.

But the westward migration of European settlers was unstoppable. In the Battle of Tippecanoe in 1811, Harrison destroyed the Shawnee Prophet's religious center, killing many Indians, and breaking the power of Tenskwatawa.*

* See especially Laurence A. Marschall: "A Tale of Two Eclipses," *Sky & Telescope*, volume 57, February 1979, pages 116–118.

## NOTES AND REFERENCES

1.    Epigraph: John Milton: *Paradise Lost*, book 1, lines 594 and 597–599. See John Milton: *The Complete Poems* (New York: Crown Publishers, 1936), page 24.
2.    Herodotus: *The History*, volume 1, translated by George Rawlinson; Everyman's Library, volume 405 (London: J. M. Dent, 1910), book 1, chapter 74, pages 36–37.
3.    A similar battle-stopping eclipse occurred over southern Japan on November 11, 1183 CE. Two clans, the Minamoto and the Taira, were in the midst of a 5-year war. They were preparing for battle when an annular eclipse intervened. The Minamoto soldiers were scared out of their wits and fled. The Taira won the day. Unlike the war between the Medes and the Lydians, no treaty was signed. Hostilities continued. Two years later the Minamoto won the war. Shigeru Nakayama: *A History of Japanese Astronomy—Chinese Background and Western Impact* (Cambridge, Massachusetts: Harvard University Press, 1969), page 51. Thomas Crump also mentions this story in *Solar Eclipse* (London: Constable, 1999), page 192.
4.    Robert R. Newton lists three annular eclipses seen in the region during a 50-year period which he feels are equally likely to have given rise to this story, although an annular eclipse is not nearly as spectacular as one that is total. *Ancient Astronomical Observations and the Accelerations of the Earth and Moon* (Baltimore: Johns Hopkins Press, 1970), pages 94–97.
5.    Clemency Montelle: *Chasing Shadows: Mathematics, Astronomy, and the Early History of Eclipse Reckoning* (Baltimore: Johns Hopkins University Press, 2011), page 78.

6. Herodotus: *The History*, volume 2, translated by George Rawlinson; Everyman's Library, volume 406 (London: J. M. Dent, 1910), book 7, chapter 37, page 136.

7. Plutarch: *The Rise and Fall of Athens: Nine Greek Lives*, translated by Ian Scott-Kilvert (Baltimore: Penguin Books, 1960), pages 201–202. Ironically, Pericles' raid was disastrous for the Athenian forces; they fell victim to the plague. Pericles was fined and temporarily stripped of power.

8. Thucydides: *History of the Peloponnesian War*, translated by Richard Crawley (New York: E. P. Dutton, 1910), book 2, paragraph 28. Stars would not have been visible at Athens. Perhaps Thucydides heard reports from where the eclipse was annular and incorporated them into his account.

9. F. Richard Stephenson: *Historical Eclipses and Earth's Rotation* (Cambridge: Cambridge University Press, 1997), page 401. Thomas Crump: *Solar Eclipse* (London: Constable, 1999), page 185.

10. From the chapter "Second Soir" in Bernard Le Bovier de Fontenelle: *Entretiens sur la pluralité des mondes* (Paris: Chez la veuve C. Blageart, 1686), cited in François Arago: *Popular Astronomy*, volume 2, translated by W. H. Smyth and Robert Grant (London: Longman, Brown, Green, Longmans, and Roberts, 1858), pages 359–360.

11. Mabel Loomis Todd: *Total Eclipses of the Sun*, revised edition (Boston: Little, Brown, 1900), pages 141–142.

12. F. Richard Stephenson and David H. Clark: *Applications of Early Astronomical Records*, Monographs on Astronomical Subjects, number 4 (New York: Oxford University Press, 1978), page 9. Also F. Richard Stephenson: *Historical Eclipses and Earth's Rotation* (Cambridge: Cambridge University Press, 1997), page 397.

13. F. Richard Stephenson and David H. Clark: *Applications of Early Astronomical Records*, Monographs on Astronomical Subjects, number 4 (New York: Oxford University Press, 1978), page 14.

14. François Arago: *Popular Astronomy*, volume 2, translated by W. H. Smyth and Robert Grant (London: Longman, Brown, Green, Longmans, and Roberts, 1858), page 359.

15. François Arago: *Popular Astronomy*, volume 2, translated by W. H. Smyth and Robert Grant (London: Longman, Brown, Green, Longmans, and Roberts, 1858), page 362.

16. William J. S. Lockyer: "The Total Eclipse of the Sun, April 1911, as Observed at Vavau, Tonga Islands" in Bernard Lovell, editor: *Astronomy*, volume 2, *The Royal Institution Library of Science* (Barking, Essex: Elsevier Publishing, 1970), pages 190–191.

17. Alice Marriott and Carol K. Rachlin: *Plains Indian Mythology* (New York: Thomas Y. Crowell, 1975), pages 131–132. There is an inconsistency in this story: The Sun Dance is held in the summer and Marriott and Rachlin

introduce the story as occurring in the summer. The total solar eclipse occurred on November 30, 1834. The Kiowa also recorded another terrifying event that had occurred a year earlier—the night when the stars fell. That was the great Leonid meteor storm of November 13, 1833. Marriott and Rachlin recount that story too, pages 129–131.

## A MOMENT OF TOTALITY

### Fashionably Late to an Eclipse

The great astronomy popularizer Camille Flammarion, quoting astronomer François Arago, tells of the eclipse of May 22, 1724, visible from Paris. Supposedly, Jacques Cassini, director of the Paris Observatory, invited a marquis and his aristocratic lady friends to observe the eclipse with him at the observatory. However, the ladies fussed so long with their gowns and hair styles that the party arrived late, a few minutes after totality had ended. "Never mind, ladies," said the marquis, "we can go in just the same. M. Cassini is a great friend of mine and he will be delighted to repeat the eclipse for you."*

---

* Camille Flammarion: *The Flammarion Book of Astronomy*, edited by Gabrielle Camille Flammarion and André Danjon, translated by Annabel and Bernard Pagel (New York: Simon and Schuster, 1964), page 147.

# 6

◄○►

# The Sun at Work

*As long as the Sun shall shine,*
*As long as the rivers shall flow,*
*As long as the Moon shall rise*
*As long as the grass shall grow.*

Anonymous, North American Indian expression
for how long a promise should last[1]

## At the Core

At this moment, deep in the core of the Sun, the nucleus of a hydrogen atom—a proton—is colliding and fusing with another hydrogen nucleus, and the collisions and fusions proceed until four hydrogen nuclei have become the nucleus of one helium atom. In this nuclear reaction, a tiny amount of mass has been destroyed: not lost, but converted into energy. It is this reaction that powers the Sun and all stars, that creates their light and heat for more than 90% of their active lives.

In the Sun, tens of trillions of these reactions take place every *second*. Every second 600 million tons of hydrogen become 596 million tons of helium, and 4 million tons of mass become energy, in accordance with Einstein's famous equation $E = mc^2$. A little bit of mass yields a vast amount of energy.[2]

Even though the Sun is actually losing mass at the rate of 4 million tons every second, this weight-reduction plan is far from a crash diet. At 4 million tons a second, it would take the Sun 14.8 trillion years to consume itself entirely, if it could. But it can't. The heat to run this nuclear fusion comes from the gravitational force of all the mass of the Sun pressing inward on the core, and it is only within this central 25% of the Sun's diameter that the temperatures—up to 27 million degrees Fahrenheit (15 million degrees Celsius)—are hot enough to generate and sustain this reaction.[3]

Over billions of years, the hydrogen at the core is converted into helium until the core is too clogged with helium for hydrogen fusion to continue. It is then that stars begin to die. Our Sun has been shining for 4.6 billion years and has enough hydrogen at its core to continue shining much as it does now for about another 5 billion.

Cross-section of the Sun and its atmosphere. [Drawing by Josie Herr after William K. Hartmann: Cosmic Voyage Through Time and Space, © 1990 Wadsworth Publishing, Inc., by permission of the publisher]

The fiercely hot reaction at the heart of our Sun is concealed from our view by 432,500 miles (696,000 km) of opaque gases. And a good thing too. The principal radiation generated at the Sun's core is not visible light but gamma rays. There would be no life on Earth if the Sun radiated large quantities of this high-energy radiation in our direction.

Fortunately, it does not. High-energy photons radiate outward from the Sun's core but are absorbed in the crush of other atomic particles, which in turn reradiate this energy. But the energy emitted is not the same. With each absorption and emission, some of the photon's original energy is lost. A photon that started as a highly energetic gamma ray, if tracked from absorption to absorption, would gradually become an x-ray, then an ultraviolet ray, and then visible light as it bounced around randomly inside the Sun.

At about 130,000 miles (200,000 km) below the Sun's surface, the temperature and density have fallen enough so that energy is conveyed upward less by radiation than by convection—the rise of gases heated by this energy from below.

If energy created at the Sun's core could leave the Sun directly traveling at the speed of light, it would emerge from the Sun's surface in 2⅓ seconds. But because of the countless absorptions and reemissions in random directions, a typical photon requires 10 million years to reach the

Sun's surface. There, at last, it is free to move directly away from the Sun at the speed of light. At that pace, it travels the distance from the Sun to the Earth in 8⅓ minutes.

## Layers Above

The Sun's internal layers of radiation and convection are hidden from our eyes. What we see of the Sun, when we or the atmosphere provide adequate filters, is an apparent disk, glaring white-hot at a temperature of about 10,000 °F (5,500 °C). This disk is an optical illusion since the Sun is gaseous throughout. We are really seeing the layer of the Sun in which the density and ionization of atoms is so great that the gas becomes opaque. This region that provides the Sun with the appearance of a surface is

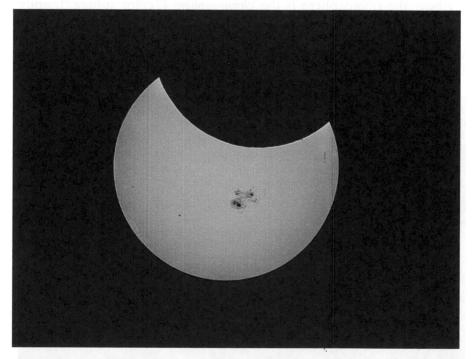

The photosphere of the Sun has sunspots that vary in number and size over an 11-year period. The biggest sunspot to grace the face of the Sun in more than two decades just happened to be visible during the partial solar eclipse of October 23, 2014. The sunspot was as big as the planet Jupiter. The enormous amount of energy stored in its twisted magnetic fields was responsible for producing several major solar flares. [Nikon D600, Borg 100ED refractor (100 mm, f/12.8, fl = 1280 mm), 1/2000 second, f/12.8, ISO 200, Thousand Oaks solar filter, © 2014 Dave Kodama]

called the *photosphere* ("light sphere"). It is only about 200 miles (300 km) deep. It is there that we notice sunspots, areas of magnetic disturbance on the Sun the size of Earth or Jupiter or even larger, appearing and disappearing and riding along in the photosphere with the Sun's rotation. The sunspots increase and decrease in number over a period of about 11 years.

If we examine the photosphere more closely, we see that it has a mottled appearance, created by rising columns of hot, bright gas surrounded by darker haloes where the gas has cooled and is descending, to be heated again. The photosphere is boiling. These *granulations* of upwelling and downfalling gases are typically 500 miles (800 km) in diameter and last for 5 or 10 minutes. The gases in the granulations, the topmost layer of the convection zone, carry with them magnetic fields from deep within the Sun. As these gases tumble up and down, the magnetic lines of force twist and snap, and this magnetic turbulence governs the behavior of gases in the photosphere and in the solar atmosphere above it.

Above the photosphere is the *chromosphere* ("color sphere"), aptly named for its vibrant reddish color. Seen with a telescope at the rim of the Sun, it looks like a fire-ocean.[4] Its lower levels are cooler than the white-hot photosphere, with temperatures of about 7,200 °F (4,000 °C). Its upper layers, however, are much hotter than the photosphere—about

Earth to Scale

On August 31, 2012 an enormous solar prominence erupted out into space. The event triggered a coronal mass ejection, or CME, which traveled at over 900 miles per second (1400 km per second). Several days later the plasma reached Earth's magnetosphere, causing the aurora to appear in the night sky. [NASA/SDO/AIA/GSFC]

18,000 °F (10,000 °C). This temperature causes hydrogen atoms to emit a red wavelength that is primarily responsible for the chromosphere's fiery color.[5] The chromosphere is a thin atmospheric layer, only about 1,600 miles (2,500 km) thick, although there are no sharp boundaries above or below.

Seen under high magnification along the rim of the Sun, the chromosphere is not a smooth layer of gas. It looks like an erratic forest of thin, topless trees. These spiked features that compose the chromosphere are known as *spicules*. They are less than 400 miles (700 km) in diameter but may tower thousands of miles high, reaching out of the chromosphere and into the corona. These spicules may be not only columns of rising gas but also the outlines of magnetic flux tubes that transfer magnetic fields from the photosphere to the corona.

It is in the photosphere and chromosphere that prominences and flares are rooted and stretch upward into the corona. They too are transient features of the Sun that vary with the rhythm of the sunspots below them. *Prominences* are the same temperature as the upper chromosphere and glow with the same red color of excited hydrogen. They are condensed clouds of solar gas, but much cooler and denser than the surrounding corona. The prominences are bent and twisted by local magnetic fields. Magnetic forces keep the gases in the prominences from all falling back to the surface, which would take only about 15 minutes if gravity were the only force acting. Most often these clouds, the prominences, do slowly rain material back toward the surface, but occasionally they erupt outward.

*Flares* are much stronger eruptions, also triggered by the magnetic activity of the Sun, that launch great torrents of mass and energy from the Sun at millions of miles an hour.

The chromosphere is also the birthplace of *solar tornados*, almost the size of Earth, whirling at speeds up to 300,000 miles per hour (500,000 km per hour), spiraling up and through the corona, probably contributing fast-moving atomic fragments to the solar wind of particles flowing spaceward from the Sun.[6] How these solar tornados form and function is not certain, but, as with virtually every feature seen on the Sun's surface and in its atmosphere, the explanation almost certainly involves twisted magnetic fields.

## Upward and Outward

In the upper reaches of the chromosphere and extending outward into the corona is a realm called the *transition region*, so named because the temperature there suddenly climbs to about 2 million °F (1 million °C).

Why should the temperature of the *corona* ("crown"), the outer atmosphere of the Sun, be so high? It is much hotter than the photosphere and chromosphere, yet those layers are closer to the core where the Sun generates its energy. The answer again lies in the Sun's magnetic fields. The Sun not only behaves like a giant bar magnet (like the Earth with its magnetic poles), but the Sun is also pocked by many local magnetic regions with intensities greater than the Sun's polar magnetism. These surface sites of magnetic activity, the largest marked by sunspots, are induced by varying magnetic fields in the Sun's interior. These magnetic fields are borne to the surface by the columns of hot gases seen as granulations in the photosphere. The random motion of the rising and falling gases contort the lines of magnetic force. These magnetic fields stretch into the chromosphere and corona as loops and arches, as if they were invisible

Solar flares are violent explosions in the Sun's atmosphere that release huge amounts of energy and subatomic particles. This close-up image from NASA's Solar Dynamics Observatory (SDO) shows the eruption of a flare on July 6, 2012. The flare caused a radio blackout as well as a solar energetic particle event in which fast particles traveling behind the flare impacted Earth's magnetosphere. [NASA/SDO/AIA]

electric wires attached to terminals in the photosphere. The twisted and coiled magnetic fields induce electrical currents into the corona, which heat the gases there. But explaining exactly how this heating is accomplished in the near vacuum of the corona remains a challenge to solar astronomers.[7]

The gases rising from the surface of the Sun are all so hot that their atoms are missing some of or all their electrons. This plasma (ionized gas) expands as it rises from the surface of the Sun. As it expands, its density declines and these charged gases are more easily warped by the Sun's magnetic fields. The patterns of this magnetic control can be seen in the loops and arches of the prominences that extend high into the corona. Here and there a magnetic loop in the corona is stressed so greatly that it snaps and perhaps 10 billion tons of million-degree plasma is slung into space as a huge expanding bubble traveling as fast as 4.5 million miles per hour (2,000 km per second). Such outpourings are called *coronal mass ejections*.[8]

Coronal mass ejections, eruptive prominences, and flares are probably different parts of the same phenomenon. They involve the expulsion of hot gases and twisted ropes of magnetic fields from the Sun, they often occur together, and all are more frequent at sunspot maximum.[9]

Coronal mass ejections enhance and create shock waves in the normal solar wind with charged particles (primarily protons and electrons) that are escaping the Sun's gravitational hold at speeds of 900,000 miles per hour (400 km per second) or more. Coronal mass ejections also carry with them some of the coronal magnetic field, and distort the existing field in the solar wind. When the altered magnetic fields and these especially energetic subatomic particles strike the Earth, they create colorful displays of the northern and southern lights—the aurora.

But solar windstorms can also damage electronic equipment on satellites in Earth orbit; disrupt telephone, radio, and television transmission; and overload electric power lines, causing blackouts (as one did for 9 hours to 6 million Canadians in Quebec on March 13, 1989).

Scientists now think that coronal mass ejections, rather than flares, are the principal cause of the aurora and other geomagnetic events.

In visible light the corona shows graceful, delicate streamers and brush-like features.[10] However, in x-ray pictures, which better capture the activity of high-temperature gases, the corona is a riot of everchanging loops, plumes, eruptions, and contrasting light and dark regions. These dark regions of lesser activity are called *coronal holes*. It is primarily through these coronal holes, where magnetic fields are weaker, that the solar wind escapes into space. Slower and more variable solar wind flows from coronal streamers.

The coronal holes and streamers seen during a total eclipse mark the departure of solar wind from the Sun. The Sun loses about 6 million tons of material a second in the solar wind.[11] But such a loss is insignificant compared to the total mass of the Sun.

Coronal loops are fountains of multimillion-degree electrified gas in the atmosphere of the Sun that are 300 times hotter than the Sun's visible surface. The shape and structure of the coronal loops map out the magnetic field in the lower corona. [NASA Transition Region and Coronal Explorer (TRACE) spacecraft]

The solar wind blows outward in all directions. The Earth is a small target 93 million miles (150 million km) away, so only two out of every billion particles in the solar wind will reach the Earth, to cause mischief here. But they are enough. Scientists are studying "space weather" with the hope of predicting when the Earth will be struck by a stormy blast of enhanced solar wind.

To the eye, the white flame brushes of the corona can extend outward from the Sun 2 million miles (3 million km) or more until they are so tenuous that they are no longer visible. But the corona is still measurable out to the Earth and beyond in the form of the solar wind. The Earth orbits the Sun within the Sun's rarefied outer atmosphere.

## The Gift of Totality

We do not normally see the chromosphere or the corona. They are concealed from our view by the overwhelming glare of the photosphere, half a million times brighter than the corona. To study the Sun, early scientists had only its surface to rely on—just the photosphere and sun-spots. Progress was slow.

We do not see the chromosphere and the corona, the prominences, and the coronal mass ejections unless something blocks the glare of the photosphere so that the faint atmosphere of the Sun is revealed. In the 19th century, astronomers discovered that the Moon was their scientific collaborator, obscuring the Sun's surface from time to time so that they might see a part of the Sun that had never been studied before.

In less than a century, total solar eclipses, and the scientific instruments and theories they helped to stimulate, revealed the composition of the Sun, the structure of the Sun's interior, and the wonder of how the Sun shines.

A coronal mass ejection blasts immense bubbles of hot plasma into the solar system. This January 8, 2002 image shows a widely spreading coronal mass ejection (right half of picture) hurling more than a billion tons of electrons and protons into space at millions of kilometers per hour. [ESA/NASA SOHO LASCO]

# NOTES AND REFERENCES

1. Epigraph listed as Anonymous: North American Indian in *Bartlett's Familiar Quotations*.

2. Solar physicists Charles Lindsey, Joseph Hollweg, and Jay Pasachoff calculated or reviewed the statistics and information in this chapter. Personal communication, October 20–November 3, 1998. Lindsey calculates that 4.26 million metric tons per second (4.69 million English tons per second) of mass are converted to energy in the core of the Sun.

3. The figure of 14.8 trillion years is based on the Sun converting *all* its mass to energy, which, of course, it cannot do. Only 0.7% of the hydrogen mass is converted into energy; the rest becomes helium. If only the total amount of hydrogen available is considered, the Sun's life span falls to about 77 billion years. Of course, only the hydrogen close to the center of the Sun is under sufficient pressure so that the temperatures are great enough for fusion to occur. Only about 10% of the Sun's hydrogen will undergo fusion, reducing the Sun's lifetime to 8–10 billion years.

   To generate and sustain a hydrogen-to-helium fusion reaction, the core of a star must have a temperature of at least 18 million °F (10 million °C). To have enough gravity to generate sufficient pressure to obtain this high a core temperature, a star must have at least 8% the mass of the Sun.

4. The expressions "fire-ocean" to describe the chromosphere and "flame-brushes" to describe features in the corona were used by Agnes M. Clerke: *A Popular History of Astronomy during the Nineteenth Century*, 4th edition (London: A. & C. Black, 1902), pages 68, 175.

5. Spectroscopically, the red of the chromosphere is produced by the hydrogen-alpha line.

6. Solar tornados were discovered in 1998 by David Pike and Helen Mason using the Solar and Heliospheric Observatory (SOHO), a collaboration of the European Space Agency (ESA) and NASA (United States).

7. The temperature of the corona can be misleading. The Sun's magnetic fields cause the electrically charged atoms of the corona to move at great speeds (high temperature), but the density of these ions is so low that the corona has relatively little heat (energy in a given volume). The corona is so rarefied that if you had a box there 100 miles (160 kilometers) on each side (1 million cubic miles; 4.1 million cubic kilometers), you would entrap less than a pound (0.4 kilogram) of matter. The corona is a good vacuum by laboratory standards on Earth.

   Solar physicist Joe Hollweg points out that it is also hard to explain the heating of the chromosphere, which requires as much energy as the corona. Personal communication, August 3, 1998.

8. Coronal mass ejections were discovered using NASA's Orbiting Solar Observatory 7 between 1971 and 1973 and confirmed using NASA's Solar

Maximum Mission satellite in 1980 and after 1984, following repair in orbit by Space Shuttle astronauts. In the 1990s, a new generation of solar spacecraft studied corona mass ejections: Yohkoh (Japan), SOHO (*Solar* and *H*eliospheric *O*bservatory—ESA and NASA), and TRACE (*T*ransition *R*egion *a*nd *C*oronal *E*xplorer—NASA). In the 2000s, the next generation of solar observatories continues the quest: Hinode (Japan) and STEREO (*Solar-T*errestrial *R*elations *O*bservatory—NASA).

9.   At sunspot minimum, about one coronal mass ejection a week is observed. Near sunspot maximum, two or three coronal mass ejections are observed each day on the average.

10.  Coronal structures trace out the magnetic field lines, like iron filings trace out the field around a bar magnetic.

11.  Peter D. Noerdlinger: "Solar Mass Loss, the Astronomical Unit, and the Scale of the Solar System," arXiv0801.3807 [astro-ph], January 24, 2008.

# A MOMENT OF TOTALITY

## The Audio of the Video

Gary Ropski and Barbara Schleck make a video recording of each eclipse, complete with sounds of people's reactions as they watch the approach of totality and the sudden revelation of the Sun's corona as totality begins. They were playing an eclipse video for friends one evening, when one said, "Play it again, but this time close your eyes and just listen." So they did. The "oohs" and "aahs" built to a crescendo at the start of totality. "Eclipse?" the friend said. "Sounds to me like an orgy."*

* Gary Ropski and Barbara Schleck live in Chicago. He is a lawyer specializing in intellectual property. She is a journalist. Interviewed May 22, 2015.

# A MOMENT OF TOTALITY

## The Audio of the Video

# 7

—◄○►—

# The First Eclipse Chasers

> *I did not expect, from any of the accounts of preceding eclipses that I had read, to witness so magnificent an exhibition as that which took place.*
>
> Francis Baily (1842)[1]

## An Unlikely Beginning

Francis Baily, the man who might be said to have founded the field of solar physics, received only an elementary education, was not trained in science, and did not get around to astronomy until the age of 37. Like his father, a banker, he entered the commercial world as an apprentice when he was 14. But adventure called. When his seven years of apprenticeship expired, he sailed for the New World and spent the next two years, 1796–1797, exploring unsettled parts of North America, narrowly escaping from a shipwreck, flatboating down the Ohio and Mississippi Rivers from Pittsburgh to New Orleans, and then hiking nearly 2,000 miles back to New York through territory inhabited mostly by Indians. He liked the United States so much that he planned to marry and become a citizen, but he finally abandoned those plans and returned home in 1798.

Back in England, he began efforts to mount an expedition to explore the Niger River in Africa. He could not raise enough money, however, so he became a stockbroker. To dedication and enthusiasm he quickly added a reputation for intelligence and integrity, and he made a fortune. He exposed stock exchange fraud and helped clean it up. He published a succession of explanations of life insurance methods and comparisons of insurance companies which became wildly popular. He also published a chart of world history that was equally popular, confirming the nickname given to him in his apprentice days: the Philosopher of Newbury (his birthplace).

Francis Baily. [Royal Astronomical Society]

His first astronomical paper (1811) tried to identify the solar eclipse allegedly predicted by Thales. In 1818, he called attention to an annular eclipse of the Sun coming in 1820, and he observed it from southeastern England. That same year, he became one of the founders of the Astronomical Society of London, later the Royal Astronomical Society.

In 1825, he retired from the stock market to devote all his time to his new profession. He was 51 years old. His revisions of a series of old star catalogs were considered so valuable that the Royal Astronomical Society twice awarded him its Gold Medal and four times elected him president. Although he was not renowned as an observer, he had an abiding fascination with eclipses, a good eye for detail, and the ability to express what he saw.

Thus it was in 1836 that a few words from Francis Baily sparked the immediate, intense, and unending study of the physical properties of the Sun that had been generally ignored or discounted until then. He traveled to an annular eclipse of the Sun in southern Scotland and watched on May 15, 1836 as mountains at the Moon's limb occulted the face of the Sun but allowed sunlight to pour through the valleys between them so that the ring of sunlight around the rim of the Moon was broken up into "a row of lucid points, like a string of bright beads."[2] With those words, Baily generated fervor for solar physics and founded the industry of eclipse chasing.

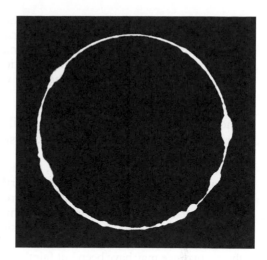

Baily's beads during the annular eclipse of April 28, 1930. [Lick Observatory]

## The Surprise of Totality

At the next accessible eclipse, July 8, 1842, a high percentage of the astronomers of Europe migrated to southern France and northern Italy to see "Baily's beads." Baily, now 68 years old, went too.

This was not an annular eclipse, as Baily had seen twice before. It was total. No European astronomer then alive had ever seen a total eclipse.

Baily set up his telescope at an open window in a building at the university in Pavia, Italy. Again Baily's beads were visible—to Baily at least. George Airy, England's astronomer royal, observing from Turin, Italy, did not see them. Baily was just jotting down the time of appearance and duration of the beads

> . . . when I was astounded by a tremendous burst of applause from the streets below, and at the *same moment* was electrified at the sight of one of the most brilliant and splendid phenomena that can well be imagined. For, at that instant the dark body of the moon was *suddenly* surrounded with a *corona*, or kind of bright *glory*, similar in shape and relative magnitude to that which painters draw round the heads of saints, and which by the French is designated an *auréole*.[3]

Baily was not the first to use the word *corona* to designate the glowing outer atmosphere of the Sun visible during a total eclipse, but his striking description of it caught everyone's attention as it never had before and forever united the word with the phenomenon.

> When the total obscuration took place, which was *instantaneous*, there was an universal shout from every observer . . . I had indeed anticipated the

appearance of a luminous circle round the moon during the time of total obscurity: but I did not expect, from any of the accounts of preceding eclipses that I had read, to witness so magnificent an exhibition as that which took place. . . . It riveted my attention so effectually that I quite lost sight of the string of *beads*, which however were not completely closed when this phenomenon first appeared.

There was so much to see, said Baily, that in future eclipses, each observer should be assigned a single observing task.

Splendid and astonishing, however, as this remarkable phenomenon really was, and although it could not fail to call forth the admiration and applause of every beholder, yet I must confess that there was at the same time something in its singular and wonderful appearance that was appalling: and I can readily imagine that uncivilised nations may occasionally have become alarmed and terrified at such an object, more especially in times when the true cause of the occurrence may have been but faintly understood, and the phenomenon itself wholly unexpected.

It was the last eclipse Francis Baily was to see. Two years later he died. Of him, historian Agnes M. Clerke wrote: "He was gentle as well as just; he loved and sought truth; he inspired in an equal degree respect and affection. . . . Few men have left behind them so enviable a reputation."[4]

Drawings of corona and prominences at different eclipses. Left: July 8, 1842; right: July 28, 1851. [François Arago: *Popular Astronomy*, edited and translated by W. H. Smyth and Robert Grant]

## Before Baily

Never again would a total eclipse over an inhabited land mass go unattended by professional and amateur astronomers, even when the observers had to travel to remote sites at the farthest ends of the world. The corona, Baily's beads, prominences, shadow bands—all had been seen before, many times, throughout the world. But now they commanded attention and explanation.

The first written record of the corona may be Chinese characters inscribed on oracle bones from about 1307 BCE that say "three flames ate up the Sun, and a great star was visible."[5] Or were these flames prominences instead?

The first unequivocal description of the corona comes from a chronicler named Leo more than 2,000 years later, observing from Constantinople the eclipse of December 22, 968 CE:

> Everyone could see the disc of the Sun without brightness, deprived of light, and a certain dull and feeble glow, like a narrow headband, shining around the extreme parts of the edge of the disc.[6]

The great astronomer Johannes Kepler did not see a total eclipse himself, but from the reports he read, he concluded that the corona must be material around the Sun and not the Moon. Giacomo Filippo Maraldi, an Italian-born French astronomer, provided evidence that the corona is part of the Sun because the Moon traverses the corona during a solar eclipse; the corona does not move with the Moon but stays fixed around the Sun.

The Spanish astronomer José Joaquín de Ferrer, well ahead of his time, traveled to the New World to observe total eclipses in Cuba in 1803 and Kinderhook, New York in 1806. He was probably the first to use the word "corona" to describe the glow of the outer atmosphere of the Sun seen during a total eclipse.

> The disk [of the Moon] had round it a ring of illuminated atmosphere, which was of a pearl colour . . . From the extremity of the ring, many luminous rays were projected to more than 3 degrees distance. —The lunar disk was ill defined, very dark, forming a contrast with the luminous corona . . .[7]

Ferrer quite correctly attributed the corona to the Sun. If this glow belonged to the Moon, he calculated, the lunar atmosphere would extend upward 348 miles (560 km)—50 times more extensive than the atmosphere of Earth. Thus, he concluded, the corona "must without any doubt belong to the Sun." Baily, too, at the 1842 eclipse, attributed the corona to the Sun.

The first sure report of solar prominences came from Julius Firmicus Maternus in Sicily, who noticed them during the annular eclipse of July 17, 334 CE.[8] Edmond Halley, the great English astronomer, saw them clearly as bright red protrusions during the total eclipse of May 3, 1715.

Baily had pointed the way. But what was the significance of the corona and the prominences? New techniques—photography and spectroscopy—were emerging with the capability to record and explore these features of the Sun. In photography and spectroscopy, observational astronomy gained its two most powerful tools to augment the telescope.

The stage was set for the eclipse of July 28, 1851. The astronomical world gathered along its path of totality through Scandinavia and Russia. They solved one mystery and uncovered another.

## The Debut of Photography

The first successful photograph of the Sun in total eclipse was a daguerreotype taken on July 28, 1851 by a professional photographer named Berkowski, assigned to the task by August Ludwig Busch, director of the observatory in Königsberg, Prussia.[9] The inner corona and prominences are clearly visible. Yet for now, photography was just a curiosity, a promising experiment. Hardcore science was still carried out by visual observations made through telescopes.

At the 1851 eclipse, two teams of astronomers provided proof of what most astronomers suspected: that prominences were part of the Sun. Robert Grant and William Swan from the United Kingdom and

First photograph of the Sun in total eclipse, July 28, 1851. Only the last name of the photographer is known: Berkowski. [Courtesy of Dorrit Hoffleit]

Karl Ludwig von Littrow from Austria documented how the eastward motion of the Moon across the face of the Sun covered prominences along the east rim of the Sun as totality began and then uncovered prominences along the west rim as totality ended, demonstrating that the prominences belonged to the Sun and that the Moon was only passing in front of them.

Often in science, observations that lead to the solution of one problem simultaneously reveal a fascinating new problem. As George B. Airy, England's astronomer royal, was observing this eclipse, he noticed a jagged edge to the solar atmosphere just above the edge of the Moon. He called it the *sierra*, thinking that he might be looking at mountains on the Sun.

Airy thus became the first to call attention to the chromosphere, the lowest level of the transparent solar atmosphere immediately above the opaque photosphere that creates the appearance of a surface for the Sun. The jaggedness that Airy observed was actually innumerable small jets of rising gas called spicules.

Drawing by Lilian Martin-Leake from a telescopic view of the chromosphere and the corona, May 28, 1900, showing red spicules in the chromosphere that George Airy had thought were mountains. [Annie S. D. Maunder and E. Walter Maunder: *The Heavens and Their Story*]

By 1860 in the field of photography, daguerreotypes were obsolete, superseded by the faster wet-plate (collodion) photographic process. Before photography, astronomers could only describe or sketch what they saw. The early photographic emulsions were not as sensitive to detail as desired, but they were more objective than the human eye.

On July 18, 1860, a total eclipse was visible from Europe and found astronomers waiting with improved cameras. Prominences remained a matter of high priority and they encountered the resourcefulness of British astronomer Warren De La Rue and Italian astronomer and Jesuit priest Angelo Secchi.

De La Rue's well-to-do family provided him with a solid education, after which he entered his father's printing business. There he demonstrated a great affinity for machinery. He could make any instrument or device run better. He was also a fine draftsman, and it was this talent that lured him into astronomy. He could produce drawings of the planets, Moon, and Sun that were better than those of astronomers. But no sooner had he produced his first excellent drawings than he found out about a new invention called photography. He never saw a mechanical device that did not fascinate him, so he was soon making improvements in cameras, inventing specialized cameras for solar photography, and photographing the Moon and Sun stereographically so that lunar features appeared in relief and sunspots revealed themselves as depressions, not mountains, in the photosphere.[10]

Angelo Secchi. [Mabel Loomis Todd: *Total Eclipses of the Sun*]

Observing the 1860 eclipse 250 miles (400 km) away from De La Rue was Angelo Secchi. He too had a remarkable aptitude for inventing instruments and coaxing results from them. Secchi was from a poor family and received his education through the Catholic Church in the demanding Jesuit tradition. From the beginning, he showed brilliance in mathematics and astronomy. When the Jesuits were expelled from Italy by a liberal, anticlerical government in 1848, Secchi spent a year in the United States as assistant to the director of the Georgetown University observatory. When the ban against Jesuits was lifted in 1849, he returned to Italy to become director of the Pontifical (now Vatican) Observatory of the Collegio Romano (the Gregorian University). He transformed it into a modern, well-equipped center for research in the new field of astrophysics. Secchi was one of the pioneers in applying spectroscopy to astronomy. He surveyed more than 4,000 stars and realized that stellar spectra could all be grouped into a handful of classifications.

De La Rue and Secchi ambushed the 1860 eclipse with improved cameras using the wet-plate process, which greatly reduced exposure time and thereby increased the clarity with which objects in motion could be seen. They captured the prominences and compared them. The prominences looked the same from the photographers' widely separated sites, so Secchi and De La Rue could conclude that they were indeed part of the Sun. If they had been features on the Moon, so much closer than the Sun, the difference in viewing angles (parallax) from Secchi's and De La Rue's separate sites would have given them a different appearance.

## The Debut of Spectroscopy

On August 18, 1868, a great eclipse touched down near the Red Sea and swept across India and Malaysia. Once again, the international scientific community had assembled.

From the United Kingdom to the path of the eclipse had come, among others, James Francis Tennant and John Herschel (son of John F. W. Herschel, grandson of William Herschel, both renowned astronomers). Norman Pogson, born in England, represented India and the observatory he directed there. The French delegation included Georges Rayet and Jules Janssen. Each of them carried a new weapon just added to the scientific arsenal for prying secrets from the Sun during an eclipse. That tool was a spectroscope. It would prove as indispensable to eclipse studies of the structure of the Sun as it was rapidly demonstrating itself to be in other realms of astronomy and in all other physical and biological sciences. By passing the light of the corona or the prominences through a prism, it could be broken down into a spectrum of lines and colors. From this spectrum, a scientist could identify the chemical elements present and even their temperature and density.

As the eclipse sped along its course, spectroscopes pointed upward toward the prominences. They showed that the prominences emitted bright lines, and most of these were quickly identified with hydrogen. More and more it seemed that the Sun must be composed primarily of gas and that hydrogen must be a major constituent.[11] The spectroscopists, properly pleased with their results, packed their equipment and headed home.

All but one. His name was Jules Janssen and the brightness of the prominences and the strength of their spectral lines had given him an idea. He wanted to look for them again when the eclipse was over, when the Moon was not blocking the intense glare of the Sun from view. Might it be possible for him to see the prominences and their spectrum in broad daylight?

The weather was cloudy the rest of that day. He would have to wait until tomorrow.

## That Extra Step

Pierre Jules César Janssen was 12 years old when Baily called attention to the beads visible during the annular eclipse of 1836. A childhood accident had left Janssen lame and he never attended elementary or high school. His family was cultured, but his father was a struggling musician, so Jules had to go to work at a young age.

While employed at a bank, he earned his college degree in 1849. He then went on to gain a certificate as a science teacher and served as a substitute teacher at a high school. In 1857, he traveled to Peru as part of a government team to determine the position of the magnetic equator. There he became severely ill with dysentery and was sent home. At age 33, he seemed destined for a quiet life in teaching, if he could get a job. He became the tutor for a wealthy family in central France. At their steel mills, he noticed that the eye could watch molten metal without fatigue or injury, while the skin had to be protected from the heat. He wrote a careful study of how the eye protects itself against heat radiation, which earned him his doctorate in 1860.

The glow of molten metal had led him to spectroscopic analysis, which he then applied to the Sun and, in 1859, he identified several Earth elements present in the Sun. He moved to Paris in 1862 to dedicate himself to solar physics and scientific instrument-making. His work had already made him a leader in solar spectroscopy. He used the changing spectrum of the Sun in its daily journey from horizon to horizon to separate spectral

## Jules Janssen (1824–1907)

Two years after his breakthrough at the 1868 eclipse, Jules Janssen again planned to apply spectral analysis to a solar eclipse, this one on December 22, 1870 in Algeria. When it came time for departure, however, the Franco-Prussian War was in progress and Paris was under siege. Colleagues in Britain had obtained from the Prussian prime minister safe passage for Janssen from Paris, but Janssen wouldn't accept favors from his country's enemy. He had a different plan: "France should not abdicate and renounce taking part in the observation of this important phenomenon. . . . An observer would be able, at an opportune moment, to head toward Algeria by the aerial route . . ."*

Although he had never been in a balloon before, on December 2, with a sailor as an assistant and himself as pilot, he ascended from Paris and headed west. Despite violent winds, he landed safely near the Atlantic coast. He reached Algeria in time—only to have the eclipse clouded out. From that experience, however, Janssen designed an aeronautical compass and ground speed indicator and prophesied methods of air travel that would "take continents, seas, and oceans in their stride." In 1898 and subsequently, he used balloons to study meteor showers from above the clouds, pioneering high altitude astronomy and foreseeing the advantages of space observations.

Janssen was also a leader in astrophotography. "The photographic plate is the retina of the scientist," he wrote. It was for spectroscopy, however, that Janssen was most renowned. His methods opened up the Sun's atmosphere to continuous study. The French government tried to find him an observatory position, but the director of the Paris Observatory did not want him. So Janssen was allowed to pick a site near Paris for a new observatory dedicated to astrophysics. He chose Meudon and directed the observatory from its founding in 1876 until his death in 1907.

"There are very few difficulties that cannot be surmounted by a firm will and a sufficiently thorough preparation," he wrote. But he was too modest in his self-assessment. To everything he investigated, he brought imagination and insight. Jules Janssen was one of the most creative scientists of any era.

* Quotations are from the sketch of Janssen by Jacques R. Lévy in the *Dictionary of Scientific Biography*.

lines caused by the Earth's atmosphere from those originating in the Sun. From those spectral lines, he demonstrated the composition and density of the Earth's atmosphere through which the sunlight passed. Janssen then applied spectroscopy to the other planets and, in 1867, discovered water in the atmosphere of Mars.

He traveled to Guntur, India for the total eclipse of August 18, 1868 to use his spectroscope on solar prominences. The great contemporary English spectroscopist Norman Lockyer spoke ringingly of that pivotal moment: "Janssen—a spectroscopist second to none . . . was so struck with the brightness of the prominences rendered visible by the eclipse that, as the sun lit up the scene, and the prominences disappeared, he exclaimed, '*Je reverrai ces lignes la!*'"[12] [I will see those lines again!] The next morning he succeeded. He had found a way to study the atmosphere of the Sun without waiting for a total eclipse, traveling halfway around the world to see it, and hoping for good weather at the critical moment.

For two weeks Janssen continued to map solar prominences by this technique and continued to perfect it on his circuitous way home, with a stop in the Himalayas to observe at high altitude. He proved that prominences change considerably from one day to the next.

A standard spectroscope breaks down the light of a glowing object into the characteristic colors of its spectrum. Janssen modified the spectroscope by blocking unwanted colors so that the observer could view an object in the light of one spectral line at a time. He had invented the spectrohelioscope. The Sun could now be analyzed in detail on a daily basis.

A month after the eclipse, on his way home to France, Janssen wrote up his findings and sent them to the Academy of Sciences in Paris. His paper arrived a few minutes after one from England that reported precisely the same discovery.

Jules Janssen. [Mary Lea Shane Archives of the Lick Observatory]

# Coincidence

Joseph Norman Lockyer came from a well-to-do family with scientific interests. He received a classical education, traveled in Europe, and then entered civil service. So wide were his interests that he wrote on everything from the construction dates and astronomical purposes of Egyptian pyramids and temples to Tennyson to the rules of golf. "The more one has to do, the more one does," was his motto.[13]

When Gustav Kirchhoff and Robert Bunsen showed in 1859 how spectroscopy could be used to determine the chemical composition of objects in space, Lockyer saw the discovery as a key to what had seemed the locked door of the universe. Had not French philosopher Auguste Comte confidently asserted only 24 years earlier that never, by any means, would we be able to study the chemical composition of celestial bodies, and every notion of the true mean temperature of the stars would always be concealed from us.[14] It was clear that Comte was wrong. Lockyer bought a spectroscope, attached it to his 6¼-inch (16-cm) refracting telescope, and began his observations.

Although he had never seen a solar eclipse, it occurred to him that, since prominences were probably clouds of hot gas, he should be able to use a spectroscope to analyze prominences without waiting for an eclipse. This idea struck Lockyer two years before the 1868 eclipse that inspired Janssen to the same realization. Lockyer tried the experiment in 1867 but found his spectroscope inadequate to the task. So he ordered a new spectroscope to his specifications. Because of construction delays, however, it did not arrive until October 16, 1868, two months after the eclipse that Janssen saw in India. Lockyer rapidly and excitedly calibrated

Medallion created by the Academy of Sciences in Paris to honor Janssen and Lockyer for their independent discovery of how to observe prominences without waiting for an eclipse. The front of the medal shows the heads of the two scientists. The reverse shows the Sun god Apollo pointing to prominences on the Sun.

his new instrument and on October 20, 1868, he trained it on the rim of the Sun and recorded bright lines typical of hot gases under very little pressure. He wrote up his findings and sent them to the Academy of Sciences in Paris for presentation by his friend Warren De La Rue.

Just minutes before De La Rue was to speak, Janssen's letter arrived and both papers were read at the same session of the Academy to great acclaim for both scientists. A special medal was struck to honor them. It showed the heads of Janssen and Lockyer side by side.[15]

## A New Element

Lockyer continued to examine the spectrum of the gases at the rim of the Sun. He recognized that the lower atmosphere of the Sun, what Airy had called the sierra, was decidedly reddish in color, so he named it the *chromosphere*, and it has been known by that name ever since.

### J. Norman Lockyer (1836–1920)

In 1869, the year after he independently showed how the atmosphere of the Sun could be analyzed without the benefit of an eclipse and discovered the element helium, Norman Lockyer founded the scientific journal *Nature*. He edited it for 50 years, until just before his death, keeping it alive through many crises.

While the French government was establishing a special observatory for Janssen, the British government likewise recognized the importance of Lockyer's contribution and set about creating a solar physics observatory for him. For its opening in 1875, Lockyer collected old and modern instruments and placed them on display. The display became permanent and grew. Lockyer had founded London's renowned Science Museum.

Lockyer was not shy in interpreting his findings to form startling theories. Often he was wrong, but always he provided useful data, and often there was a nucleus of truth in his grand speculations. He thought that all atoms shared certain spectral lines and were therefore made of smaller common constituents. He was wrong about the spectra but right about the composition of atoms.

He offered dates for the construction of ancient Egyptian temples based on their alignments with the rising and setting positions of the Sun and certain stars. His dates were wrong, but he was right that many of the temples had astronomical orientations and his work helped to establish the field of archeoastronomy.

When he died, a colleague wrote of him: "Lockyer's mind had the restless character of those to whom every difficulty is a fresh inspiration. His enthusiasm never failed him, despite repeated disappointments and opposition."*

---

* Alfred Fowler: "Sir Norman Lockyer, K.C.B., 1836–1920," *Proceedings of the Royal Society of London*, series A, volume 104, 1923, pages xiv.

J. Norman Lockyer in 1895, the year helium, the element he discovered on the Sun, was finally found to exist on Earth as well.

Lockyer was not done yet. In examining the spectrum of the prominences, he noticed a yellow line that he could not identify. It did not seem to belong to any element known on Earth. So he announced the existence of a new element and proposed the name *helium* for it, because it had been found in the Sun—*helios* in Greek. Most scientists rejected the idea of a new element, suggesting that this line was produced by a known element under unusual physical conditions. But Lockyer clung tenaciously to his interpretation.

Finally, in 1895, William Ramsay found trapped in radioactive rocks on Earth an unknown gas that exhibited the mysterious spectral line that Lockyer had discovered on the Sun. Helium was an element. Lockyer had been right.

In 1869, just after he discovered helium, Lockyer had urged: "Let us . . . go on quietly deciphering one by one the letters of this strange hieroglyphic language which the spectroscope has revealed to us—a language written in fire on that grand orb which to us earth-dwellers is the fountain of light and heat, and even of life itself."[16]

The total eclipse of 1868 had raised the strong possibility of the existence of a new element—and, for the first time, one discovered not on Earth but in the heavens. One year later, on August 7, 1869, the United States

# Some Past Eclipses of Historic Interest

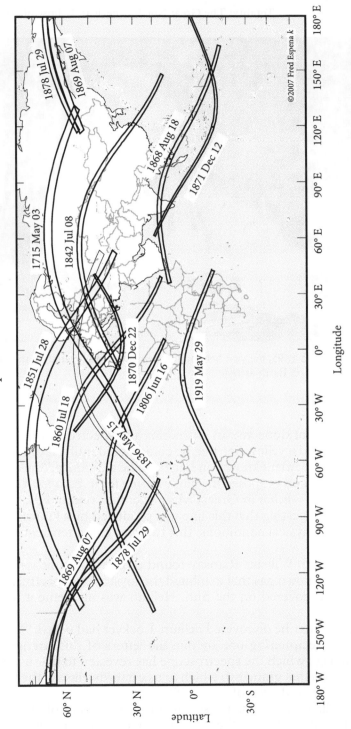

Paths of totality for the solar eclipses of 1715, 1806, 1836, 1842, 1851, 1860, 1868, 1869, 1870, 1871, 1878, and 1919. [Map and eclipse calculations by Fred Espenak]

lay on the path of a total eclipse. Two American astronomers, Charles A. Young and William Harkness, working separately, observed the event with spectroscopes. Each noticed a green line in the spectrum of the corona that defied identification with known elements. This suspected new element was called *coronium*.

In 1895, Lockyer's helium was identified in rocks on Earth, but coronium remained a spectral presence seen only on the Sun during total eclipses. As time passed, more elements were discovered on Earth until the Periodic Table of stable chemical elements was nearly complete. There was no room left for coronium to be an element. What could it be?

Might it be an already known element under such unusual conditions that it emitted a spectrum never before seen in a laboratory? Walter Grotrian of Germany in 1939 pointed the way and Bengt Edlén of Sweden in 1941 identified the green line of coronium as the element iron with 13 electrons missing—a "gravely mutilated state."[17] To ionize iron so greatly, the temperature of the corona had to be about 2 million °F (1 million °C) and its density had to be less than a laboratory vacuum. Because the conditions necessary for the production of such lines cannot be achieved in a laboratory, they are known as forbidden lines.

## The Reversing Layer

Astronomy was a family tradition for Charles Augustus Young. His maternal grandfather and his father had been professors of astronomy at Dartmouth College. Charles entered Dartmouth at age 14 and four years later graduated first in his class. He immediately began teaching— classics!—at an elite prep school, and commenced studies at a seminary to become a missionary. But in 1856 he changed his plans and became professor of astronomy at Western Reserve College, with a break in his duties to serve in the Civil War. He returned to Dartmouth in 1866 at the age of 32 as professor of astronomy, holding the same chair as his father and grandfather before him. There he pioneered in spectroscopy, especially applied to the Sun.

At the eclipse of December 22, 1870, which he observed at Jerez, Spain, Young noticed that the dark lines in the Sun's spectrum become bright lines for a few seconds at the beginning and end of totality. He had discovered the *reversing layer*, the lowest 600 miles (1,000 km) of the chromosphere, which is cooler than the photosphere and thus absorbs radiation of specific wavelengths, producing the ordinary dark-line spectrum of the Sun. However, when the Moon blocks the photosphere from view and the reversing layer of the chromosphere can be seen momentarily before

Charles A. Young. [Mary Lea Shane
Archives of the Lick Observatory]

it too is eclipsed, the bright-line spectrum of its glowing gases briefly
flashes into view. Here at last was the layer responsible for the dark-line
spectrum of the Sun seen on ordinary days. A long-missing piece in the
puzzle of the structure, composition, and density of the solar atmosphere
was fitted into place.

In 1877, Young was lured away from Dartmouth by the offer of more
equipment and research time at the College of New Jersey, now Princeton
University. Not only was he a great researcher but he was a revered teacher
and an acclaimed writer. His textbooks were the standard of his day.

## The Legacy of Eclipses

Throughout the final three decades of the 19th century, Janssen, Lockyer,
and Young led expeditions to the major total eclipses and, weather permitting, always contributed useful data and often new discoveries.

Many scientists had noticed that the corona changes its appearance
from one eclipse to the next. But it was Jules Janssen who first spotted
a pattern to those variations. He compared the coronas of the 1871 and
1878 eclipses and concluded that the shape of the corona varies according

Bolivia, November 12, 1966     Mexico, March 7, 1970     India, February 16, 1980

Siberia, July 31, 1981          Java, June 11, 1983      Philippines, March 18, 1988

The shape of the corona changes with sunspot activity. Near sunspot minimum, the corona is elongated at the Sun's equator. Near maximum, the corona is more symmetrical. These photographs were taken with a radial density filter developed by Gordon Newkirk that captures faint detail in the outer corona without overexposing the inner corona—similar to what the eye sees. The corona never appears the same twice. [High Altitude Observatory/ National Center for Atmospheric Research] 1) Bolivia, November 12, 1966; 2) Mexico, March 7, 1970; 3) India, February 16, 1980; 4) Siberia, July 31, 1981; 5) Java, June 11, 1983; 5) Philippines, March 18, 1988]

to the sunspot cycle. In 1871, the Sun was near sunspot maximum and the corona was round. In 1878, near sunspot minimum, the corona was more concentrated at the Sun's equator.

As the 20th century began, solar eclipses were still the principal means of gathering information about the workings of the Sun. Every total eclipse over land was attended by scientists willing to travel great distances, endure hostile climates, and risk complete failure because of clouds for a few minutes' view of the corona—vital for the systematic study of the Sun launched more than half a century earlier by Francis Baily in his report on the annular eclipse of 1836.

One could see the Sun best when it was obscured.

## What Eclipses at Jupiter Taught Us

by Carl Littmann

As the moons in the solar system revolve around their planets, they too create and undergo periodic solar and lunar eclipses. These events allowed the Danish astronomer Ole Römer in 1676 to prove, contrary to prevailing opinion, that light travels at a finite speed. He even succeeded in making the first good estimate of the speed of light. Such luminaries as Aristotle, Kepler, and Descartes had been certain that the speed of light was infinite. Gian Domenico Cassini, director of the Paris Observatory where Römer made his discovery, refused to believe the results, and Römer's triumph was not fully appreciated for half a century.

In observations of Io, innermost of Jupiter's four large moons, Römer noticed discrepancies between the observed times of its disappearance into the shadow of the planet and the calculated times for these events. He correctly explained that these discrepancies were due to the travel time of light between Jupiter and the Earth. When the Earth is approaching Jupiter, the interval between satellite eclipses is shorter because the distance light must travel is decreasing. When the Earth is moving farther from Jupiter, the interval between eclipses lengthens because the distance light must travel is increasing. Römer determined that light from Io took about 22 minutes longer to reach the Earth when the Earth was farthest from Jupiter than when it was closest. Thus light required about 22 minutes to cross the orbit of Earth. The diameter of the Earth's orbit was not known at the time. Modern measurements show that light actually requires about 16⅔ minutes to make this journey.

Left: Solar eclipse on Jupiter. The black dot left of center is the shadow cast by Io. Io, slightly larger than Earth's Moon, is visible over Jupiter's clouds in the upper right center. [NASA Hubble Space Telescope—WFPC 2]

Right: Solar eclipse on Saturn caused by its ring system and a moon. At the bottom are moons Tethys and Dione. Tethys' shadow on Saturn can be seen at the upper right, just below the rings. [NASA/Jet Propulsion Laboratory Voyager 1]

When Römer returned to Denmark, the king gave him a succession of appointments, including master of the mint, chief judge of Copenhagen, chief tax assessor (everyone said he was fair!), mayor and police chief of Copenhagen, senator, and

Saturn's shadow eclipses its rings [NASA/Jet Propulsion Laboratory—Cassini mission]

head of the state council of the realm—all this and more while he served as director of the Copenhagen observatory and astronomer royal of Denmark. He discharged all his duties with distinction.

Carl Littmann is a physicist and historian.

## NOTES AND REFERENCES

1.  Epigraph: Francis Baily: "Some Remarks on the Total Eclipse of the Sun, on July 8th, 1842," *Memoirs of the Royal Astronomical Society*, volume 15, 1846, page 4.
2.  Francis Baily: "On a Remarkable Phenomenon That Occurs in Total and Annular Eclipses of the Sun," *Memoirs of the Royal Astronomical Society*, volume 10, 1838, pages 1–42. Baily, searching back through the records, realized that Edmond Halley in 1715 and many other observers had seen and reported this bead-like apparition before him. Among the previous observers of the beads, he named José Joaquin de Ferrer, who also described the corona and first called it by that name. Perhaps it was Ferrer's account and his use of the word corona that came to Baily's mind when he saw the eclipse of 1842.

3. This and the following Baily quotations are from Francis Baily: "Some Remarks on the Total Eclipse of the Sun, on July 8th, 1842," *Memoirs of the Royal Astronomical Society*, volume 15, 1846, pages 1–8.

4. Agnes M. Clerke: "Baily, Francis," *The Dictionary of National Biography*, volume 1 (London: Oxford University Press, 1921), page 903.

5. Joseph Needham and Wang Ling: *Science and Civilisation in China, volume 3, Mathematics and the Sciences of the Heavens and the Earth* (Cambridge: Cambridge University Press, 1959), page 423.

6. F. Richard Stephenson: *Historical Eclipses and the Earth's Rotation* (Cambridge: Cambridge University Press, 1997), page 390.

7. José Joaquín de Ferrer: "Observations of the Eclipse of the Sun, June 16th, 1806, Made at Kinderhook, in the State of New-York," *Transactions of the American Philosophical Society*, volume 6, 1809, pages 264–275.

8. J. M. Vaquero and M. Vázquez: *The Sun Recorded Through History* (Heidelberg, Germany: Springer, 2009), page 203.

9. Dorrit Hoffleit: *Some Firsts in Astronomical Photography* (Cambridge, Massachusetts: Harvard College Observatory, 1950). The first successful photograph of the uneclipsed Sun, also a daguerreotype, was achieved by the French scientists Hippolyte Fizeau and Léon Foucault in 1845.

10. De La Rue made that discovery in 1861. Earlier evidence for sunspots as depressions had come from observations by Scottish astronomer Alexander Wilson in 1774. He noted that the geometry of sunspots seemed to change as they were seen from different angles as the Sun's rotation carried them across the solar disk and toward the limb.

11. The Sun as a sphere of hot gas had been proposed independently by Angelo Secchi and John Frederick William Herschel (William's son) in 1864.

12. J. Norman Lockyer: "On Recent Discoveries in Solar Physics Made by Means of the Spectroscope" in Bernard Lovell, editor: *Astronomy*, volume 1, The Royal Institution Library of Science (Barking, Essex: Elsevier Publishing, 1970) page 90.

13. Alfred Fowler: "Sir Norman Lockyer, K.C.B., 1836–1920," *Proceedings of the Royal Society of London*, series A, volume 104, December 1, 1923, pages i–xiv.

14. Auguste Comte: *The Essential Comte, Selected from Cours de philosophie positive*, translated by Margaret Clarke (London: Croom Helm, 1974), pages 74, 76.

15. A. J. Meadows: *Science and Controversy: A Biography of Sir Norman Lockyer* (London: Macmillan, 1972), page 53.

16. J. Norman Lockyer: "On Recent Discoveries in Solar Physics Made by Means of the Spectroscope" in Bernard Lovell, editor: *Astronomy*, volume 1, The Royal Institution Library of Science (Barking, Essex: Elsevier Publishing, 1970), pages 101–102.

17. "A gravely mutilated state" is the picturesque description found in Gabrielle Camille Flammarion and André Danjon, editors: *The Flammarion Book of Astronomy*, translated by Annabel and Bernard Pagel (New York: Simon and Schuster, 1964), page 227.

# A MOMENT OF TOTALITY

## Stumbling onto a Total Eclipse

Sheridan Williams and his fellow observers had set up their equipment on Antigua for the total eclipse of 1998. A young couple walking by approached him. Why the telescope? they asked. They were on their honeymoon and were utterly unaware that a total eclipse of the Sun would be visiting the island that day. Antigua had done little to inform its citizens and tourists about this rare event.

"Stay with us for the next hour and a half and you will see the most amazing sight you've ever seen," said Williams. He gave them solar filters and shared views through his telescope. He talked them through what was happening and what was going to happen as the Moon blocked the Sun. They were awed and grateful. It gave new meaning to the word *honeymoon.*[*]

---

[*] Sheridan Williams is a computer scientist who lives in London. Interviewed May 23, 2015.

# A MOMENT OF TOTALITY

## Stumbling onto a Total Eclipse

# 8

———◄o►———

# The Eclipse That Made
# Einstein Famous

*Oh leave the Wise our measures to collate.*
*One thing at least is certain, LIGHT has WEIGHT*
*One thing is certain, and the rest debate—*
*Light-rays, when near the Sun, DO NOT GO STRAIGHT.*

Arthur S. Eddington (1920)[1]

Of all the lessons that scientists learned from eclipses, the most profound and momentous was the confirmation of Einstein's general theory of relativity. That lesson was provided by the total eclipse of the Sun on May 29, 1919.

## The Birth of Relativity

In 1905, an obscure Swiss patent examiner third class named Albert Einstein published three articles in the same issue of the leading German scientific journal *Annalen der Physik* that utterly changed the course of physics. One proved the existence of atoms. The second laid the cornerstone for quantum mechanics. The third was a revolutionary view of space and time known as the special theory of relativity. It required only high-school algebra yet its implications were so profound that this theory baffled many of the leading scientists of the day.

In the decade that followed the publication of the special theory of relativity, Einstein labored mightily to expand his concept to accelerated systems. In 1907, he formulated his principle of equivalence: there is no way for a participant to distinguish between a gravitational system and an accelerated system. For an observer in a closed compartment, does a ball fall to the floor by gravity because the compartment is resting on a planet or does the ball fall because the compartment is accelerating

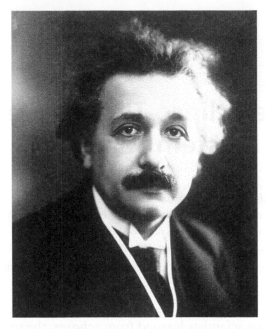

Albert Einstein in 1922. [Albert Einstein Collection, The Huntington Library, San Marino, California]

Einstein's principle of equivalence. A ball falls in a compartment. Does it fall by gravity because the compartment is resting on a planet or does it fall because the compartment is accelerating upward? Einstein realized that an observer in the compartment could not distinguish whether gravity or acceleration caused the ball to fall.

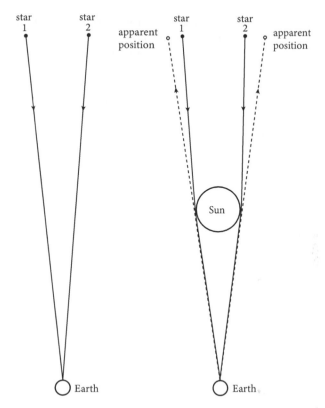

In a perfect vacuum with no gravity, light from distant stars travels in a straight line (left). The gravitational field of the Sun bends light (right), making the stars appear slightly farther apart than they actually are.

toward the ball? In both cases, the ball falls to the floor. In both cases, the observer feels weight. It is impossible to tell whether that weight is from gravity or acceleration.

Einstein then realized from his principle of equivalence that relativity required gravity to bend light rays, much as it bends the paths of particles. In a closed accelerating compartment, a light on one wall is aimed directly at the opposite wall, across the line of motion. The beam travels at a finite speed, so in the time it takes to traverse the compartment, the opposite wall has moved upward. For an observer in the compartment, the light beam has struck the opposite wall below where it was aimed. The observer concludes that light has been bent.

What about that same experiment performed in a compartment at rest on a planet? According to Einstein's principle of equivalence, there can be no difference between phenomena measured in the two compartments.

Therefore, for the observer in the gravitational environment, light must be bent by gravity. But the effect is very small. It takes a lot of mass to bend light enough to be measured. In 1911, Einstein realized that this peculiar idea might be tested during a total solar eclipse.

The Sun is sufficiently massive that light from distant stars passing close to its surface would be deflected just enough to be measurable. This bending of the light from such stars causes them to appear displaced slightly outward from the Sun. The displacement could be recognized by comparing a photograph of the star field around the Sun with a photograph of the same star field when the Sun was not present.

When the Sun is visible, however, the stars are much too faint to be seen. If only the Sun's brightness could be shut off for just a few minutes—as happens during a total solar eclipse! Using his special theory of relativity, Einstein initially predicted that starlight just grazing the Sun would be bent 0.87 second of arc.[2]

## The Search for Proof

A talented young scientist named Erwin Freundlich was the first to attempt to test this aspect of relativity.[3] Fascinated by the theory, he obtained solar eclipse photographs from observatories around the world that might show the displacement of stars. But photographs of previous eclipses were not adequate for the purpose.[4] The theory would have to be tested at a future eclipse, such as the one that would occur in southern Russia on August 21, 1914.[5] Freundlich was an assistant at the Royal Observatory in Berlin and tried to interest his colleagues in mounting a scientific expedition to test the theory. His superiors, however, were uninterested. Freundlich was allowed to go if he took unpaid leave and raised his own money. With youthful enthusiasm, he made his plans and informed Einstein of his intentions.

Einstein had just moved from Switzerland to Berlin to take a distinguished position created for him at the Kaiser Wilhelm Institute, the most prestigious research center in the world's science capital. Even among the giants there, Einstein stood out. Physicist Rudolf Ladenburg recalled, "There were two kinds of physicists in Berlin: on the one hand was Einstein, and on the other all the rest."

It was at this time, the spring of 1914, that Einstein was becoming increasingly withdrawn and oblivious to conventions of social behavior. He was confident in his new general theory of relativity but was deeply engrossed in its final formulation and its implications. Freundlich's wife Käte told of inviting Einstein to dinner one evening. At the conclusion of the meal, as the two scientists talked, Einstein suddenly pushed back his plate, took out his pen, and began to cover their prized tablecloth with equations. Years later, Mrs. Freundlich lamented, "Had I had kept it unwashed as my husband told me, it would be worth a fortune."[6]

That summer Freundlich took his scientific equipment and headed for the Crimea and the eclipse. On August 1 Germany declared war on Russia: the World War I had begun. Freundlich was a German behind Russian lines. He and his team members were arrested and their equipment impounded. Within a month, Freundlich and crew were exchanged for high-ranking Russian officers, but they had missed the eclipse.[7]

Einstein deplored the war and German militarism, an attitude that drew the ire of many of his colleagues in Berlin. He ignored the hostility and concentrated all his energies on his research. In 1915 Einstein announced the completion of his general theory of relativity, a radically new theory of gravitation. It is perhaps the most prodigious work ever accomplished by a human being. Not only were its implications profound but the mathematics needed to understand it were formidable. "Compared with [the general theory of relativity]," said Einstein, "the original theory of relativity is child's play."[8]

Einstein offered three tests to confirm or reject his theory. The first was the peculiar motion of Mercury. The entire orbit of Mercury was turning (precessing) more than Newton's law of universal gravitation could explain. It was a tiny but measurable amount: 43 seconds of arc a *century*. This unexplained advance in Mercury's *perihelion* (its closest point to the Sun) had given rise to a suspicion that one or more planets lay between Mercury and the Sun. The suspicion was so strong that this suspected planet, scorched by the nearby Sun, had already been given

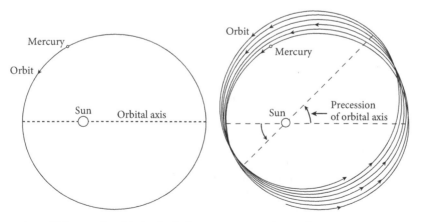

Precession of Mercury's orbital axis. *Left*: According to Newtonian theory, a planet orbiting the Sun follows a fixed elliptical path. *Right*: Einstein's general theory of relativity predicts that the axis of the ellipse will gradually rotate.

a name: Vulcan, after the Roman god of fire. Many observers had tried to spot it as a tiny dot passing across the face of the Sun, and some even claimed success. But when an orbit for the planet was calculated and its next passage across the face of the Sun was due, the planet never kept the appointment. It did not exist. But it was not until Einstein formulated the general theory of relativity that the 43 seconds per century anomaly could be explained.

Einstein was more than just pleased when he realized that his theory could account for this discrepancy in the motion of Mercury. "I was beside myself with ecstasy," he wrote.[9] This explanation of a perplexing problem gave general relativity significant credibility. But the power of the general theory would be even more evident if it could predict something never before contemplated or detected.

Einstein offered two such predictions: that starlight passing close to the Sun would be bent, and that light leaving a massive object would have its wavelengths stretched so that the light would be redder. This *gravitational redshift* was so small an effect that it could not be detected in the Sun with the equipment available at the time, so this proof of general relativity had to wait many years. It was finally detected in 1959 by Robert V. Pound and Glen A. Rebka, Jr., using the recently discovered Mössbauer effect in which the gamma rays emitted by atomic nuclei serve as the most precise of clocks. One of these atomic clocks was placed in the basement of a building and moved up and down so that its depth in the Earth's gravitational field varied minutely. The deeper in the basement the clock was, the longer the wavelength of its radiation. It had taken 45 years, but the gravitational redshift predicted by Einstein had at last been confirmed.[10]

In contrast, the gravitational deflection of starlight predicted by Einstein could be tested at most total eclipses of the Sun. Between 1911 and 1915, Einstein revised his calculation, using his new general theory of relativity. He found the deflection to be twice the initially assigned value. Starlight passing near the Sun would be bent 1.75 arc seconds. (Einstein and the world were fortunate that his initial prediction was not tested before it was revised; otherwise his later figure, although rigorously honest, might have seemed to be a manipulation to make the numbers come out right. There would have been far less drama in the confirmation of relativity.)

In 1916 Einstein published his complete general theory of relativity. But the world hardly noticed. For two years, nations had been locked in the World War I. Feelings against Germans ran strong in France and Great Britain, just as the Germans hated the French and British. Einstein sent his paper to a friend, scientist Willem de Sitter in The Netherlands. De Sitter, in turn, forwarded the paper to Arthur Eddington in England. At the age of 34, Eddington was already famous for his pioneering work in how stars emit energy. Eddington instantly recognized the significance of

A stunning view of the March 20, 2015 total eclipse from Svalbard was created from 29 individual exposures combined with custom software. In addition to the wealth of fine structure seen in the corona, lunar surface features are also revealed. [Nikon D810, TS Photo Line refractor, fl = 800mm, f/6.9, exposures: 1/1600 to 4 seconds. © 2015 Miloslav Druckmüller, Shadia Habbal, Peter Aniol, Pavel Starha]

Eclipse cruises are a popular way to chase solar eclipses. Comfort, good food, and entertainment are all pluses. One negative is that ship stability can make high magnification photography a challenge. Total solar eclipse of July 11, 2010 shot from South Pacific Ocean, near French Polynesia. [Canon SD940 IS, auto-exposure, video frame capture, © 2010 Alson Wong]

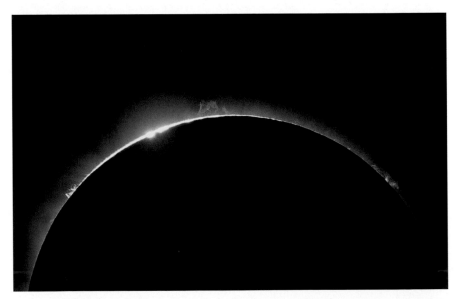

An enormous, red solar prominence is joined by Baily's beads as totality ends during the total solar eclipse of Aug, 21, 2017, and shot near Casper, Wyoming. [Canon 5DS R, CFF Telescopes 160mm refractor, ISO100, f/6.5, 1/1000, © 2017 Catalin Beldea]

A scenic wilderness foreground graces this time-lapse image of the 2017 total solar eclipse from Green River Lakes, WY. [Nikon D750, 17mm, f/8, photos taken every five minutes, filter removed at totality. © 2017 Ben Cooper]

A two-image composite of the Nov. 14, 2012 total solar eclipse from Queensland, Australia. The diamond ring combined with a longer exposure of the corona taken during totality. [Canon 60Da, Astro-Physics Traveler 105mm refractor, f/6, 1/1000 and 1/60, © 2017 Alan Dyer]

The rising sun in partial eclipse is captured against the iconic skyline of New York City on Nov. 3, 2013. Careful planning placed the photographer about 8 miles west of the Empire State Building. [Canon EOS 40D, Canon EF 400mm f/5.6 L USM, 1/4000 sec. © 2013 Chris Cook]

Seventy-two separate exposures (1/1000 to 2.5 seconds) were combined and processed using Photoshop CC 2017 and Photomatix Pro 6 to produce this HDR (High Dynamic Range) image of the corona. [Nikon D850, Vixen 90mm Fluorite Refractor (90mm, f/9, fl=810mm), 1/1000 to 2.5 seconds, f/9, ISO 250, Mamalluca Observatory, Vicuña, Chile, © 2019 Fred Espenak]

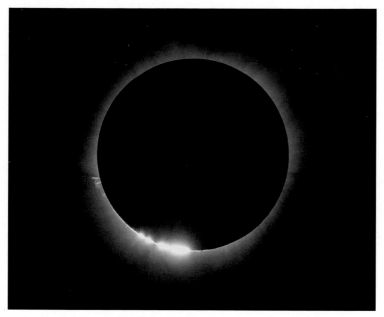

Bailey's beads form in the last 3 seconds before second contact while a ruby-red loop prominence appears along the Moon's limb. This dramatic capture was done from the deck of a cruise ship in Indonesian's Makassar Strait during the total solar eclipse of Mar. 9, 2016. [Canon 5D, EF100-400mm +2x, (fl=800mm), f/11, 1/1000, ISO 100, © Tommy Tat-fung Tse]

This time-lapse sequence captures the entire total solar eclipse of July 3, 2019 over ESO's La Silla Observatory, Chile. [Credit: ESO/P. Horálek]

Before leaving for his remote observing site, Catalin Beldea set up a small camera on the balcony of his hotel room to capture 2019 totality and the crowds on the beach of La Serena, Chile. [Fujifilm X30, 28mm f/5.6, ISO100, 1/4, © 2019 Catalin Beldea]

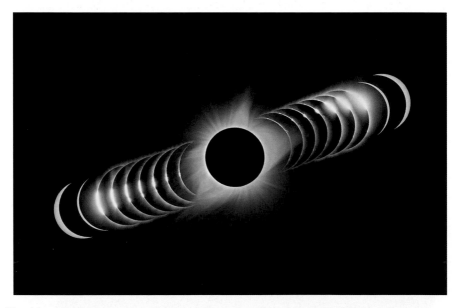

Here's a variation on a time-sequence composite of the August 21, 2017 total solar eclipse. Time runs from left to right and captures the diamond ring and Baily's beads both before and after totality. The corona image is a blend of seven exposures exposures, from 1/1600 to 1/15. Shot from Driggs, Idaho [Canon 6D MkII, Astro-Physics Traveler refractor, f/5, ISO 100, © 2017 Alan Dyer]

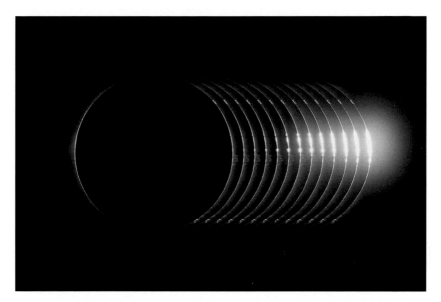

A time sequence shot at 2 frames per second captures the formation of Baily's beads just after third contact (i.e., as totality ends) during the 2017 total eclipse. The bright red arc along the edge of the Moon is the Sun's chromosphere. Several large solar prominences are also visible. [Nikon D810, Vixen ED100SF Refractor (100mm, f/9, fl=900 mm), 1/1000 sec, f/9, ISO 200, Casper College, Casper, WY, © 2017 Fred Espenak]

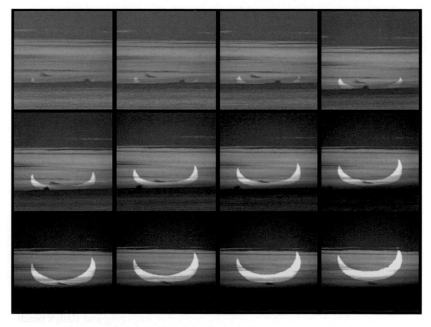

A sequence captured at sunrise from the Atlantic shore of New Jersey shows the Sun in deep parial eclipse on June 10, 2021. This eclipse was annular from central Canada. [Nikon Z7ii, Nikkor 500mm f/4. © 2021 Michael Zeiler]

Time lapse of the total solar eclipse of Mar. 20, 2015 captures the spectacle above a frozen landscape from Svalbard, Norway. [Nikon D800, 50mm, f/11, 1/800 and solar filter for partial phases, 0.6 sec for totality, © 2015 Thanskrit Santikunaporn]

The Sun's chromosphere and large prominences peek above the Moon's disk during the short total eclipse visible from Pokwero, Uganda on Nov. 3, 2013. [Nikon D800, TeleVue 102 Refractor, Focal Length 880mm, F8.6, ISO800, 1/400 sec. © 2013 Jaime Vilinga]

Einstein's discoveries and was deeply impressed by its intellectual beauty. He shared the paper with other scientists in Britain.

## The 1919 Test

Frank Dyson, the astronomer royal of England, began planning for a British solar eclipse expedition in 1919 to test relativity. It was the perfect eclipse for the purpose because the Sun would be standing in front of the Hyades, a nearby star cluster, so there would be a number of stars around the eclipsed Sun bright enough for a telescope to see. But the timing could hardly have been worse. In 1917 Britain was in the midst of a terrible war whose issue was still very much in doubt. Yet somehow Dyson managed to persuade the government to fund the expedition, despite the fact that its purpose was to test and probably confirm the theory of a scientist who lived in Germany, the leader of the hostile powers.

Meanwhile, US astronomers had an earlier opportunity to verify or disprove relativity. In the final months of the World War I, a total eclipse passed diagonally across the United States from the state of Washington to Florida. On June 8, 1918, a Lick Observatory team led by William Wallace Campbell and Heber D. Curtis, observing from Goldendale, Washington, pointed their instruments skyward to render a verdict on relativity. The weather was mostly cloudy, but the Sun broke through for 3 minutes during totality. Measuring and interpreting the plates had to wait several months, however, until comparison pictures could be taken of the same region of the sky at night, without the Sun in the way. By this time, Curtis was in Washington, D.C. working on military technology for the government as the war came to a close. Curtis returned to Lick in May 1919 and the results were announced in June. The star images were not as point-like as desired. Curtis could detect no deflection of starlight. By this time, word was spreading about the findings of the British expeditions, and the Lick paper was never published.[11]

Four months after the armistice, two British scientific teams were poised for departure. Andrew C. D. Crommelin and Charles R. Davidson, heading one party, were to set sail for Sobral, about 50 miles (80 km) inland in northeastern Brazil. Eddington, Edwin T. Cottingham, and their team were headed for Principe, a Portuguese island about 120 miles (200 km) off the west coast of Africa in the Gulf of Guinea. On the final day before sailing, the four team leaders met with Dyson for a final briefing. Eddington was extremely enthusiastic and confident that Einstein was right. A deflection of 1.75 arc seconds would confirm relativity. Half that amount—a deflection of starlight by 0.87 arc seconds—would reconfirm Newtonian physics.[12]

The total solar eclipse of May 29, 1919 featured an enormous prominence. Arthur Eddington's observation of this eclipse from Principe provided the first observational confirmation of Einstein's general theory of relativity. [Royal Astronomical Society]

"What will it mean," asked Cottingham, "if we get double the Einstein deflection?" "Then," said Dyson, "Eddington will go mad and you will have to come home alone!"

On Principe, worrisome weather conditions greeted Eddington and Cottingham. Every day was cloudy. However, May was the beginning of the dry season and no rain fell—until the morning of the eclipse. The fateful day, May 29, 1919, dawned overcast, and heavy rain poured down. The thunderstorm moved on about noon, but the cloud cover remained. The Sun finally peeped through 18 minutes before the eclipse became total but continued to play peekaboo with the clouds. Said Eddington: "I did not see the eclipse, being too busy changing plates, except for one glance to make sure it had begun and another half-way through to see how much cloud there was."

Eddington had much cause for worry. He was not interested in prominences or the corona. He needed to see the region around the Sun clearly. With great care, the plates were developed one at a time, only two a night.

Arthur S. Eddington. [AIP Emilio Segrè Visual Archives, gift of S. Chandrasekhar]

Eddington spent all day measuring the plates. Clouds had interfered with the view, but five stars were visible on two plates. When he had finished measuring the first usable plate, Eddington turned to his colleague and said, "Cottingham, you won't have to go home alone."

Eddington measured the displacement in the stars' positions, extrapolated to the limb of the Sun, to be between 1.44 and 1.94 arc seconds, for a mean value of 1.61 ± 0.30 arc seconds. The deflection agreed closely with Einstein's prediction. In later years, Eddington referred to this occasion as the greatest moment of his life.[13]

The expedition to Sobral had equally threatening weather but there was a clear view of totality except for some thin fleeting clouds in the middle of the event. Crommelin and Davidson stayed in Brazil until July to take reference pictures of the star field without the presence of the Sun and then brought all their photographic plates back to Britain before measuring them. They found that their largest telescope had failed because heat caused a slight change in focus, which spoiled the pinpoint images of the stars. But the other instrument had worked well, and its plates also supported Einstein's prediction. Their mean value was 1.98 ± 0.12 arc seconds.

English astronomers photographed the total solar eclipse of 1919 from two sites to test Einstein's general theory of relativity. Andrew Crommelin and Charles Davidson took this photograph from Sobral, Brazil. The white bars mark the location of stars recorded during totality. Careful measurements of their positions showed shifts because the stars' light, passing close to the Sun, was bent by the Sun's gravity, just as Einstein predicted. [Royal Society of London]

It was now September and no news about eclipse results had been released. Einstein was curious and inquired of friends. Hendrik Antoon Lorentz, the Dutch physicist, used his British contacts to gather news and telegraphed Einstein: "Eddington found star displacement at rim of Sun . . ."[14] He also announced the favorable results at a scientific meeting in Amsterdam on October 25, with Einstein in attendance. But no reporters were present, and no word of the discovery was published.

Finally, on November 6, 1919, the Royal Society and the Royal Astronomical Society held a joint meeting to hear the results of the eclipse expeditions. The hall was crowded with observers, aware that an age was ending. From the back wall, a large portrait of Newton looked down on the proceedings. Joseph John Thomson, president of the Royal Society and discoverer of the electron, chaired the meeting, praising Einstein's work as "one of the highest achievements of human thought." He called upon Frank Dyson to summarize the results and introduce the reports of the eclipse team leaders. The astronomer royal concluded his presentation by saying: "After careful study of the plates I am prepared to say that there

Albert Einstein and Arthur S. Eddington in Eddington's garden, 1930. [Royal Astronomical Society]

can be no doubt that they confirm Einstein's prediction. A very definite result has been obtained that light is deflected in accordance with Einstein's law of gravitation."[15]

Einstein awoke in Berlin on the morning of November 7, 1919 to find himself world famous. Hordes of reporters and photographers converged on his house. He genuinely did not like this attention but he found a way to turn the disturbance to the benefit of others. He told the reporters about starving children in Vienna. If they wanted to take his picture, they first had to make a contribution to help those children. Suddenly Einstein was a celebrity.

## More Tests

For the next half century, scientists from many nations traveled the world to measure and re-measure the bending of starlight around the Sun at every total eclipse. And always the results confirmed Einstein's general theory of relativity.[16]

In the 1970s, a new method emerged for testing relativity by the deflection of radiation around the Sun. Radio astronomers were using widely separated radio telescopes to measure the positions of celestial objects with an accuracy greater than a single optical telescope.

Quasars had been discovered and nearly all astronomers interpreted them to be the most distant objects in the universe—so distant that they were essentially fixed markers. Their angular distance from one another provided a new coordinate system against which the positions of all other objects could be referred. Because some of the quasars lay near the ecliptic (the Sun's apparent path through the heavens in the course of a year), the Sun's gravity annually would deflect the quasars' radiation, making them seem to shift slightly in position. Only this time, the radiation would be radio waves rather than visible light. And no eclipse was necessary because radio telescopes do not require darkness. In 1974 and 1975, Edward B. Fomalont and Richard A. Sramek used a 22-mile (35-km) separation between radio telescopes to measure the deflection of light at the limb of the Sun as $1.761 \pm 0.016$ arc seconds, a result that not only confirmed relativity, but also favored Einsteinian relativity over slightly variant relativity formulations by other scientists.[17]

Solar eclipses may no longer be the only or most accurate way to determine the relativistic deflection of starlight. But when the general theory of relativity predicted that the gravity of the Sun would bend starlight, a phenomenon never before observed, it was an eclipse that provided proof. The 1919 total eclipse of the Sun supplied the first and most dramatic demonstration of Einstein's masterpiece and brought relativity and Einstein to the attention of the entire world.

## NOTES AND REFERENCES

1.  Epigraph: Arthur S. Eddington as cited in Allie Vibert Douglas: *The Life of Arthur Stanley Eddington* (London: T. Nelson, 1956), page 44.
2.  The angular displacement of a star is inversely proportional to the angular distance of that star from the Sun's center.
3.  In 1911, Einstein had asked Freundlich to investigate another aspect of his emerging general theory of relativity: the motion of Mercury. Freundlich reviewed the anomaly in Mercury's motion, recognized since the work of Le Verrier in 1859, and reconfirmed that Mercury was indeed deviating slightly from the law of gravity as formulated by Newton. Freundlich's results, which matched Einstein's relativistic recalculation for the precession of Mercury's orbit, were published in 1913 over the objections of his superiors.
4.  To make certain that the telescope optics and the camera system introduced no unknown deflection in the positions of stars, it was important to have for comparison with the eclipse plate a picture of that same region taken with the same equipment when the Sun present was not present. No such plates were available.

5.  Charles Dillon Perrine, an American who was directing the Argentine National Observatory, prepared to test the bending of starlight during the October 10, 1912 eclipse in Brazil, but was rained out.

6.  Ronald W. Clark: *Einstein, the Life and Times* (London: Hodder and Stoughton, 1973), page 176.

7.  Freundlich's team was not alone in the Crimea. An American team from the Lick Observatory also journeyed to Russia to test relativity, but they were clouded out. In 1918, Freundlich left the Royal Observatory to work full time with Einstein at the Kaiser Wilhelm Institute. In 1920, he was appointed observer and then chief observer and professor of astrophysics at the newly created Einstein Institute at the Astrophysical Observatory, Potsdam. Freundlich continued to be plagued by miserable luck on his eclipse expeditions to measure the deflection of starlight. He returned empty-handed in 1922 and 1926 because of bad weather. He finally got to see an eclipse in Sumatra in 1929, although he obtained a deflection (now known to be erroneous) considerably greater than Einstein predicted. When Hitler came to power, Freundlich left Germany and eventually settled in Scotland, where he changed his name to Finlay-Freundlich, based on his mother's maiden name, Finlayson. He built the first Schmidt-Cassegrain telescope, the prototype for almost all large photographic survey telescopes today.

8.  Banesh Hoffmann with the collaboration of Helen Dukas: *Albert Einstein, Creator and Rebel* (New York: Viking Press, 1972), page 116.

9.  Banesh Hoffmann with the collaboration of Helen Dukas: *Albert Einstein, Creator and Rebel* (New York: Viking Press, 1972), page 125.

10.  Robert V. Pound and Glen A. Rebka, Jr.: "Resonant Absorption of the 14.4-kev Gamma Ray from 0.10-microsecond $Fe^{57}$," *Physical Review Letters*, volume 3, December 15, 1959, pages 554–556. The gravitational redshift was detected previously, based on Einstein's suggestion, in light emitted from extremely dense white dwarf stars, but to nowhere near the same degree of certainty.

11.  The Lick team was working with mediocre equipment and improvised mounts. Their excellent regular equipment had stood ready under cloudy skies in Russia in 1914, but it was too cumbersome to transport home after the outbreak of the First World War. After the war, it was delayed in shipment home and missed the eclipse of 1918.

12.  The "Newtonian prediction" is calculated on the basis that light energy has a mass equivalent (Einstein's $E = mc^2$). That mass is then treated as ordinary matter using Newton's equations for gravity.

13.  Allie Vibert Douglas: *The Life of Arthur Stanley Eddington* (London: T. Nelson, 1956), pages 40–41.

14.  Ronald W. Clark: *Einstein, the Life and Times* (London: Hodder and Stoughton, 1973), page 226.

15.  Frank W. Dyson, Andrew C. D. Crommelin, and Arthur S. Eddington: "Joint Eclipse Meeting of the Royal Society and the Royal Astronomical Society," *The Observatory*, volume 42, November 1919, pages 389–398. The article

includes dissenting comments from scientists in the audience. The article based on eclipse results was Frank W. Dyson, Arthur S. Eddington, and Charles R. Davidson: "A Determination of the Deflection of Light by the Sun's Gravitational Field, From Observations Made at the Total Eclipse of May 29, 1919," *Philosophical Transactions of the Royal Society of London*, series A, volume 220, 1920, pages 291–333. A summary of the eclipse results had appeared the week after the joint meeting: Andrew C. D. Crommelin: "Results of the Total Solar Eclipse of May 29 and the Relativity Theory," *Nature*, volume 104, November 13, 1919, pages 280–281. Eddington had yet another reason to be proud: "By standing foremost in testing, and ultimately verifying, the 'enemy' theory, our national observatory kept alive the finest traditions of science; and the lesson is perhaps still needed in the world today" (Clark: *Einstein*, page 284). British physicist Robert Lawson made a similar observation: "The fact that a theory formulated by a German has been confirmed by observations on the part of Englishmen has brought the possibility of cooperation between these two scientifically minded nations much closer. Quite apart from the great scientific value of his brilliant theory, Einstein has done mankind an incalculable service" (Clark: *Einstein*, page 297). The world soon forgot this lesson.

16.  For a description of other expeditions, see Mark Littmann, Fred Espenak, and Ken Willcox *Totality: Eclipses of the Sun* (New York: Oxford University Press, 3rd Edition: 2008; updated 3rd Edition: 2009), pages 101–102.

17.  Edward B. Fomalont and Richard A. Sramek: "A Confirmation of Einstein's General Theory of Relativity by Measuring the Bending of Microwave Radiation in the Gravitational Field of the Sun," *Astrophysical Journal*, volume 199, August 1, 1975, pages 749–755. The result reported in their article, $1.775 \pm 0.019$ arc seconds, was later revised to $1.761 \pm 0.016$ arc seconds. (Personal communication, March 1990.) They used three 85-foot (26-meter) antennas and a distant 45-foot (14-meter) antenna, instruments of the National Radio Astronomy Observatory at Green Bank, West Virginia. The idea of using radio waves and radio telescopes to test the bending of light was first proposed by Irwin I. Shapiro: "New Method for the Detection of Light Deflection by Solar Gravity," *Science*, volume 157, August 18, 1967, pages 806–807.

# A MOMENT OF TOTALITY

## The Difference Between Partial and Total

Canadian filmmaker and eclipse veteran Jean Marc Larivière is always frustrated when people confuse a partial eclipse of the Sun with a total eclipse. "Oh yeah," they say, "I saw an eclipse of the Sun when I was young. It was no big deal."

A total eclipse of the Sun is utterly different from a partial eclipse.

For his film *Shadow Chasers*,* Jean Marc interviewed a busload of French tourists who had traveled to the west coast of Australia in February 1999 to see an annular eclipse of the Sun that was almost total. The Moon just a bit too far from Earth to completely cover the disk of the Sun.

After the eclipse, he interviewed a woman who told him: "It was a wonderful annular eclipse, very tight. But you know, despite so much of the Sun being eclipsed, it was still a partial eclipse. A partial eclipse is like a first kiss. A total eclipse is like a night of passionate love."

---

* The 2000 documentary *Shadow Chasers* was written and directed by Jean Marc Larivière and produced and distributed by the National Film Board of Canada: <https://www.nfb.ca/film/shadow_chasers>.

# 9

—◄o►—

# Observing a Total Eclipse

A total eclipse of the Sun . . . is the most sublime<br>and awe-inspiring sight that nature affords.

Isabel Martin Lewis (1924)[1]

A total eclipse of the Sun. What is this sight that lures people to travel to the ends of the Earth for a brief view at best, a substantial possibility of no view at all, and with no rain check? And how do you get the most out of the experience of a total eclipse?

In these pages, eclipse veterans of today plus some eclipse seekers from earlier times share their experiences with you. Among them they have witnessed over 500 total eclipses.

"It's like a religious experience," says Jay Anderson, "the anticipation as the time until totality is counted in days, then hours, then minutes. It's the perfect buildup. Spielberg couldn't do it better. It's an intensely moving event."

Steve Edberg agrees: "It is the intensity of the event. You grab as much as you can. I like action in the heavens and you can't get much better than this."

Mike Simmons starts his countdown to the eclipse four weeks in advance when the new moon becomes visible after sundown in the west. He watches the Moon night by night edging eastward across the sky, increasing from crescent to first quarter to gibbous to full, then waning from gibbous to last quarter to crescent. It seems like an ordinary month—but it isn't. This month the Moon is headed for a perfect alignment between the Sun and Earth.

Roger Tuthill had a ritual he followed the day before an eclipse. He walked around inside the path of totality observing the people who live there. They have no idea what they are about to experience and there are no words to adequately prepare them.

## Panel of Eclipse Veterans

Jay Anderson, Meteorologist, noted for eclipse weather predictions (Canada)
Paul and Julie (O'Neil) Andrews, Information technologist; Businesswoman (United Kingdom)
John Beattie, Eclipse researcher (New York)
Richard Berry, Astronomy author and former Editor-in-Chief, *Astronomy* (Oregon)
Joe Buchman, Marketing consultant (Utah)
Kristian Buchman, University student (Utah)
Dennis di Cicco, Senior Editor (retired), *Sky & Telescope* (Massachusetts)
Stephen J. Edberg, Astronomer, NASA Jet Propulsion Laboratory (California)
Alan Fiala, Astronomer, US Naval Observatory (Washington, D.C.) (He died in 2010)
George Fleenor, Planetarium director (retired) (Florida)
Ruth S. Freitag, Senior Science Specialist, Library of Congress (Washington, D.C.) (She died in 2021)
Joseph V. Hollweg, Professor of Astronomy (retired), University of New Hampshire, Durham
Xavier Jubier, Computer scientist; created Google interactive eclipse maps (France)
Jean Marc Larivière, Filmmaker (Canada)
Dawn Levy, Science writer (Tennessee)
Charles Lindsey, Astronomer, Southwest Research Institute, Boulder (Colorado)
George Lovi, Astronomy author/columnist, Lakewood, New Jersey (He died in 1993)
David Makepeace, Filmmaker (Canada)
Larry Marschall, Professor Emeritus of Physics, Gettysburg College (Pennsylvania)
Frank Orrall, Professor of Physics & Astronomy, University of Hawaii (He died in 2000)
Jay M. Pasachoff (He died in 2022), Field Memorial Professor of Astronomy, Williams College (Massachusetts)
Patrick Poitevin, Technology scientist (Belgium, United Kingdom)
Luca Quaglia, Physicist and financial engineer (New York)
Joe Rao, Meteorologist (New York)
Leif J. Robinson, Editor-in-Chief (emeritus), *Sky & Telescope* (Massachusetts) (He died in 2011)
Michael Rogers, Writer and futurist (New York City)
Gary and Barbara (Schleck) Ropski, Attorney; Journalist (retired) (Illinois)
Walter Roth, Travel agency owner (Florida) (He died in 2002)
Glenn Schneider, Astronomer, Stewart Observatory, University of Arizona
Mike Simmons, Founder, Astronomers Without Borders (California)
Roger W. Tuthill, Electrical engineer and solar filter inventor (New Jersey) (He died in 2000)
Ken Willcox, Geologist, Phillips Petroleum Company (Oklahoma) (He died in 1999)
Sheridan Williams, Rocket scientist (retired) (United Kingdom)
Michael Zeiler, Computer scientist, eclipse mapmaker (New Mexico)
Jack B. Zirker, Astronomer (retired), National Solar Observatory (New Mexico)

## First Contact

There is a special feeling at the instant when the Moon begins to slide in front of the Sun. In less than a minute, observers with small telescopes see the first tiny "bite" out of the western side of the Sun. That's when the magic starts.

First contact remains a very special moment for Jay Pasachoff. As an astronomer, he could appreciate more readily than most all the factors that go into predicting precisely when an eclipse will occur and exactly where on Earth it will be seen. And when the call "first contact" came right on schedule, he always found this accomplishment of mankind astounding.

Famed nature writer Annie Dillard saw the 1979 solar eclipse. "It began," she said, "with no ado. It was odd that such a well-advertised public event should have no starting gun, no overture, no introductory speaker. I should have known right then that I was out of my depth. Without pause or preamble, silent as orbits, a piece of the sun went away."[2]

Even two centuries ago, the commencement of this rare event was already exerting a powerful effect on its beholders. French astronomer François Arago observed the 1842 eclipse from Perpignan in southern France amid townspeople and farmers who had been educated about the eclipse and who were watching the sky intently. "We had scarcely, though provided with powerful telescopes, begun to perceive a slight indentation in the sun's western limb, when an immense shout, the commingling of twenty thousand different voices, proved that we had only anticipated by a few seconds the naked eye observation of twenty thousand astronomers equipped for the occasion, and exulting in this their first trial."[3]

Yet as George B. Airy, Astronomer Royal of England, learned when he saw his first total eclipse in 1842: "No degree of partial eclipse up to the last moment of the sun's appearance gave the least idea of a total eclipse . . ."[4]

## The Crescent Sun

The partial phase of a total eclipse has a power all its own, as the Moon steadily encroaches upon the Sun, covering more and more of its face. A partial eclipse close to total visited a Russian monastery in medieval times, near sunset on May 1, 1185. A chronicler recorded: "The sun became like a crescent of the moon, from the horns of which a glow similar to that of red-hot charcoals was emanating. It was terrifying to men to see this sign of the Lord."[5]

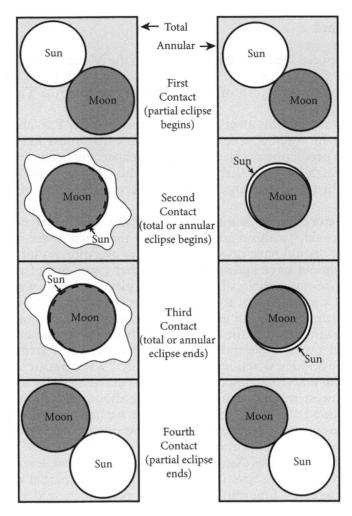

Points of contact in a total solar eclipse (*left*) and an annular eclipse (*right*). In a partial eclipse, there are only first and fourth (last) contacts.

As the partial phase proceeds, it's a good time to use your cardboard pinhole camera to project the Sun in crescent phase on a sheet of white cardboard. Alternatively, use a strainer or cheese grater and get dozens of crescent Suns. You can also project images of the crescent Sun with just your hands. Lace your fingers and let the sunlight pass between them onto the ground or a piece of cardboard.

Check nearby trees and bushes to see if they are casting a thousand tiny images of the crescent Sun on the ground beneath them. Small gaps between the leaves create natural pinhole cameras to focus the crescent image of the waning Sun.

During the early partial stages of a total eclipse, be sure to notice the shadows cast by the Sun. For now they will be the usual solar shadows: a little fuzzy at the edges. But as the crescent of the Sun narrows, they will be quite different.

If you are observing the partial phases with binoculars or a telescope— with a solar filter covering the front of the binoculars or telescope, of course—make note of any sunspots you see, then watch as the Moon steadily occults them. If there are sunspots visible at the limb of the Sun, remember their positions and, when totality is beginning or ending, see if those positions are marked by vivid reddish prominences—magnetic storms in the Sun's atmosphere associated with the sunspots. When the corona flares into view, see if it shows lengthened rays or shock waves near where you saw the sunspots at the edge of the Sun. You may be seeing evidence of eruptions from the Sun—coronal mass ejections—rushing through the corona and out into space at more than a million miles per

Pinhole images of the crescent Sun are formed through the leaves of a tree during the solar eclipse of October 3, 2005 from St Juliens, Malta. [Canon PowerShot A20, f/2.8, 1/40 second. Wikimedia Commons]

hour (1.6 million km/h). But that's too slow for you to notice this motion within the corona during a few minutes of totality.

## The Changing Environment

Every eclipse veteran urges newcomers to pause as totality approaches to look around at the landscape and notice the changes in light levels and color. Many people are surprised how little the landscape darkens until the last 15 minutes before the eclipse becomes total. All the better for the drama, because once the light begins to fade, it sinks quite noticeably. Usually, everyone becomes silent. You can feel the tension and rising emotion. A primitive portion of your brain tugs at you to say that something peculiar is going on, that it ought not to be growing dark in the midst of the day.

During the 1976 eclipse in Australia, Dennis di Cicco remembers that the birds responded to the fading light by raising a racket and going to roost. In 1973, he witnessed an annular eclipse in Costa Rica that took place shortly after sunrise. The cows grazing around him ignored the partial phases of the eclipse, but as the eclipse reached maximum, the cows ceased grazing, formed a line, and marched back to their barn.

Walter Roth watched the 1973 total eclipse from a game preserve in Africa. As the light faded in the minutes just before totality, the birds flocked into the trees, complaining madly. Elephants, which had been grazing peacefully, milled around, nervous and confused.

"As the light fades, the Sun is a thin crescent and shrinking quickly," says Richard Berry. "The quantity and quality of the light have begun to change noticeably. The temperature is dropping; the air feels still and strange."

## The Sharp Edge of Shadows

On ordinary days and in the early partial phase of the eclipse, sunlight falling on an object casts a shadow with fuzzy edges. Now, as the crescent of the Sun narrows and the daylight dims, look again at the shadows cast by trees and leaves and buildings—and you. Spread your fingers and let your hand cast a shadow on the ground. Look how sharp the edges of the shadow are.

On an ordinary day, every part of the disk of the Sun—top, bottom, left side, right side—is casting your shadow from a slightly different angle, creating a fuzzy edge. Now the Sun is reduced to a thin crescent, more like a narrow slit than a disk. Your shadow is fuzzy in one direction but razor sharp in the other. This is no ordinary day.

## Color Change

It's now about 15 minutes before totality. What's happening to the colors around you: the sky, the landscape, the buildings, the water, the trees, and leaves? The colors are fading. The sky is taking on a steely blue-gray, even purplish cast. You've never seen the sky that color, even with the approach of a storm. Yet today the sky is clear, or nearly so. The Sun is still there in the sky. But the light is going out. The shadows are sharp, crisp.

## Temperature Drop

As you waited for first contact, for the eclipse to begin, the Sun was warming the Earth. The temperature was rising, as it does on a usual day. But now—what's happening? Is it how weird nature is beginning to look? Is it how weird you feel? The temperature is no longer rising. It's maybe even a little cooler. It will be getting noticeably cooler soon as more and more of the Sun is blocked by the Moon.

If small cumulus clouds have been puffing up, maybe their buildup has stopped. Maybe some clouds have begun to thin and disappear because less warm, moist air is rising from the ground and because the upper atmosphere cools more rapidly than the lower air. The cooling upper air sinks, taking the clouds lower, where the slightly higher atmospheric pressure causes the water droplets that form the clouds to heat up and vaporize, turning the droplets of the clouds into transparent water gas scattered in the atmosphere.

The cooling as the Sun is increasingly blocked from view is not sufficient to break up a developing thunderstorm or a frontal system, but the approach of totality can substantially reduce clouds and improve observing conditions.

## The Moon's Shadow Visible

Now look toward the western horizon. Looming there, growing ever larger, is the Sun-cast dark shadow of the Moon. And it is coming toward you. Alan Fiala described its appearance as the granddaddy of thunderstorms, but utterly calm. If you are observing from a hill with a view to the west, the approach of the Moon's shadow can be quite dramatic, even chilling.

In the words of astronomer Mabel Loomis Todd more than a century ago, it is "a tangible darkness advancing almost like a wall, swift as imagination, silent as doom."[6]

## Shadow Bands

As the eclipse nears totality and shortly after it emerges from totality, shadow bands—faint undulations of light rippling across the ground at jogging speed—are sometimes visible. They are one of the most peculiar and least expected phenomena in a total eclipse. Many eclipse veterans have never seen them. Some do not want to take time to try because they occur in the last moments before totality as Baily's beads and other beautiful sights are visible overhead. Other veterans consider shadow bands one of the true highlights of a total eclipse. They resemble the graceful patterns of light that flicker or glide across the bottom of a swimming pool.

### Catching Shadow Bands

*by Laurence A. Marschall*

Even though shadow bands are only visible for a few fleeting minutes, it is possible to catch them if you prepare in advance. Get a large piece of white cardboard or white-painted plywood to act as a screen—the bands are subtle and can be more easily seen against a clean, white surface. A large white sheet staked to the ground may be more portable and will serve in a pinch, but ripples in the sheet can mask the faint gradations of the shadow bands.

Lay out on the screen one or two sticks marked with half-foot intervals (yardsticks will do nicely). Orient the sticks at right angles to one another so that at the first sign of activity you can move one stick to point in the direction that the shadow bands are moving. Then, using the marks on that stick, make a quick estimate of the spacing between the bands (typically 4–8 inches; 10–20 centimeters). Finally, using a watch, make a quick timing of how long a bright band takes to go a foot or a yard. Jot down the figures or, better yet, dictate your measurements into a small audio recorder. If you practice this procedure before the eclipse, you will be able to see the shadow bands and then swing your attention back to the sky to catch the diamond ring, Baily's beads, and the onset of totality.

The second stick, by the way, is reserved for marking the direction of the shadow bands *after* totality, if they are visible.

After the eclipse, you can take stock of your data. Do the shadow bands seem to move at all? At some eclipses, especially when the air is very still, they just shimmer without going anywhere. If they move, how fast? Typical speeds are about 5–10 miles per hour (8–16 kilometers per hour). Do they change directions after eclipse? Usually they do, unless you happen to be standing directly along the central line of totality.

Dr. Laurence A. Marschall is Professor of Physics (Emeritus) at Gettysburg College and Deputy Press Officer, American Astronomical Society.

Shadow bands occur when the crescent of the remaining Sun becomes very narrow so that only a thin shaft of sunlight enters the atmosphere of Earth overhead. There it encounters currents of warmer and cooler air which have slightly different densities. The different densities act as weak prisms to bend the light passing from one parcel to the next.[7] It is the slight bending and re-bending of light by these ever-present air currents in motion that causes stars to twinkle. The Sun would twinkle too if it were a starlike dot in the sky. And so it does, near the total phase of a solar

## The Personalities of Eclipses

*by Stephen J. Edberg*

Each total eclipse is different from all others and these differences continue to lure eclipse veterans. They use them in planning their observations.

One factor is the magnitude of the total eclipse, the degree to which the disk of the Moon more than covers the disk of the Sun. When the angular size of the Moon is great, the Moon at mid-totality will mask not only the Sun's photosphere but also its chromosphere and lowermost corona. Except at the beginning and end of totality (or near the eclipse path limits), the stunning fluorescent pink of the prominences will be hidden from view, unless an absolutely gigantic prominence happens to be present on the Sun's limb (as there was for the 1991 eclipse).

However, because a large magnitude eclipse blocks from view the relatively bright lower corona, a large magnitude eclipse is the best time to observe the full extent and detail in the corona. Because the Moon appears larger, its shadow is wider. Thus the total eclipse lasts longer and the darkness is deeper, allowing the full majesty of the corona to shine through, as well as any planets and bright stars that happen to be above the horizon.

The Moon's apparent motion over the Sun's disk is easier to photograph during long eclipses. One photo taken at the onset of totality and a second taken just before the end maximizes the change in the Moon's position with respect to the Sun. When viewing these pictures as a stereo pair, the Moon appears to actually hang in front of the corona and prominences.

By contrast, in a smaller magnitude total eclipse, the Moon's disk is not big enough to mask the lower corona, and prominences may be seen all the way around the disk of the Sun during most of totality. But this eclipse will be comparatively brief and the full extent of the corona may not be evident.

A total eclipse with less obscuration often provides a better display of Baily's beads and the diamond ring effect. Because the Moon's disk is nearly the same apparent size as the Sun's disk, as totality nears, the crescent of the Sun is long and narrow, allowing Baily's beads to glimmer like jewels on a necklace. When the Moon's apparent disk is large, the length of the Sun's crescent is greatly shortened as it thins. There may be only one or two Baily's beads.

eclipse. When the Sun has narrowed from a disk to a sliver, the light from the sliver twinkles in the form of shadow bands rippling across the ground.

In 1842, George B. Airy, the English Astronomer Royal, saw his first total eclipse of the Sun and recalled shadow bands as one of the highlights:

As veteran observers plan for upcoming eclipses, they also take into consideration the sunspot cycle. At sunspot maximum, the corona is brighter, rounder, and larger. Prominences also tend to be more numerous. At sunspot minimum, the corona is fainter and broader at the equator than at the poles. Brush-like coronal features projecting from the poles are more noticeable.

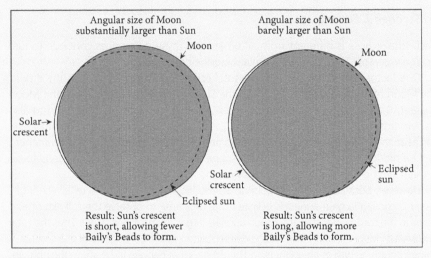

*Left: When the Moon's angular size is large, the Sun's crescent is short, reducing the span of Baily's beads. Right: When the Moon just barely covers the Sun, so that the disks appear nearly the same size, the Sun's crescent is much longer and the span of Baily's beads is greater.*

A third factor used by eclipse followers is the proximity of the shadow path to the Earth's equator, where the rotation of the Earth causes the ground speed of the Moon's shadow to be slowest, making the duration of totality longest. Eclipses with 6 to 7 minutes of totality are almost always found between the Tropic of Cancer and the Tropic of Capricorn. The rotation of the Earth extends not just the period of totality but all aspects of the eclipse, so that near the equator, the partial phases of the eclipse last longer. The last crescent of the Sun is covered more slowly, so Baily's beads and the diamond ring effect, while still brief, last a little longer, and the view of the prominences at the limb of the Sun is prolonged.

Stephen J. Edberg is a Jet Propulsion Laboratory scientist, astronomy author, and executive director of the RTMC Astronomy Expo, a major annual conference of amateur astronomers.

Shadow bands on an Italian house in 1870. [Mabel Loomis Todd: *Total Eclipses of the Sun*]

"As the totality approached, a strange fluctuation of light was seen . . . upon the walls and the ground, so striking that in some places children ran after it and tried to catch it with their hands."[8]

## The Approach of Totality

During the final one to two minutes before totality, the Sun is about 99% covered and the light is fading fast. Now everything happens at once. In these brief moments, the Moon's shadow on the horizon rushes at you, while at the rim of the vanishing Sun, the corona looms into view and the sliver of sunlight becomes a brilliant diamond ring, then breaks up into Baily's beads.

This is how astronomer Isabel Martin Lewis in 1924 described the onrush of the Moon's shadow as totality begins: "When the shadow of the moon sweeps over us we are brought into direct contact with a tangible presence from space beyond and we feel the immensity of forces over which we have no control. The effect is awe-inspiring in the extreme."[9]

Glenn Schneider went to Roy, Montana for the 1979 eclipse. He watched the approaching shadow of the Moon growing on the western horizon dark and vast. Suddenly it flew at him. Instinctively, he ducked.

As the shadow races toward you, it seems to accelerate. Ten seconds before totality, you feel as if you are being swallowed by a gigantic whale.

At this moment, above you—there it is. The soft white glow of the corona begins to silhouette the Moon's dark disk while a last dazzling beacon of sunlight clings to one edge of the Moon—the diamond ring effect.[10] "The remaining light seems to scintillate," says Michael Zeiler. "Your surroundings shimmer—eerie, sharp, electric."

A few seconds before totality begins, the ends of the remaining sliver of the Sun fracture into Baily's beads. The last rays of sunlight are shining through the lunar valleys. Each bead lasts only an instant and flickers out until only one remains. "For one fleeting moment this last bead lingers, like a single jewel set into the arc that is the lunar limb," says John Beattie. And then it is gone.

The eclipse is total.

Diamond ring effect at the total solar eclipse of July 11, 2010 from Easter Island. [Nikon D700, Borg 100ED & 2× teleconverter, fl = 1280 mm, ISO 800, f/12.8 at 1/4000 second. © 2010 Dave Kodama]

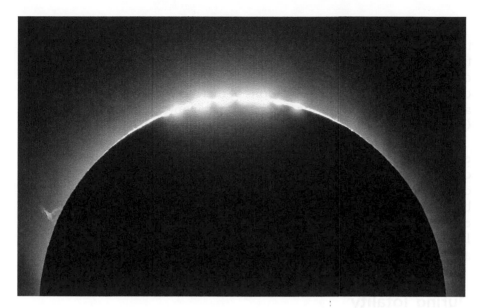

Baily's beads are seen just before and after the total phase of a solar eclipse. They are formed by sunlight passing through deep valleys along the edge of the Moon. This image was shot from Jinta, China on August 1, 2008. [Nikon D300, Vixen 90 mm f/9 Fluorite Refractor, 1/1000 second, ISO 200. © 2008 Fred Espenak]

## The Corona

It is difficult to find words that do justice to the corona, the central and most surprising of all features of a total eclipse. The Sun has vanished, blocked by the dark body of the Moon. A black disk surrounded by a white glow. "It's the blackest black I've ever seen," says Michael Zeiler. "It is the eye of God," says Jack Zirker.

In the bitter cold of the high plateau in Bolivia, astronomer Frank Orrall was part of a research team studying the total eclipse of 1966. In reviewing his observing notes, he realized that he had written "The heavens declare the glory of God." "My notebooks normally do not say such things," he mused.

The color of the corona is difficult to capture. "I prefer 'pearly,'" said Ruth Freitag, "because it conveys a luminous quality that 'white' lacks." "A silver gossamer glow," suggests Steve Edberg. To Michael Zeiler "it appears to be a living object with delicate filamentary structure." "Remember," said George Lovi, "the human eye is the only instrument that can see the corona in all its splendor."

Take time to examine the corona. Is it symmetrical or bunched up at the Sun's equator? Do you see bristle-like rays extending outward from the poles of the Sun—"polar brushes"? Are there streamers stretching outward from the equator? Do you see loops, arcs, or plumes in the corona? Any swirls or concentrations of material? You might be looking at a coronal mass ejection—the Sun hurling million-degree gases into space. Only during a total eclipse can you see the corona with your unaided eyes.

How far can you see the corona extending away from the Sun? You can estimate it by comparing it to the diameter of the eclipsed Sun, which is conveniently and just barely covered by the Moon. The angular diameter of the Moon—and Sun—is ½°. Does the corona extend away from the black rim of the Moon by two solar diameters? Three? More? The diameter of the Sun is 865,000 miles (1.4 million km)—almost a million miles. If you can see the corona extending outward three solar diameters, that's about 3 million miles (about 4¾ million km) into space.

## During Totality

In the midst of totality, it is particularly impressive to look around at the nearby landscape and the distant horizon. Depending on your position within the eclipse shadow, the size of the shadow, and cloud conditions away from the Sun's position, the light level during totality can vary greatly. It also varies from eclipse to eclipse and from place to place within an eclipse. On some occasions, totality brings the equivalent of twilight soon after sunset. On others, it is dark enough to make reading difficult. Yet it is never as black as night.

The color that descends upon everything is hard to specify. Most veteran observers describe it as an eerie bluish gray or slate gray, and then apologize for the inadequacy of the description. The Sun's corona contributes some brightness, but only about as much as a full moon. Primarily, the light that brightens your location within the shadow of the Moon is light reflected from miles away where the Sun is not totally eclipsed.

There in the distance in every direction, it is twilight—painted with sunset oranges and yellows. It surrounds you. The colors are familiar. Yet it's all so strange. This twilight does not tower high into the sky. It is cut off by something dark overhead—a lid that confines the colors to a band around the horizon. You are seeing beyond the Moon's shadow. Out there, far away, the eclipse is not total. Where you stand is very special.

Here, in the midst of totality, looking around at the colors on the horizon and the darkness overhead, you have the fullest impression of standing in the shadow of the Moon. You also have a renewed appreciation for the power of the Sun. There it is, its face completely blocked by the Moon.

How big in the sky is that body whose overwhelming brightness creates the day and banishes the stars from view? Stretch out your arm to full length toward the eclipsed Sun, just as you did to measure the Moon. Squeeze the darkened Sun between your index finger and your thumb until it just fits between them. It is the size of a pea. Yet that little spot in the sky, darkened now, is usually enough to blanket the Earth in light and warmth and completely dazzle the eye. For just a few moments its face is hidden. Daytime has become night and the temperature is falling. Do you feel something of what ancient people must have felt as they watched the Sun, upon which they depended for warmth and light and life itself, disappear?

Leif Robinson emphasized that a total eclipse cannot be experienced vicariously. Even the best photography and video recordings are pale reflections of the event. Taking pictures during an eclipse is fine, but be sure you *look* at what is truly a visual spectacle. Do not miss the ambience of the moment either, he advised. Look at other people to see how they are reacting. Notice the changing colors in the sky.

Robinson did not recommend trying to see planets or stars in the sky during a total eclipse. "Time during totality is too precious to spend straining to see stars when you can see those same objects much better in the nighttime sky." Steve Edberg feels differently: "One of my strongest memories from eclipses is seeing stars during the day."

The glorious corona is visible for just a few short minutes during the total solar eclipse of July 3, 2019 near Punta Colorada, Chile. Details in the Sun's corona are revealed in this HDR composite made from 50 separate exposures. The longest exposure reveals features on the dark face of the Moon (these were not visible to the naked eye). [Panasonic Lumix S1R, CFF Telescopes 80 mm refractor, f/6, ISO 100, 20 exposures: 1/2000 to 2 seconds, © 2019 Catalin Beldea and processed by Alson Wong]

The transformation in the appearance of the Sun from a bright crescent to a dark disk surrounded by ghostly light was recorded by astronomer François Arago as he watched the eclipse of July 8, 1842 with other astronomers and nearly 20,000 townspeople in southern France.

> When the sun, being reduced to a narrow filament, began to throw only a faint light on our horizon, a sort of uneasiness took possession of every breast; each person felt an urgent desire to communicate his emotions to those around him. Then followed a hollow moan resembling that of the distant sea after a storm, which increased as the slender crescent diminished. At last, the crescent disappeared, darkness instantly followed, and this phase of the eclipse was marked by absolute silence . . . The magnificence of the phenomenon had triumphed over the petulance of youth, over the levity affected by some of the spectators as indicative of mental superiority, over the noisy indifference usually professed by soldiers. A profound calm also reigned throughout the air: the birds had ceased to sing.
>
> After a solemn expectation of two minutes, transports of joy, frenzied applauses, spontaneously and unanimously saluted the return of the solar rays. The sadness produced by feelings of an undefinable nature was now succeeded by a lively satisfaction, which no one attempted to moderate or conceal. For the majority of the public the phenomenon had come to a close. The remaining phases of the eclipse had no longer any attentive spectators beyond those devoted to the study of astronomy.[11]

How do you take it all in? It is not possible. There is too much to see—and feel. "Everybody, myself included, tries to do too much," says Steve Edberg. "Save time near the beginning, middle, and end of totality just to stare," urges Jay Anderson. "Make a deliberate effort to store the sights in your mind."

More than a century ago, Mabel Loomis Todd wrote that "[w]hen Dr. Peters of Hamilton College was asked what single instrument he would select for observing an eclipse, he replied, 'A pillow.'"[12]

## The End of Totality

All too soon, it's over. Totality can never last longer than 7 minutes 32 seconds.[13] Even that is not long enough. The Moon moves on in its orbit, beginning to uncover the Sun. The western rim of the Moon's black disk brightens as the Sun's crimson prominences and chromosphere (lower atmosphere) peek out from behind the Moon. A dazzling dot of white light appears, then two, three—a diamond necklace, Baily's beads—joining to form the thinnest of crescents, sunlight bursting from the western edge of the Moon. A second diamond ring.

Too bright to look at. Filters on.

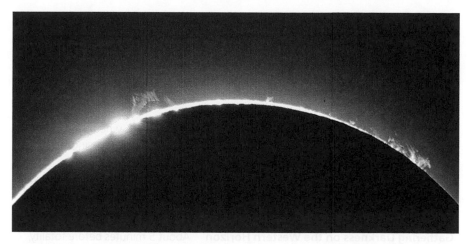

Brilliant red prominences and the chromosphere are joined by Baily's beads at third contact during the total solar eclipse of 2017 from Redfish Lake, Stanley, ID. [Canon EOS 6D, Meade 8 LX10 SCT, f/10, 1/2000 seconds. © 2017 Tunç Tezela]

The corona fades as daylight returns. The dark lunar shadow rushes off to the east, bringing totality to observers farther along the path.

Rebecca Joslin and her Smith College astronomy teacher, Mary Emma Byrd, traveled from the United States to Spain for an eclipse in 1905, only to be clouded out. So they shifted their attention to nature around them until light pierced the clouds to tell them that the unseen total phase of the eclipse was over.

> But we hardly had time to draw a breath, when suddenly we were enveloped by a palpable presence, inky black, and clammy cold, that held us paralyzed and breathless in its grasp, then shook us loose, and leaped off over the city and above the bay, and with ever and ever increasing swiftness and incredible speed swept over the Mediterranean and disappeared in the eastern horizon.
>
> Shivering from its icy embrace, and seized with a superstitious terror, we gasped, "WHAT WAS THAT?" . . . The look of consternation on M's face lingered for an instant, and then suddenly changed to one of radiant joy as the triumphant reply rang out, "THAT was the SHADOW of the MOON!"[14]

"I doubt if the effect of witnessing a total eclipse ever quite passes away," wrote Mabel Loomis Todd. 'The impression is singularly vivid and quieting for days, and can never be wholly lost. A startling nearness to the gigantic forces of nature and their inconceivable operation seems to have been established. Personalities and towns and cities, and hates and jealousies, and even mundane hopes, grow very small and very far away.'[15]

## Stages of a Total Eclipse

**First Contact**   The Moon begins to cover the western limb of the Sun.

**Crescent Sun**    Over a period of about an hour, the Moon obscures more and more of the Sun, as if eating away at a cookie. The Sun appears as a narrower and narrower crescent.

**Light and Color Changes**    About 15 minutes before totality, when 80% of the Sun is covered, the light level begins to fall noticeably—and with increasing rapidity. The landscape takes on a metallic gray-blue hue.

**Animal, Plant, and Human Behavior**    As the level of sunlight falls, animals may become anxious or behave as if nightfall has come. Some plants close up. Notice how the people around you are affected.

**Gathering Darkness on the Western Horizon**    About 5 minutes before totality, the shadow cast by the Moon causes the western horizon to darken as if a giant but silent thunderstorm is approaching.

**Temperature**    As the sunlight fades, the temperature may drop perceptibly.

**Shadow Bands**    A minute or so before totality, ripples of light may flow across the ground and walls as the Earth's turbulent atmosphere refracts the last rays of sunlight.

**Thin Crescent Sun**    Only a sliver of the Sun remains, then thinner still until . . .

**Corona**    Perhaps 15 seconds before totality begins, as the Sun becomes the thinnest of crescents, the corona begins to emerge.

**Diamond Ring Effect**    As the corona emerges, the crescent Sun has shrunk to a short, hairline sliver. Together they form a dazzlingly bright diamond ring. Then the brilliant diamond fades into . . .

**Baily's Beads**    About 5 seconds before totality, the remaining crescent of sunlight breaks into a string of beads along the eastern edge of the Moon. These are the last few rays of sunlight passing through deep valleys at the Moon's limb, creating the momentary effect of jewels on a necklace. Quickly, one by one, Baily's beads vanish behind the advancing Moon as totality begins.

**Shadow Approaching**    While all this is happening, the Moon's dark shadow in the west has been growing. Now it rushes forward and envelops you.

**Second Contact**    Totality begins. The Sun's disk (photosphere) is completely covered by the Moon.

**Prominences and the Chromosphere**    For a few seconds after totality begins, the Moon has not yet covered the lower atmosphere of the Sun and a thin strip of the vibrant red chromosphere is visible at the Sun's eastern limb. Stretching above the chromosphere and into the corona are the vivid red prominences. A similar effect occurs along the Sun's western limb seconds before totality ends.

**Corona Extent and Shape**    The corona and prominences vary with each eclipse. How far (in solar diameters) does the corona extend? Is it round or is it broader at the Sun's equator? Does it have the appearance of short bristles at the poles? Look for loops, arcs, and plumes that trace solar magnetic fields.

**Planets and Stars Visible**    Venus and Mercury are often visible near the eclipsed Sun, and other bright planets and stars may also be visible, depending on their positions and the Sun's altitude above the horizon.

**Landscape Darkness and Horizon Color**    Each eclipse creates its own level of darkness, depending mostly on the Moon's angular size. At the far horizon all around you, beyond the Moon's shadow, the Sun is shining and the sky has twilight orange and yellow colors.

**Temperature**    Is it cooler still? A temperature drop of about 10 °F (6 °C) is typical. The temperature continues to drop until a few minutes after third contact.

**Animal, Plant, and Human Reactions**    What animal noises can you hear? How are other people reacting? How do you feel?

**End of Totality Approaching**    The western edge of the Moon begins to brighten and vividly red prominences and the chromosphere appear. Totality will end in seconds.

**Third Contact**    One bright point of the Sun's photosphere appears along the western edge of the Moon. Totality is over. The stages of the eclipse repeat themselves in the reverse order.

**Baily's Beads**    The point of light becomes two, then several beads, which fuse into a thin crescent with a dazzling bright spot emerging, a farewell diamond ring.

**Diamond Ring Effect and Corona**    As the diamond ring brightens, the corona fades from view. Daylight returns.

**Shadow Rushes Eastward**

**Shadow Bands**

**Crescent Sun**

**Recovery of Nature**

**Partial Phase**

**Fourth Contact**    The Moon no longer covers any part of the Sun. The eclipse is over.

"Beware of post-eclipse depression," warns Steve Edberg. One minute after third contact, with the passage of the Moon's shadow, the disappearance of the shadow bands, and the reappearance of the crescent Sun, you feel exhausted: worn out by totality and the excitement of getting ready for it. Most observers are too tired to watch after third contact, and what follows is anticlimactic anyway. The cure for blue sky blues? "Socialize," says Edberg. "Ask people what they saw. Share war stories."

"The end of totality is a time for celebration," says Jay Anderson. "In the wake of such a powerful shared experience, conversations become animated and casual acquaintances tend to become lifelong friends."

George Fleenor watched the 1991 eclipse from the beach at San Blas, Mexico. Not everyone there had come to see the eclipse. As the Moon covered more and more of the Sun, people continued to play in the ocean. And then the Sun went out. The people in the water stopped where they were and started screaming and cheering. When the Sun emerged from totality, the people in the water started dancing.

As Mike Simmons watched the 2006 total eclipse in Turkey, there was a video camera recording the reactions of the couple near him. The husband was observing with binoculars. His wife was standing behind him. As the eclipse became total, the wife began hammering on the husband's back with both fists and screaming. Afterwards, the wife had no recollection of her reaction to totality. But her husband did. He remembers ducking.

Charles Piazzi Smyth, Astronomer Royal of Scotland, saw his first total eclipse in 1851 and recognized its ability to overwhelm an observer, distracting him from his carefully planned research. "Although it is not impossible but that some frigid man of metal nerve may be found capable of resisting the temptation," he wrote, "yet certain it is that no man of ordinary feelings and human heart and soul can withstand it."[16]

## Confessions of Eclipse Junkies

What if you are clouded out, as happens even with good planning about one out of every six times? Unlike a rocket countdown, which can be "T minus 30 minutes and holding" while the clouds clear, an eclipse countdown is inexorable, and it is not always possible to race to another location where there is a break in the clouds.

"My first eclipse was in Maine in the summer of 1963," Joe Hollweg recalls. Totality was about one minute long. "I was working in New York City, but I made a trip to Maine just for the eclipse. The sky was partly cloudy, with lots of small cottony fairweather clouds. What a disappointment when one of those little puffs moved over the Sun just at the start of totality! I got to see Baily's beads, and that was all. A few hundred yards away people were in the clear, and they were cheering with excitement. However, I do remember seeing the Moon's shadow racing across those puffy clouds, both at the start of the eclipse and at the end. The motion of the shadow seemed incredibly rapid, perhaps because the clouds weren't very high. That was the first time I had some sense of how fast 1,000 miles per hour really is."

All the veterans agree: plan your eclipse trip so that you see things, go places, and meet people that will shine in your memory even if the corona does not. "An eclipse is the perfect excuse to go traveling," says Jay Anderson.

Crucial to the enjoyment of a solar eclipse is eye safety. You wouldn't stare directly at the Sun during a normal day, so you shouldn't stare at the Sun when it is partially eclipsed without proper eye protection (described in Chapter 10). However, when the Sun is totally eclipsed—no portion of its disk is showing—it is perfectly safe to look at the Sun without any eye protection. A filter would hide the view.

The eerie twilight of totality silhouettes astronomers as they quickly make their measurements during the total solar eclipse of February 16, 1980 near Hyderabad, India. [Nikon FE SLR, 24 mm lens, f/2.8, 1 second, ISO 200 film. © 1980 Jay M. Pasachoff]

## Equipment Checklist for Eclipse Day

Checklist of your intended activities during eclipse

Solar filters for your eyes

Portable seat or ground covering

Binoculars and/or telescope

Solar filters for binoculars and/or telescope

Camera equipment and tripod

Video camera and tripod

Flashlight with new batteries and a piece of red gel to filter flashlight during totality

Straw hat, kitchen pasta colander, or cooking spoon with small holes to project pinhole images of partially eclipsed Sun on a white piece of cardboard

Suitable clothing and hat (you will be outside for several hours)

Sunglasses (not for direct viewing of partial phases)

Bug repellent, sunscreen lotion, basic first aid kit

Snacks and a canteen of water

Audio recorder for your comments and impressions or to capture reactions of people or wildlife near you

Audio recorder with earphones and prerecorded message timed to cue you about what you want to do next (to run from about 5 minutes before totality until 5 minutes after totality)

Pencil and paper to record impressions or to sketch (also to take down the names and addresses of fellow observers)

Jay Pasachoff was irritated by governments and news media in foreign lands—and the United States—that mislead the public by implying that the Sun gives off "special rays" at the time of an eclipse. Instead of teaching simple safety procedures, they frighten people, thereby depriving them of a rare and magnificent sight.

Alan Fiala recalled the 1980 eclipse in India where pregnant women were instructed to remain indoors during the eclipse. For the 1992 eclipse, Roger Tuthill rented a jumbo jet in Brazil to meet and chase the shadow over the Atlantic, using the speed of the aircraft to expand three minutes of totality into six. To serve the 50 passengers aboard the DC-10, the airline assigned 12 flight attendants. There were plenty of windows for them to watch the eclipse, but almost all the stewardesses refused to look. They feared that if they saw the Sun in eclipse, they could never become pregnant.

The pilot, however, was not afraid to look. He was wildly enthusiastic. Even after the Moon's shadow outran the jet and totality was over, the pilot pressed on eastward toward Africa because he was so fascinated by the fast-moving pillar of darkness ahead of him. "It's even more exciting than an engine fire," he explained.

George Lovi, Alan Fiala, and others remember the 1983 eclipse in Indonesia. Before the eclipse, the streets were teeming with people. On eclipse morning, however, there was scarcely anyone outdoors. Fire sirens wailed like an air raid warning for people to take cover. School children, on government orders, were kept indoors, forbidden to see the eclipse, except perhaps on television. A stunning natural event that comes to you, if you are lucky, once in five lifetimes, was denied to them.

On Java for that 1983 eclipse, Jay Anderson recalls soldiers patrolling the streets to discourage unauthorized observers, although the citizens were so frightened by government warnings that almost all stayed indoors. As the Sun was gradually disappearing, the soldiers near him began to be caught up in the excitement. Anderson offered them a view through his telescope, but the warnings they had received overtook them and they refused. Several observers gave the officers a few minutes of careful explanation before the officers nervously took a peek. "After their first look all apprehension disappeared, and they participated in the event as fully as we did. Ten minutes before totality, the officer in charge was interrupted by a call on his walkie-talkie from his superior at headquarters:

'How are things going out there?'

'OK. There are no problems.'

'It's time to come in now. Collect your men and bring them back to the barracks. We are supposed to have all troops back before the eclipse begins.'

'There are no problems here; everyone is safe; the tourists are all working on their equipment. Are you sure you want us in? I see no problems.'

'Orders are to bring everyone in, and leave the tourists to themselves. Bring the men in.'

'I'm sorry, I can't hear you. I'm having trouble with the radio. What did you say?'

At this point the officer turned off the radio and put it back in his pocket. The entire troop stayed to watch the eclipse with us."

Even more memorable to Anderson were two dozen Indonesian students from a local college. They were studying English and attached themselves to Anderson's group to practice their speaking skills. As totality neared, the students became extremely nervous. To reassure them, Anderson and his fellow observers held hands with them during totality—the most magical moment in all the eclipses he has seen.

## NOTES AND REFERENCES

1.   Isabel Martin Lewis: *A Handbook of Solar Eclipses* (New York: Duffield, 1924), page 3.
2.   Annie Dillard: "Total Eclipse" in *An Annie Dillard Reader* (New York: Harper Perennial, 1995).
3.   François Arago: *Popular Astronomy*, volume 2, translated by W. H. Smyth and Robert Grant (London: Longman, Brown, Green, Longmans, and Roberts, 1858), page 360.
4.   George B. Airy: "On the Total Solar Eclipse of 1851, July 28" in Bernard Lovell, editor: *Astronomy*, volume 1, The Royal Institution Library of Science (Barking, Essex: Elsevier Publishing, 1970), page 1.
5.   Anton Pannekoek: *A History of Astronomy* (London: G. Allen & Unwin, 1961), page 406.
6.   Mabel Loomis Todd: *Total Eclipses of the Sun*, revised edition (Boston: Little, Brown, 1900), page 21.
7.   Johana L. Codona: "The Enigma of Shadow Bands," *Sky and Telescope*, volume 81, 1991, page 482.
8.   George B. Airy: "On the Total Solar Eclipse of 1851, July 28" in Bernard Lovell, editor: *Astronomy*, volume 1, The Royal Institution Library of Science (Barking, Essex: Elsevier Publishing, 1970), page 4. This speech was given on May 2, 1851, prior to the total eclipse on July 28.
9.   Isabel Martin Lewis: *A Handbook of Solar Eclipses* (New York: Duffield, 1924), page 62.
10.  On rare occasions, depending on the terrain at the rim of the Moon, a double diamond ring is possible—and predictable.
11.  François Arago: *Popular Astronomy*, volume 2, translated by W. H. Smyth and Robert Grant (London: Longman, Brown, Green, Longmans, and Roberts, 1858), pages 360–361.
12.  Mabel Loomis Todd: *Total Eclipses of the Sun*, revised edition (Boston: Little, Brown, 1900), page 19.
13.  Jean Meeus, "The Maximum Possible Duration of a Total Solar Eclipse," *Journal of the British Astronomical Association*, volume 113, number 6 (December 2003), pages 343–348.
14.  Rebecca R. Joslin: *Chasing Eclipses: The Total Solar Eclipses of 1905, 1914, 1925* (Boston: Walton Advertising & Printing, 1929), pages 14–15.
15.  Mabel Loomis Todd: *Total Eclipses of the Sun*, revised edition (Boston: Little, Brown, 1900), page 25.
16.  Mabel Loomis Todd: *Total Eclipses of the Sun*, revised edition (Boston: Little, Brown, 1900), page 174.

# A MOMENT OF TOTALITY

## The Accidental Eclipse Tourist

The total solar eclipse of 1992 would arc across the South Atlantic Ocean so Glenn Schneider and his friends chartered a jet to fly from Rio de Janeiro out over the ocean to intercept the eclipse, using Glenn's calculations. The problem was timing. The eclipse wouldn't wait for a late-arriving aircraft and the Rio airport was notorious for ground delays. So the captain was delighted when the Rio control tower gave him clearance a little early to push back from the gate, start his engines, and begin to taxi for takeoff. He accepted. The cabin doors closed, the jetway retracted, the pushback truck eased the aircraft away from the terminal, and the engines began to roar.

Suddenly there was a frantic woman pounding on the cockpit door. She was a member of the ground crew still onboard cleaning the plane. Let me off, she demanded. The plane wasn't scheduled to leave yet. Go back to the gate. Extend the jetway. She had to get off. No, the captain said. The eclipse would wait for no one. He couldn't risk a delay waiting for another clearance. The woman was incensed. You have to let me off, she shouted. She had another plane to clean, and more after that. Her boss would be furious if she wasn't there. She would lose her job. She was being kidnapped.

You are coming with us, the captain said, but I will radio the airline and let your boss know what happened and where you are, that it's not your fault, and that you'll be with us for the next four hours.

The flight crew found her a small window to view from and the eclipse chasers gave her eclipse glasses and instructions on how to use them. So, without intending to, and even against her will, the cleaning crew woman saw the eclipse—and was flabbergasted. After the eclipse she thanked everyone aboard, grateful that she had been trapped aboard an aircraft bound for a total eclipse.*

---

* Glenn Schneider is a University of Arizona astronomer. Interviewed May 5, 2015.

# 10

◄◦►

# Eye Safety During Solar Eclipses

*People expect a total solar eclipse to be a curiosity.*
*They don't expect it to move their souls.*

David Makepeace, Eclipse veteran (2015)[1]

Watching the Sun safely during an eclipse is simple. **When the Sun is *partially* eclipsed, you need eye protection. When the Sun is *totally* eclipsed, you need no eye protection at all.**

You would never think of staring at the Sun without eye protection on an ordinary day. You know the disk of the Sun is dazzlingly bright, enough to damage your eyes permanently. Likewise, any time the disk of the Sun is visible—throughout the partial phase of an eclipse—you need proper eye protection. Even when the Sun is nearing total eclipse, when only a thin crescent of the Sun remains, the 1% of the Sun's surface still visible is about 10,000 times brighter than the full moon.

Once the Sun is entirely eclipsed, however, its bright surface is hidden from view and it is completely safe to look directly at the totally eclipsed Sun without any filters. In fact, it is one of the greatest sights in nature.

Here are ways to observe the partial phases of a solar eclipse without damaging your eyes.

## Solar Eclipse Glasses

The most convenient way to watch the partial phases of an eclipse is with solar eclipse glasses. These devices consist of solar filters mounted in cardboard frames that can be worn like a pair of eyeglasses. If you normally wear prescription eyeglasses, you place the eclipse glasses right in front of them.

The filters in eclipse glasses are made from one of several types of materials. Black polymer is the most common type of filter and is composed of carbon particles suspended in a stiff plastic. It produces a natural yellow image of the Sun. Here are companies that specialize in black polymer solar eclipse glasses:

A Samburu man wears a pair of eclipse glasses in preparation for an annular eclipse in Kenya. These inexpensive glasses with cardboard frames have become very popular for safe eclipse viewing. [© 2010 Fred Espenak]

American Paper Optics www.eclipseglasses.com/

Rainbow Symphony www.rainbowsymphony.com/

Thousand Oaks Optical www.thousandoaksoptical.com/

An older alternative to black polymer is the shiny silver filter made of aluminized polyester (Mylar).[2] A minor problem with this material is that it gives the Sun an unnatural blue color. But more importantly, aluminized polyester may have small pinholes that could allow unfiltered sunlight to reach your eyes and damage them. For that reason, aluminized polyester has largely been replaced by safer materials.

Advertisements for other eclipse glasses may be found in popular astronomy and science magazines and on websites. Check the glasses for a printed message certifying that they are safe for eclipse viewing. And verify the optical density of eclipse glasses by making sure you can look comfortably at the filament of a high-intensity lamp.

When you are using a filter, do not stare for long periods at the Sun. Look through the filter briefly and then look away. In this way, a tiny hole that you miss will not cause you any harm. You know from your ignorant childhood days that it is possible to glance at the Sun and immediately look away without damaging your eyes. Just remember that your eyes can be damaged without you feeling any pain.

## Welders' Goggles

Another safe filter for looking directly at the Sun is welders' goggles (or the filters for welder's goggles) with a shade of 13 or 14. They are relatively inexpensive and can be purchased from a welding supply company. The down side is that they cost more than eclipse glasses and give the Sun an unnatural green cast.

## The Pinhole Projection Method

If you don't have eclipse glasses or a welder's filter you can always make your own pinhole projector, which allows you to view a *projected* image of the Sun. There are fancy pinhole cameras you can make out of cardboard boxes, but a perfectly adequate (and portable) version can be made out of two thin but stiff pieces of white cardboard. Punch a small clean pinhole in one piece of cardboard and let the sunlight fall through that hole onto the second piece of cardboard, which serves as a screen, held behind it. An inverted image of the Sun is formed. To make the image larger, move the screen farther from the pinhole. To make the image brighter, move the screen closer to the pinhole.

Do not make the pinhole wide or you will have only a shaft of sunlight rather than an image of the crescent Sun.

Remember, a pinhole projector is used with your back to the Sun. The sunlight passes over your shoulder, through the pinhole, and forms an image on the cardboard screen behind it. Do **not** look through the pinhole at the Sun.

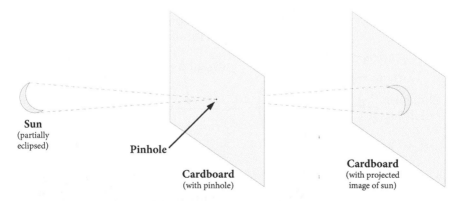

A pinhole projector can be used to watch the partial phases of a solar eclipse safely. It is easily fashioned from two stiff pieces of cardboard. One piece serves as the projection screen. Make a pinhole in the second piece and hold it between the Sun and the first piece. If the two cardboards are held 2 feet apart, the projected image of the Sun will appear about 1/4-inch in size. [Drawing by Fred Espenak]

## Solar Filters for Cameras, Binoculars, and Telescopes

Many telescope companies provide special filters that are safe for viewing the Sun. Black polymer filters are economical but some observers prefer the more expensive metal-coated glass filters because they produce sharper images under high magnification. Baader Planetarium AstroSolar Safety Film is another alternative. It's a metal-coated resin with excellent optical quality and high contrast. The company even offers instructions on how to make an inexpensive cardboard cell to mount the filter on your telescope, binoculars, or camera.

### Eye Damage from a Solar Eclipse

*by Lucian V. Del Priore, MD, PhD*

The dangers of direct eclipse viewing have not always been appreciated, despite Socrates' early warning that an eclipse should only be viewed indirectly through its reflection on the surface of water. A partial eclipse in 1962 produced 52 cases of eye damage in Hawaii, and a total eclipse along the eastern seaboard of the United States produced 145 cases in 1970. As many as half of those affected never fully recovered their eyesight.

There is nothing mysterious about the optical hazards of eclipse viewing. No evil spirits are released from the Sun during a solar eclipse, and there is no scientific reason for running indoors to avoid "the harmful humors of the Sun." Eye damage

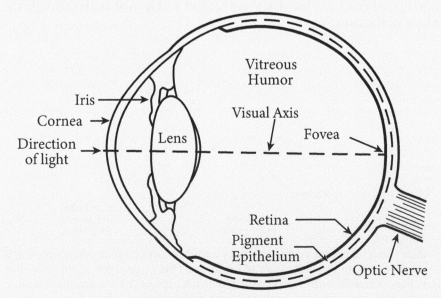

Cross-section of the eye. [Drawing by Josie Herr]

from eclipse viewing is simply one form of light-induced ocular damage, and similar damage can be produced by viewing any bright light under the right (or should I say the wrong!) conditions.

Light enters the eye through the cornea and is focused on the retina by the optical system in the front of the eye. Any light that is not absorbed by the retina is absorbed by a black layer of tissue under the retina called the retinal pigment epithelium.

The retina is the human body's video camera: it contains nerve cells that detect light and send the electrical signal for vision to the brain. Without it, we cannot see. Most of the retina is devoted to giving us low-resolution side vision. Fine detail reading vision is contained in a small area in the center of the retina called the fovea. People who damage their fovea are unable to read, sew, or drive, even though this small area measures only 1/100th of an inch across, and is less than 1/10,000th of the entire retinal area. Unfortunately, this is the precise area that is damaged if we stare directly at the surface of the Sun. Damage has been reported with less than one minute of viewing. The image of the Sun projected onto the retina is about 1/150th of an inch in size, and this is large enough to seriously damage most of the fovea.

Why is sunlight damaging to a structure which is designed to detect light? Bright sunlight focused on the retina is capable of producing a thermal burn, mainly from the absorption of infrared and visible radiation. The absorption of this light raises the temperature and literally fries the delicate ocular tissue. There is no mystery here: every schoolchild knows that sunlight focused through a magnifying glass can cause a piece of paper to burst into flames.

Yet Sun viewing seldom produces a thermal burn. All but the most intoxicated viewer would surely turn away before this occurs. Instead, most cases of eclipse blindness are related to photochemically induced retinal damage, which occurs at modest light levels that produce no burn and no pain. Two types of light lesions are recognized clinically and experimentally, and both are probably responsible for the damage observed after improper eclipse viewing. Blue light (400–500 nanometers) damages the retinal pigment epithelium and leads to secondary changes in the retina, while near ultraviolet light (340–400 nanometers) is absorbed by and directly damages the light-sensitive cells in the outer retina.

Viewing a partial eclipse recklessly is not the only way to produce light-induced retinal damage. Sungazing is a well-known cause of retinal damage even in the absence of an eclipse. Numerous cases of blindness have been reported in sunbathers, in military personnel on anti-aircraft duty, and in religious followers who sungaze during rituals and pilgrimages. The Sun is not even required to produce light damage. Other types of bright lights, including lasers and welders' arcs, will have the same effect.

The common thread here is clear: direct viewing of bright lights can damage the retina regardless of the source. A solar eclipse merely increases the number of potential victims—and brings the problem to public attention.

Dr. Lucian V. Del Priore, MD, PhD (physics) is Robert R. Young Professor of Ophthalmology & Visual Science and Chair of Ophthalmology, Yale University School of Medicine; and Chief of Ophthalmology, Yale New Haven Hospital.

Check the Internet or astronomy magazines for dealers offering these filters. Some of the major companies include Celestron, Meade, Orion Telescopes, and Thousand Oaks Optical.

Caution: Do not confuse these filters, which are designed to fit over the *front* of a camera lens or the aperture of a telescope, with a so-called solar *eyepiece* for a telescope. Solar eyepieces are still sometimes sold with small amateur telescopes. They are not safe because they absorb heat and tend to crack, allowing the sunlight concentrated by the telescope's full aperture to enter your eye.

## Sources of Safe Solar Filters

Here are some suppliers of filters that are designed for safe solar viewing with or without a telescope. The list is representative, not exhaustive. Additional sources may be found in advertisements in *Astronomy, Sky & Telescope*, and other popular astronomy magazines.

### United States
American Paper Optics: www.3dglassesonline.com/
Astronomics: www.astronomics.com/
Celestron: www.celestron.com/
DayStar Filters: www.daystarfilters.com/
Lunt Solar Systems: luntsolarsystems.com/
Meade Instruments: www.meade.com/
OPT Telescopes: www.optcorp.com/
Orion Telescopes and Binoculars: www.telescope.com/
Rainbow Symphony: www.rainbowsymphony.com/
Seymour Solar: www.seymoursolar.com/
Spectrum Telescope: www.spectrumtelescope.com
Thousand Oaks Optical: www.thousandoaksoptical.com/

### Canada
Kendrick Astro Instruments: www.kendrickastro.com/
Khan Scope Centre: www.khanscope.com/
KW Telescope: www.kwtelescope.com/

### Europe
Baader Planetarium: www.baader-planetarium.com/
First Light Optics: www.firstlightoptics.com/solar-filters.html
Rother Valley Optics: www.rothervalleyoptics.co.uk/
Solar Scope: www.solarscope.co.uk/

## Eye Suicide

Do not use standard or polaroid sunglasses to observe the partial phases of an eclipse. They are *not* solar filters. Standard and polaroid sunglasses cut down on glare and may afford some eye relief if you are outside on a bright day, but you would never think of using them to stare at the Sun. So you must not use sunglasses, even crossed polaroids, to look directly at the Sun during the partial phases of an eclipse.

Also, do not use smoked glass, medical x-ray film with images on them, photographic neutral-density filters, and polarizing filters. All these "filters" offer utterly inadequate eye protection for observing the Sun.

## Observing with Binoculars

Binoculars are excellent for observing total eclipses. Any size will do.

Astronomy writer George Lovi's favorite instrument for observing eclipses was 7 × 50 binoculars—magnification of seven times with 50-millimeter (2-inch) objective lenses. "Even the best photographs do not do justice to the detail and color of the Sun in eclipse," Lovi said, "especially the very fine structure of the corona, with its exceedingly delicate contrasts that no camera can capture the way the eye can." He felt that the people who did the best job of capturing the true appearance of the eclipsed Sun were the 19th-century artists who photographed totality with their eyes and minds and developed their memories with paints on canvas.

For people who plan to use binoculars on an eclipse, Lovi cautioned common sense. Totality can and should be observed *without* a filter, whether with the eyes alone or with binoculars or telescopes. But the partial phases of the eclipse, right up through the diamond ring effect, must be observed with filters over the objective (front) lenses of the binoculars. Only when the diamond ring has faded is it safe to remove the filter. And it is crucial to return to filtered viewing as totality is ending and the western edge of the Moon's silhouette brightens with the appearance of the second diamond ring.

After all, binoculars are really two small telescopes mounted side by side. If observing a partially eclipsed Sun without a filter is quickly damaging to the unaided eyes, it is far quicker and even more damaging to look at even a sliver of the uneclipsed Sun with binoculars that lack a filter.

If you don't have solar filters for your binoculars, there is a second way to safely view the partial phases with them. Use the binoculars to project the Sun's image onto a white piece of cardboard. Just hold the binoculars

Binoculars can be used to safely project a magnified image of the Sun onto a piece of white cardboard. Never look at the Sun directly through binoculars unless they are equipped with solar filters. [© 2000 Fred Espenak]

two or three feet from the cardboard, using the binoculars' shadow to point them towards the Sun. It takes a little practice but it works great once you get the hang of it. You have two magnified images for the price of one because each half of the binoculars projects a separate image.

If you use binoculars to *project* the image of the Sun, you don't need solar filters. So when totality begins, you have no solar filters to remove. You can quickly bring the binoculars up to your eyes and look through them.

## Observing with a Telescope

Some observers, including astronomy historian Ruth Freitag, prefer to watch the progress of the eclipse through a small portable telescope, which offers stability and is much less tiring to use for extended periods than binoculars. A telescope also provides more detail at higher powers, if this is desired.

Obviously, solar filters are required for the partial phases of the eclipse— one for the main telescope and a second for the finder scope. The solar filters are removed at totality. When Freitag wanted to see a wider view of the corona, she switched to the finder scope.

Inexpensive eclipse glasses are a great way to share the eclipse with others and make new friends. Bring extras to hand out on eclipse day. [© 1997 Fred Espenak]

Make sure, she said, that your solar filters are easy to remove without bumping the telescope.

## A Final Thought

Just remember, George Lovi said, "Don't try to do too much. Look at the eclipse visually. Don't be so busy operating a camera that you don't see the eclipse. And don't set off for the eclipse so burdened down by baggage and equipment that you are tired and stressed and too nervous to enjoy the event."

Astronomer Isabel Martin Lewis also warned of the dangers of too many things to do: "A noted astronomer who had been on a number of eclipse expeditions once remarked that he had never SEEN a total solar eclipse."[3]

## NOTES AND REFERENCES

1. Epigraph: David Makepeace, Canadian filmmaker (<http://eclipseguy.com>), interviewed May 1, 2015.
2. Mylar is a registered trademark of DuPont.
3. Isabel Martin Lewis: *A Handbook of Solar Eclipses* (New York: Duffiel, 1924), page 98.

# A MOMENT OF TOTALITY

## Unexpected Totality

In 1993, when she was 88 years old, Florence Andsager McPherson told this story to her great niece Julie Andsager about growing up with her family on a farm in Reno County, Kansas.

> One day in the summer we were outside in the yard. It became real dark—the chickens all went to roost. Dad was in the field cultivating the corn and came in. Mom got us kids together. She thought the world was coming to an end. I don't think it lasted over 20 minutes. It seemed like hours. In those days, we never had any way to know what was happening. It was an eclipse of the Sun.

The Andsagers were in the path of totality for the eclipse of June 8, 1918.*

* Florence Andsager McPherson, as told to Julie Andsager: *Florence's Memories—as Remembered at Age 89–90*. Self-published family history, 1998.

# 11

◄◉►

# The Strange Behavior of Man and Beast—Modern Times

*Henry Holiday was an artist hired to travel to India for the 1871 eclipse to draw the corona accurately for scientific purposes. Upon seeing the total eclipse, he wrote: "I could with diffi-culty control myself so as to be fit for making a decent observation." As soon as totality ended "[I] plunged my head into water, for I was in a fever of excitement."[1]*

## Beasts in Eclipse

In centuries past, eclipse observers were fascinated by the reaction of animals. In modern times, as totality approaches and as it ends, it's equally interesting to watch how animals respond to nightfall in the midst of day.

During the 2001 total eclipse in Zambia, Pat and Fred Espenak were ser-enaded by thousands of crickets during the 20 minutes surrounding totality.

David Makepeace observed the 1995 eclipse in India, not far from the Taj Mahal. In India, cows are protected animals and roam the streets freely. As the eclipse darkened the morning sky, half a dozen confused cows lay down in the middle of the street and went to sleep. When the Sun returned, it took eight men pulling and pushing with all their might to urge the cows to get back up and get out of the way to allow traffic to move.

At age 13, Kristian Buchman watched the 2008 eclipse from eastern Mongolia. A flock of 20 sheep was grazing contentedly during the partial phases. But as the light began to fade, the sheep gathered together and began a slow trek back to their enclosure.

In 1991, in Baja California, when darkness descended, Mike Simmons remembers nighthawks flying at noon. Mike was in Iran for the 1999 eclipse. As the light faded in the midst of day, ants came out of the ground in huge numbers. When the Sun returned, they vanished.

George Fleenor watched the 1998 eclipse in Aruba near the lighthouse at the north end of the island. He could see the Moon's dark shadow looming in the west and then coming at him across the Caribbean Sea at more than a thousand miles an hour. It made the hair on the back of his neck stand up, he says. He wasn't looking for shadow bands but they were everywhere, rippling vividly. A herd of goats was grazing on the lighthouse grass, bleating noisily. As the sky darkened, they fell silent. Some lay down and went to sleep.

Glenn Schneider traveled to the shores of the St. Lawrence River northeast of Quebec City, Canada for the 1972 total solar eclipse. Eclipse day delivered broken clouds, but 30 seconds before totality he and the other eclipse chasers could see the thin crescent Sun. And then it was gone, covered by clouds. The landscape darkened, but there was no view of the Sun in eclipse. In the false twilight, mosquitos awakened—clouds of mosquitoes. They fed with delight on the heartbroken eclipse goers.

Australian newspaper reporter Stephen O'Baugh reported that during totality for the 1976 eclipse, the wolves and dingoes at the Melbourne Zoo began to howl, an antelope ran into a fence, and birds went to roost.[2]

Eclipse veteran Patrick Poitevin took his wife Joanne to her first total eclipse in Antigua in 1998, an idyllic site. With the diamond ring approaching, hornets suddenly stormed out of a nearby bush and swarmed around them, perhaps startled by the darkening in the midst of the day. Patrick and Joanne watched totality surrounded by hornets, buzzing furiously. Fortunately they were not stung—but Joanne was badly bitten by the eclipse bug. Ever since, Patrick and Joanne always go to eclipses together.

Joe Buchman went to the Beach of Goats, south of Mazatlán, Mexico for the 1991 eclipse. There were no goats but there were chickens roaming the beach near a weather-beaten cinderblock bar and restaurant. As the sky brightness faded to a silvery twilight, the chickens quit clucking and pecking, formed a line, and marched back to their pen.

Glenn Schneider watched the 1976 eclipse from a mountaintop in southeastern Australia after a 14-hour overnight drive from Melbourne to escape bad weather. His observing site overlooked a large valley where a thousand cows were grazing. As the eclipse became total, Glenn and his friends were treated to a bovine symphony—a thousand cows mooing in confusion. The cacophony ended with the return of the Sun.

## Humans in Eclipse

In ancient times, lunar and solar eclipses—especially total eclipses of the Sun—caused desperate fear and bizarre behavior in humans. Gradually through the centuries we have come to appreciate the science and the beauty of eclipses and have stripped away our superstitions.

Or have we?

A group of people gather on a Paris street corner to watch a solar eclipse in 1912. [Eugène Atget, 1912]

People in southernmost India saw an annular eclipse on January 15, 2010. For the residents in Mangalore, farther up the coast, the eclipse was near annular—a very large partial. The *Times of India* reported: "Myths continue to rule. Despite assurances from scientists that the eclipse will have no harmful effects on humans, the majority here preferred to stay indoors during the eclipse hours." Few people walked the streets, few jitney buses ran. "Restaurants and hotels saw a dip in customers as many preferred not to eat during the eclipse hours." Most schools closed because students didn't show up.

But a local amateur astronomy society offered an eclipse observing party at one high school. "Teenagers said they were thrilled at the magic in the sky," the newspaper reported.[3]

Remember Hsi and Ho, the mythical Chinese astronomers who got drunk and bungled an eclipse prediction? Their negligence delayed the drum banging, pot thumping, and vocal screeching needed to scare away the celestial dragon that eats the Sun. But times change, says writer Michael Rogers. When the celestial dragon made another pass at the Sun over Cambodia in 1965 in the form of a near-total annular eclipse, "Lon Nol's troops simply aimed their American automatic rifles straight up in the air and blasted the gluttonous reptile right out of the sky—sustaining, it is reported, only scattered casualties from the bullets that the fleeing reptile spat back to earth."[4]

George Fleenor and his wife Stephanie went to Aruba to see the 1998 eclipse. After the event, they stopped by a bar to hear from native islanders what they had seen. No, they told him, they stayed inside during the eclipse. They would not watch it. "It is forbidden," they said.

Researchers from the Royal Observatory of Belgium went to Baja California for the 1991 eclipse. They set up their equipment in the garden of their hotel. During totality, while the astronomers were concentrating on their measurements, they were surprised "by the weeping and wailing of the hotel staff, who were terrified by the unexpected fall of darkness."[5]

Patrick Poitevin had a similar experience when he watched the 1991 eclipse from San Blas, Mexico. The Mexican government was quite aggressive in its warnings about the potential danger to eyesight from looking at the Sun. As the eclipse neared totality, local citizens screamed at one another: "Don't look, don't look." and they rushed under a shelter to protect themselves from the view.

Sometimes superstitions and misinformation about eclipses can be tragic. In the aftermath of the 1998 eclipse in the Caribbean, an Associated Press story[6] reported:

> They were frightened by an eclipse—and apparently, they were frightened to death.
>
> Four members of a Haitian family have been found dead in their home. Officials say the four may have been killed by an overdose of sleeping pills they took to alleviate their fears of last Thursday's eclipse. They also may have suffocated. They'd plugged all the openings to their home with rags to keep the Sun out.
>
> Radio broadcasts in Haiti say another young girl suffocated Thursday in a home that had been sealed. Thousands of Haitians were afraid that the eclipse would blind them or kill them.
>
> The government declared a national holiday, in hopes of preventing panic. Police ordered pedestrians off the street during the eclipse, yelling, "Go home! It's dangerous to be out!"

## Providing Misinformation

Fortunately, superstitions about eclipses seldom result in tragedy. More often they are colorful—and sometimes humorous. But when they deprive people of an inspiring, once-in-a-lifetime sight, they are truly unfortunate. Far worse though is when modern-day, well-educated government officials and science professionals give out information that robs tens of thousands or millions of people of the chance to see a total eclipse of the Sun.

One such occasion was a total eclipse of the Sun in Australia on October 23, 1976.[7] The headline in the Sydney *Daily Telegraph* on October 23, 1976 blared: "Television is the only safe way to watch today's eclipse of the sun." The first line of the story explained: "Doctors and scientists stress there is NO other safe way to watch the eclipse." Later, the article provided a list of "Don'ts" and "Dos" for the eclipse, including: "Do supervise children strictly as they . . . will be tempted to look. Keep them inside during the danger period."

That admonition was mild compared to the warning issued in Tony Murphy's story the day before under this headline: "Lock up children during eclipse—expert."

The article began: "Children should be locked indoors tomorrow during the solar eclipse." The advice came from C. Waldron, Queensland President of the Australian Optometrical Association, and was endorsed by the state government. Waldron continued: "Parents still have only themselves to blame if their children's sight is harmed—they are the only ones responsible."

The eclipse was not total in Sydney. It was partial with 90% of the Sun obscured. But the eclipse was total in Melbourne and eastward. And the citizens of Melbourne suffered from misinformation as well.

Notices were posted everywhere on walls and poles throughout Melbourne. In huge letters, they screamed:

WARNING!
SOLAR ECLIPSE TODAY
NEVER look directly at the Sun at any time.
You may damage your eyes. . . .
It is DANGEROUS to look at the sun through any kind of filter . . .
It is VERY DANGEROUS to look at the sun through a telescope,
binoculars, a camera lens, or viewfinder.
WATCH THE ECLIPSE ON TELEVISION—
this is the only SAFE way to see the eclipse.
Prepared by: Community Services Centre . . . Melbourne.
Authorized by: Solar Eclipse Committee, 1976 . . . Melbourne.

"Eclipse, but watch out" was the eclipse-day headline of John Rentsch's article for *The Age*, a Melbourne daily newspaper. "Authorities again warned that people looking directly at the eclipse risked permanent eye damage," Rentsch reported. As in Sydney, there was no effort by the government to teach people the simple precautions for observing a natural wonder safely—what filters to use and when to use them.

Instead, "Scientific and Government authorities have urged people wanting to observe the moon passing across the sun to watch the magnified view on television." However, for irresponsible people who insist on viewing the eclipse, thereby damaging their eyes, "[t]he Royal Victoria Eye and Ear Hospital has set up a special clinic to deal with the expected flood of people seeking treatment after the eclipse."

This road warning was photographed along a highway in France before the total eclipse of August 11, 1999. It means Speed Limit 70 km/hr during eclipse period. [Creative Commons (S. Klüsener)]

After such dire warnings, the optometrists and government officials had one concession:

"Authorities said yesterday that although people should not look directly at the sun, there was no harm in being outside during the eclipse."

Perhaps the only expression of skepticism and satire for the stories about the danger of the eclipse was sounded by William Ellis Green, a cartoonist for the Melbourne *Herald*. His cartoon shows two convicts tied to poles against a cinderblock wall riddled with bullet holes. The convicts' eyes are covered with handkerchiefs tied around their heads. They are about to be executed by firing squad. One is saying to the other: "Luckily we're blindfolded—we won't harm our eyes looking at the eclipse!"

The day after the eclipse the Australian Optometrical Association reported many phone calls and hospital visits, but no confirmed eye injuries. It credited itself for warning the public and convincing people to watch the eclipse on television.[8]

The warnings worked. Most Australians stayed inside and watched the 1976 solar eclipse on television. In Melbourne, where the eclipse was total, a daily newspaper, *The Age*, happily reported that "[t]he eclipse provided one of the biggest television audiences of any event in Australian history. In Melbourne, two million viewers were glued to their television sets to watch the once-in-a-lifetime phenomenon."[9]

Put another way, two million people could have seen a stunning, unforgettable, life-changing total eclipse of the Sun just by stepping outdoors. But because of professional and governmental misinformation, they missed it.

For more recent total solar eclipses Down Under—2002, 2012—the Australian government conducted effective public education campaigns to help people appreciate the eclipse and view it safely.

Government wrong-headedness has struck in Canada too. Steve Edberg was in Winnipeg for the eclipse of 1979. Students had to have notes from their parents to be allowed outside during the eclipse. All the others were confined to their classrooms where the shades were drawn and the students watched the eclipse on television. Jay Pasachoff recalled that "One school in Winnipeg even asked for permission to ignore fire alarms if any sounded during the eclipse, lest the students rush outside and be blinded."

On a more rational note, Jay Anderson remembers that many Winnipeg parents kept their children home from school on eclipse day and took the day off themselves so that they could witness the eclipse as a family.

## American Snippets

Canada? Australia? But surely such misinformation and superstitious behavior could never hold sway in the United States, depriving people of a rare gift of nature. Well . . .

The eclipse of 1970 was the first and most impressive of the many George Lovi was to see. It was visible in the eastern United States and he was watching from Virginia Beach. The news media had trumpeted the event, but laid great emphasis on the dangers. Around him were thousands of people, most of them casual observers, watching as the crescent Sun thinned and vanished. Instantly a cry went up, spreading from group to group: "Look away! Look away!" Many did.

Lovi knew of other people who traveled substantial distances to reach Virginia Beach and then, fearful, stayed in their motel rooms and watched the event on television.

When Californians heard that the Sun would experience an annular eclipse on May 20, 2012, some expressed their fear to the BabyCenter blog: "I'm pregnant. Today there will be a solar eclipse. Did you know I'm supposed to wear red and some sort of metal to protect the baby?"

Yes, said blogger Katherine Martin. Many Hispanic mothers and grandmothers tell their pregnant daughters and granddaughters to wear red underwear and to attach a safety pin to it over their belly to protect the fetus from birth defects. No safety pin? Any sort of metal will do. No red panties? The mother-to-be can use underwear of any color with a red ribbon attached to it.

Also, a pregnant woman must not go outside during the eclipse.

Martin said that these superstitions dated back to the Aztecs who believed that an eclipse was caused by a celestial monster taking bites out of the face of the Sun or Moon. If an expectant mother watched such an atrocity, the same thing would happen to her baby.[10]

## What a Shame

Dennis di Cicco was in a cornfield in North Carolina for the 1970 eclipse and recalls that in the midst of totality, a car came up the road with its lights on and drove right by without stopping, its passengers oblivious or impervious to the wonder taking place above them.

A cruise ship brought Joe Buchman and many eclipse chasers to Aruba for the total eclipse of 1998. He observed from the back patio of a hotel looking out over the Caribbean. It was a hot day and there wasn't any shade, but as the Moon cut off more and more sunlight, the temperature fell and it became quite comfortable for one of nature's most stirring sights. Later that day, back aboard the ship, Buchman found that many passengers had not left the boat. They spent the eclipse in the windowless casino, pulling handles on the slot machines during totality.

## The Effect of a Total Eclipse

Sometimes people's behavior during a total eclipse is out of the ordinary, but quite understandable.

The 2013 eclipse in Uganda occurred late in the afternoon. Patrick and Joanne Poitevin noticed that farm laborers continued to work in the fields until the sunlight, increasingly blocked by the Moon, began to diminish. The workers stopped, looked around, and, oblivious to the eclipse, picked up their tools and started walking home from the fields, apparently confused at the early onset of evening.

Total eclipses bring out strong emotions in both animals and humans. Xavier Jubier observed the 2008 total eclipse in northwestern China. Just before totality when the diamond ring effect appeared, a man dropped to one knee and proposed to his girlfriend, asking that she give him an answer when totality was over. (She said yes.)

At the 2024 eclipse and those to follow, make your own observations of animal—and human—behavior. During totality, says Joe Buchman, be sure to spend a moment looking at the people around you. Look at their expressions. It's like a medieval painting—the look of adoration. Eyes wide open, mouth agape. Then, as the Sun emerges from eclipse, cheers for what they have seen.

## NOTES AND REFERENCES

1.  Epigraph: Henry Holiday: *Reminiscences of My Life* (London: William Heinemann, 1914, pages 209–210). Quoted by Alex SooJung-Kim Pang: *Empire and the Sun:Victorian Solar Eclipse Expeditions* (Stanford, California: Stanford University Press, 2002), pages 72–73 plus photo insert.
2.  Stephen O'Baugh: "Eclipse Had Top Rating," *The Age* (Melbourne, Australia), October 24, 1976.
3.  "Solar eclipse: Myths continue to rule," *Times of India*, January 15, 2010.
4.  Michael Rogers: "Totality—A Report," *Rolling Stone*, October 11, 1973. Reprinted in Robert Gannon, editor: *Best Science Writing: Readings and Insights* (Phoenix: Oryx Press, 1991), pages 168–185. In this article, Rogers concentrates on the 1973 total eclipse he saw in Mauritania, with this Cambodian eclipse mentioned anecdotally. Rogers thought this eclipse happened in 1972, but no total or partial eclipse was visible from Cambodia that year, or in 1971 or 1970. The only significant solar eclipse that was visible from Cambodia close to that date was the annular eclipse of November 23, 1965.
5.  Thomas Crump: *Solar Eclipse* (London: Constable, 1999), page 191.
6.  Associated Press, March 2, 1998, reporting on the eclipse of February 26.
7.  Thanks to Glenn Schneider who provided photocopies of newspaper articles and posters from the 1976 total solar eclipse in Australia.
8.  The text of the article provided by Glenn Schneider was complete but did not include the name of the newspaper, the reporter's name, or the date.
9.  Stephen O'Baugh: "Eclipse Had Top Rating," *The Age* (Melbourne, Australia), October 24, 1976.
10. Katherine Martin: "Interesting Pregnancy and Solar Eclipse Superstitions," *BabyCenter blog*, May 20, 2012. We have not been able to verify that this superstition comes from Aztec or Maya folklore.

# A MOMENT OF TOTALITY

## Spoiled

On December 14, 2001, George Fleenor drove from Bradenton, Florida to Sarasota to photograph a partial eclipse of the Sun. The maximum of the eclipse, with about half the Sun obscured, would happen at sundown over the Gulf of Mexico. George had a perfect beach site in mind, with a lifeguard stand in the foreground, looking out over the water to the horizon. While George took pictures, people were still lounging on the beach and splashing in the waves, oblivious to the eclipse in progress. But as the Sun reached the horizon, the atmosphere filtered the Sun's brightness enough that people could look at it without eye protection and see that half the Sun was missing. As a couple walked by George on their way off the beach, the woman moaned, "This is our last day in Florida and our sunset is ruined."[*]

[*] George Fleenor, a planetarium consultant and former planetarium director, helped the couple appreciate the rarity and significance of the event they were seeing. Interviewed July 14, 2015.

# 12

———◄◦►———

# Eclipse Photography

*One picture is worth a thousand words.*

Anonymous[1]

How do you capture the spectacle of a total eclipse with a camera? Photographing an eclipse isn't difficult. It doesn't take fancy or expensive equipment. You can take a snapshot of an eclipse with a simple camera (even a smart phone) if you can hold the camera steady or place it on a tripod.

The first step in eclipse photography is to decide what kind of pictures you want. Are you partial to scenes with people and trees in the foreground and a small but distinct eclipsed Sun overhead? Or do you prefer a close-up in which the gossamer corona or vivid red prominences of the eclipsed Sun fill the frame? Your decision will determine what kind of equipment you need. Look at the photographs and captions throughout this book. They illustrate some of what can be done with a range of cameras, lenses, telescopes, and exposures.

New technologies in cameras and electronics are making eclipse photography easier than ever before. Even beginners can take great eclipse photos with some careful planning. *Planning* is the key. The day of the eclipse is *not* the time to try out a new camera, lens, or tripod. You need to be completely familiar with your camera and equipment, and you need to *rehearse* with them weeks before the eclipse. A total eclipse grants you only a few precious minutes and everything must work perfectly. Nature does not provide instant replays.

## Digital Photography

In less than a decade, digital cameras have revolutionized the way we shoot pictures. The price of electronic image sensors and memory cards has plunged while the ability to capture fine detail has leapt into many megapixels. The result: Digital cameras have completely replaced film cameras. But perhaps you are too young to remember film cameras!

Smart phones are capable of shooting great eclipse landscapes in either photo or video mode. This image is a *single* frame from an HD video made with an iPhone 5s during the total solar eclipse of March 20, 2015 from Svalbard. [Apple iPhone 5s, HD video mode. © 2015 Sarah Marwick]

Back in the old days of film, eclipse photographers had to carefully pace themselves because there were only 36 exposures on a roll of film. To capture the diamond ring effect at the beginning of totality, the corona during totality, and the second diamond ring at the end of totality, you had to take pictures sparingly to make your film last. Today, with a large memory card, you can shoot hundreds of digital eclipse images in a few minutes and see your results immediately after totality ends.

The following sections explain how you can bag some prized eclipse photos no matter what kind of digital camera you own. We'll start with some simple cameras and techniques and progress to the more challenging.

## Simple Cameras

Point-and-shoot or compact cameras are the simplest digital cameras you can buy. They usually have autofocus lenses, automatic exposure modes, and a small built-in flash. Although the most basic cameras have a single-focal-length lens, more advanced models include a zoom lens.

Nine-year-old Maggie nabbed her first total eclipse photo using a simple point-and-shoot camera from Oregon during the 2017 eclipse. [Nikon Coolpix L32, auto-exposure, © 2017 M. Delos-Reyes]

Many point-and-shoots are small enough to fit in a pocket, making them perfect for snapshots of vacations, parties, and other events. They are popular with people who do not consider themselves photographers but want an easy-to-use camera.

Smart phones are replacing point-and-shoot cameras as the most popular type of digital camera. These ingenious devices were the stuff of science fiction only a decade ago. Besides shooting photos, they have downloadable applications or apps for almost everything imaginable: texting, e-mail, web browsing, music, games, and alarm clock, just to mention a few. You can even use them to make phone calls!

Both point-and-shoots and smart phones are great for capturing simple snapshots of an eclipse. You can also use them to shoot a panorama around the horizon during totality. Although you can handhold either one, a tripod is a better choice for support, especially during the subdued twilight of totality. Point-and-shoots have a threaded socket on the bottom for tripod attachment, while smart phones will require a special bracket to securely attach them to a tripod.

After the eclipse, you can quickly review your photos. Of course, a smart phone also lets you share them immediately with friends and family via Facebook, Twitter, Instagram, and Flickr.

## Landscape Eclipse Photography

The easiest way to capture the eclipse is to take a few wide-angle shots of the sky, horizon, and landscape before, during, and after totality. This sequence might show the Moon's dark, fast-moving shadow approaching from the west or racing away over the horizon to the east. Place yourself or some other interesting subjects in the foreground to give the photos some scale. If your camera has a zoom lens, the widest-angle setting will produce the most dramatic images. Your camera's built-in auto-exposure feature should work well. Just be sure to *turn off the automatic flash feature* so you don't annoy observers nearby. If your camera has an exposure-bracketing feature, use it for insurance to get the best exposure of the total eclipse. You'll also capture any bright planets near the Sun during totality. You might even discover a new comet!

## Suggested Targets for Solar Eclipse Photographs

**Wide-Angle Photos**

People setting up telescopes

People watching eclipse while wearing solar-eclipse glasses

People watching eclipse through telescope or binoculars

Eclipse crescents on the ground (sunlight falling between tree leaves, holes between fingers, etc.)

The lunar shadow on the western horizon (before totality) or eastern horizon (after totality)

The colors at the horizon during totality

People silhouetted against twilight sky

Wide-angle view of totally eclipsed Sun with interesting foreground (people, trees, buildings, statues, etc.)

**Close-Up Photos** (with telescope or zoom lens)

Partial phases (through solar filter)

Diamond ring effect during the 15-second period before and after totality—no solar filter

Baily's beads during the 5-second period before and after totality—no solar filter

Chromosphere and prominences during totality—just after totality begins and just before totality ends—no solar filter

Totality—no solar filter—use a range of shutter speeds to get inner, middle, and outer corona

The gaps in the fronds of a palm tree produce thousands of eclipsed Sun images on the sand below during the January 15, 2010 annular eclipse from Elaidhoo, Maldives. [Canon 450D, 24 mm, f/6.3, 1/100 second, ISO 200. © 2010 Stephan Heinsius]

## Photographing Pinhole Crescents

Eclipses provide other phenomena that make interesting pictures, such as the crescent images of the partially eclipsed Sun produced by tree foliage. The narrow gaps between leaves act as "pinhole cameras" and each projects its own tiny (and inverted) image of the crescent Sun on the ground. This pinhole camera effect becomes more pronounced as the eclipse progresses.

You can make your own pinhole camera to project the crescent Sun with pinholes punched in thin cardboard, or with a wide-brimmed straw hat. You can even produce the effect in the shadow of your hands by loosely lacing your fingers together. Watch the crescents form when the light passes through the gaps between your fingers. The profusion of crescents on a white wall or on a person's face makes a nice photographic memento. Once again, you'll want to disengage the automatic flash on your camera.

## Eclipse Close-Ups

If your goal is to shoot close-ups of the Sun and Moon during the eclipse, a little more effort is required. These kinds of eclipse images are more challenging than wide-angle landscapes and pinhole crescents. You will need a more expensive camera, a powerful telephoto lens or small tele-scope, a strong tripod, and a special solar filter (for partial phases only). Successfully capturing magnified views of the eclipse also requires a basic understanding of how digital cameras work as well as careful planning and preparation before eclipse day.

So if you want more advanced eclipse photography, read on.

## Dynamic Range: RAW vs. JPEG

The sensor in your camera is a two-dimensional array of individual pixels. Each pixel is a tiny light sensor that measures and records the brightness of light falling on it in either 1 of 4096 (12-bits) or 1 of 16,384 (14-bits) values, depending on the camera sensor. When the camera's processing chip converts a RAW image into a JPEG, it compresses the range of brightness in the original image down to just 256 (8-bits) values. In other words, some of the brightness information in the original image is thrown away in order to make the JPEG image smaller than the RAW image. When you save an image as a JPEG, this extra brightness information is lost forever.

## Digital "Film"

Digital cameras use solid-state memory cards to store images. This is a type of computer storage made from silicon microchips that record your photos electronically. Most memory cards fall into one of five types: secure digital (SD), Micro SD, compact flash (CF), Memory Stick (Sony), and xD. Depending on the make and model of a digital camera, it is designed to work with one (and possibly two) of these cards. The cards themselves come in a range of capacities: 32GB to 512GB and larger. GB stands for gigabyte (1 billion bytes).

When a digital camera records an image onto a memory card, it usually writes a JPEG file.[2] JPEG *(pronounced "jay-peg")* is a clever way of com-pressing the raw image recorded by the image sensor and storing it in less space than it normally would require. JPEGs come in varying levels of compression: the higher the compression, the smaller the file. But there's no such thing as a free lunch and this is especially true of JPEGs. When you save an image as a JPEG, some of the information in the photo is lost for-ever. JPEGs are called "lossy" because the decompressed image isn't quite the same as the one you started with. The more you compress a JPEG, the *smaller* the file size and the *greater* the loss of quality in the original image.

Images can be saved at several levels of JPEG quality. Typical choices include something like "low, medium, high" or "basic, normal, fine." Although the intermediate level is sufficient for most photography, a total eclipse is an exceptional event warranting the highest-level setting. This

**Storage Chart—Approximate Number of JPEGs per Memory Card**

| Sensor Size (Megapixels) | File Size (MB) | 8GB | 16GB | 32GB | 64GB | 128GB | 256GB |
|---|---|---|---|---|---|---|---|
| 8MP | 2.4 | 2861 | 5722 | 11444 | 22888 | 45776 | 91552 |
| 10MP | 3.0 | 2288 | 4577 | 9155 | 18310 | 36620 | 73240 |
| 12MP | 3.6 | 1907 | 3814 | 7629 | 15258 | 30516 | 61032 |
| 14MP | 4.2 | 1634 | 3269 | 6539 | 13078 | 26156 | 52312 |
| 16MP | 4.8 | 1430 | 2861 | 5722 | 11444 | 22888 | 45776 |
| 24MP | 7.2 | 1135 | 2269 | 4539 | 9079 | 18157 | 36314 |

means fewer images can be stored on a memory card but they will be the best quality JPEGs the camera can deliver.

How many images can fit on a given memory card? That depends on three factors: the JPEG quality level, the number of pixels in the image, and the amount of detail in the image. The storage chart above estimates the capacity for a range of image and memory card sizes assuming a highest JPEG quality ("high" or "fine").

When you put an empty memory card in your camera, the display shows the estimated number of exposures for the camera's current settings. Change the JPEG quality and watch the numbers change. This is just an estimate but it is still a useful guide.

The size of a JPEG also depends on the amount of detail in the subject matter. An image of an outdoor landscape (with rocks, trees, grass, and clouds) produces a larger JPEG file than a partially eclipsed Sun (dark sky with a relatively featureless yellow crescent).

Professional and higher-end consumer cameras can also store images in something called camera RAW format (RAW for short—it's not an acronym; it just means raw data). Earlier, a JPEG was described as "a clever way of compressing the raw image recorded by the image sensor and storing it in less space than it normally would require." A RAW image is essentially the original image recorded by the camera sensor before any clever processing is applied to compress the image size and make a JPEG. RAW image files are about 10 times larger than high-quality JPEG files, so fewer of them will fit on a memory card. To roughly estimate the number of RAW images on a memory card, divide the number of JPEG images in the Storage Chart table by 10.

Why would anyone want to use the RAW format if it takes up more room on a memory card? Because RAW images can store a much wider range of brightness values than JPEGs, and that is what you need to best capture the beautiful corona (see vignette on Dynamic Range: RAW vs. JPEG).

One of the most remarkable things about the Sun's corona is that it exhibits an enormous range in brightness—something referred to as dynamic range. The inner parts of the corona are thousands of times brighter that the outer parts. The more brightness values your camera can record, the more successful it will be in capturing the dynamic range of brightness in the corona.

If you're a more casual photographer, you can shoot perfectly acceptable photos of the corona using the best JPEG setting in your camera. But if you consider yourself a serious eclipse photographer, and if you plan to process your images in a computer to bring out the maximum amount of detail possible, then you need to shoot your eclipse photos in camera RAW.

## Quick and Easy Eclipse Photography

*by Patricia Totten Espenak*

Is your idea of the perfect eclipse experience—"I just want to watch!"? Do your eyes glaze over when serious eclipse photographers delve into the minutiae of exposure times, f-stops, and cameras that require you to carry the manual with you? *But* do you still have just a tiny desire for some photos of your very own? Then read on.

After 16 total solar eclipses, my methods and gear have evolved to accommodate both. I can watch almost the entire eclipse while taking great photos using the following equipment: a comfortable chair, a sturdy tripod, a cable release, a right-angle finder,* and a solar filter for my telescope or telephoto lens.

I use a Bogen 3001 tripod with a Bogen 3275 (410) compact geared head. The geared head has two large knobs that allow me to make small, precise adjustments in altitude and azimuth every minute or so to easily track the Sun. With a right-angle finder, I can bend over to check the position of the image and then quickly lean back to view the eclipse. I focus manually on a sunspot and the focus ring is then taped down to secure it. After setting the camera in program mode and matrix metering, I'm ready to go. I just sit back with the cable release in one hand and my eclipse glasses in the other.

Thirty seconds before totality begins, I check to see that the Sun is centered in my viewfinder and I remove the solar filter. A dozen or more shots are quickly taken in the seconds before second contact and after third contact. This results in nice diamond ring sequences while the camera automatically adjusts the exposure.

During totality, I might take another dozen shots. I also check the Sun's position in the viewfinder, but mostly I'm gazing up at the spectacle and pressing the cable-release button.

After my third-contact sequence, I replace the solar filter and return to the leisurely pace of the partial phases.

## Sensor Sensitivity and ISO

The correct exposure of a digital photograph is determined by three factors: shutter speed, lens aperture, and sensor sensitivity. The shutter speed controls how long the shutter is open during an exposure. The lens aperture controls how much light passes through the camera lens. The sensor sensitivity controls how much light is needed to obtain a well-illuminated exposure of the subject.

The camera setting used to adjust and change the sensor sensitivity is called the ISO value. A powerful feature of digital cameras is the ability to dial in the ISO sensitivity of your choosing. This was not possible with

Eclipse photography doesn't have to be difficult or interfere with your view of totality. Just use a 500 mm lens and set your camera on "program." This image was shot in Libya during the 2006 total solar eclipse. [Nikon D200 DSLR, Sigma 170–500 mm zoom at 500 mm, f/5.0, 0.8 second, ISO 400. © 2006 Patricia Totten Espenak]

Capture the moment with your eyes and your camera. You *can* get the best of both!

\* If your digital camera has a fold-out LCD screen, it will work fine in place of a right-angle finder.

Patricia Totten Espenak is a retired chemistry teacher who dotes on her granddaughters Valerie and Maggie—and chases eclipses around the world with her husband Fred.

film cameras because each roll of film had to be exposed and processed for a fixed ISO value, but with digital cameras you can change the ISO value any time you wish. In bright Sun, you might pick ISO 100, while indoor photography with the available light might call for ISO 800. The downside of higher ISO speeds is the increase in digital "noise," which appears as grainy specks in images shot at higher ISO values. In modern cameras, digital noise usually becomes significant by ISO 1600. You can study the noise in your camera by shooting the same scene using a range of ISO settings and comparing the results.

Fortunately, eclipses are relatively bright, so an ISO value of 400 is generally a good compromise. It's fast enough to minimize blurring from vibrations without sacrificing image quality caused by digital noise.

The greatest threat to image sharpness in eclipse photos, especially among beginners, is camera vibration caused by wind, flimsy tripods, and nervous hands. Other causes of blurry images may include shooting handheld without a tripod, shooting from a rocking ship, and shooting at a very long focal length without adequate support. If any of these conditions are true on eclipse day, you might consider boosting the camera sensitivity

The diamond ring effect precedes the beginning of totality during the total solar eclipse of August 21, 2017 from Riverton, Wyoming. [Canon 5D, EF100–400 mm +2×, (fl = 800 mm), f/11, 1/2000 second, ISO 800, © 2017 Tommy Tat-Fung Tse]

to ISO 800 or even ISO 1600. This will let you use faster shutter speeds, which can help minimize blurring from vibrations and camera motion. A little extra noise is a small price to pay for a sharper image.

## Digital Single-Lens Reflex Cameras

Close-ups of the eclipsed crescent Sun (through a solar filter) and detailed portraits of the solar corona require the use of either a digital single-lens reflex (DSLR) camera or a mirrorless interchangeable-lens camera (MILC). Both types feature interchangeable lenses, from extreme wide-angle to high-power telephoto. You can also remove the lens and hook the camera body up to a telescope. All DSLRs and MILCs have a continuous shooting mode (or burst mode) that allows you to shoot two or more frames per second. Their electronic shutters have sophisticated metering and exposure modes while lenses autofocus on their target in a split second.

Most consumer DSLRs and MILCs use something called a crop sensor chip measuring about $16 \times 24$ mm. The more expensive professional models use a full-size sensor chip of $24 \times 36$ mm. So the sensor dimensions of the consumer DSLR/MILC (or crop DSLR/MILC) are about two-thirds the size of the sensor in the full-frame models and 44% of the area.

Although a camera lens focuses the same size image of the Sun on both full frame and crop sensor models, it fills a larger fraction of the crop camera's imaging area. This so-called digital magnification factor means an image shot with a crop camera appears to be made with a lens having 1.5 times the focal length compared to the image shot with a full-frame camera using the same lens. The increased image scale on crop camera has important implications when shooting eclipses.

## Super Telephotos and Telescopes

The size of the Sun's image is determined solely by the focal length of the lens you use. The larger the focal length, the larger is the Sun's image. To shoot close-ups of an eclipse, you'll need a telephoto lens with a focal length of 400 mm or 600 mm or more.

Professional sports and nature photographers typically use these types of lenses. They are big, heavy, and expensive. Fortunately, mirror lenses are an economical alternative. They use a combination of mirrors and lenses to focus the light into the camera. The folded light path allows a long focal length (500–600 mm) to fit within a short lightweight tube, making small mirror lenses easily portable—ideal for most eclipse photography.[3]

With crop DSLRs, the 1.5× magnification factor (actually 1.6× for Canon crop DSLRs) offers a free ride for eclipse photographers. A 500-mm lens

Full Frame: 200 mm

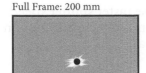

Crop: 135 mm

Full Frame: 400 mm

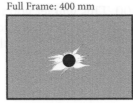

Crop: 270 mm

Full Frame: 500 mm

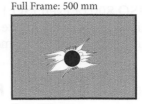

Crop: 330 mm

Full Frame: 1000 mm

Crop: 670 mm

Full Frame: 1500 mm

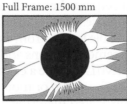

Crop: 1000 mm

Full Frame: 2000 mm

Crop: 1330 mm

The image sizes of the eclipsed Sun and corona are shown for a range of focal lengths on both professional (full-frame) and consumer (crop-sensor) DSLRs or MILCs. The same lens produces an image appearing 1.5 times larger on a crop DSLR than on a full-frame camera.
(Adapted from Fred Espenak and Jay Anderson: *Eclipse Bulletin: Total Solar Eclipse of 2017 August 21* [Astropixels Publishing, 2015].).

captures the same field of view on a crop DSLR that a 750 mm does on a full-frame DSLR. You can easily determine the size of the Sun's disk in any DSLR/MILC, whether crop or full-frame. Take a coin from your pocket and measure its diameter. Now place the coin 110 times its diameter from the lens and shoot a picture of the coin, which now appears the same size as the Sun or Moon. The full moon also makes a great target for evaluating the magnification of your lens for shooting eclipses.

The recommended focal length for close-up photos of solar eclipses ranges from 500 mm to 2000 mm for a full-frame DSLR/MILC (350–1350 mm for crop cameras), depending on whether you concentrate on the corona or the prominences. Again allowing one solar radius of corona on either side of the Sun, you can calculate that lenses with focal lengths longer than about 1400 (900 mm for crop cameras) may clip the corona, and focal lengths longer than 2500 mm (1700 mm for crop cameras) may not capture the entire solar disk. The longer the lens, the more expensive and less stable a telescope/telephoto system will be, which means that you will need a heavy-duty tripod or mount, adding to your expense and weight.

Modern refracting telescopes are excellent for eclipse photography. The new extra-low-dispersion (or apochromatic) lenses eliminate the color halos around images that handicapped older refractors. Apochromatic refractors (apos) have wider objective lenses, yet shorter tubes, making them fast lenses.

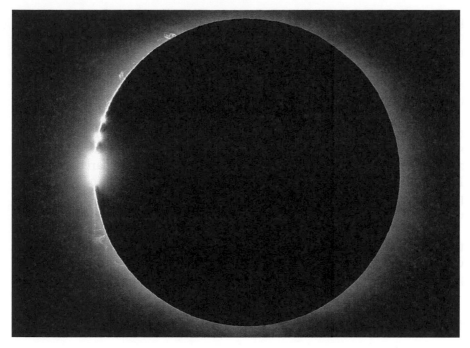

Baily's beads were photographed at the end of totality during the total eclipse of August 1, 2008 from Jinta, China. The beads are formed by sunlight passing through deep valleys along the edge of the Moon. [Nikon D300, Vixen 90 mm fluorite refractor, fl = 810mm, f/9, 1/1000 second, ISO 320. © 2008 Fred Espenak]

Apos are expensive and they are not as portable as mirror lenses. Nevertheless, advanced eclipse photographers prize apos for their image quality.[4]

Using a telescope means that you lose the autofocus function (and possibly autoexposure) of your camera. This handicap isn't as serious as it might seem because most cameras don't autofocus well for eclipse close-ups. The rapidly changing light conditions confuse many autofocus cameras, causing them to hunt for the correct focus, so the best solution is to turn autofocus off and focus manually.

## Rise of the Superzoom

A less expensive alternative to the DSLR and MILC for high-magnification eclipse photography is the superzoom or bridge camera. This digital camera is positioned between the simple point-and-shoot camera and the DSLR. While it does not have an interchangeable lens like the DSLR or MILC, it is equipped with a built-in zoom lens with an amazing range of magnifications: 20×, 40×, 60×, and higher. This means you can zoom into the Sun and get good close-ups of the partial phases (through a solar filter). You can also capture the diamond ring and the corona with no telescope or expensive telephoto lens. Of course, a good tripod helps eliminate camera shake.

Superzooms are available from all the major camera manufacturers including Canon, Fujifilm, Nikon, Panasonic, Pentax, and Samsung, and they typically cost hundreds of dollars less than a consumer DSLR.

So why would anyone choose a DSLR instead of a superzoom? DSLRs (even crop DSLRs) have larger sensors than superzoom cameras. A large sensor has less noise and better image quality. And the all-in-one super-zoom lens makes some compromises in image quality compared to a good DSLR lens. But it's hard to beat the superzoom for size, convenience, portability, and expense. It's a great choice for first-time or more casual eclipse photographers, and it also makes an excellent vacation camera.

To determine the size of the Sun and Moon in your superzoom, simply take a photo of the full moon. If you don't want to wait for full moon, you can use the coin trick described earlier. Just take a picture of the coin after placing it 110 times its diameter from the superzoom.

## Camera Tripods and Cable Releases

Flimsy tripods are the main reason that eclipse photographs come out fuzzy and blurred. Small portable tripods that are convenient for airline travel are not sturdy enough to hold your camera and heavy telephoto lens steady for sharp eclipse pictures. Because you must touch your camera to adjust exposures, your tripod must also dampen any vibrations quickly.[5]

To maximize stability and minimize vibrations, don't extend the tripod's legs more than halfway and do not extend the center column. Adjust the tripod height so that you can reach the camera controls while kneeling or sitting on a chair. Test your setup by tapping on the camera or tripod while viewing the Sun (through a solar filter) in your camera. Vibrations should be small and should damp out quickly.

You can help reduce vibrations by suspending some weight under the tripod. Put rocks or sand in a sack or in plastic zip-lock bags and hang them from the center of the tripod using string or duct tape. Setting up the tripod on sand or grass is better than concrete or asphalt because the softer surfaces dissipate vibrations faster.

Before you travel to an eclipse, put your camera and lens on the tripod and make sure they can be pointed at the Sun's predicted altitude for the eclipse and that all controls work smoothly. You don't want to discover that your equipment becomes unstable when you try to view the crucial portion of the sky on eclipse day.

You cannot take crisp close-ups of a total solar eclipse without using an electronic cable release. Cable releases have a button at one end and

Five images of totality and the diamond rings are combined into a single mosaic using Photoshop. The images were taken during the total solar eclipse of August 21, 2017 from Driggs, Idaho [Canon 6D MkII, Astro-Physics Traveler refractor, f/5, © 2017 Alan Dyer]

a connection to the camera on the other end. When you press the cable release button, it triggers the camera's shutter without jostling it. Check with your owner's manual or camera store to choose a cable release that fits your camera.

## Solar Filters

***When viewing or photographing the partial phases of any solar eclipse, you must always use a solar filter.*** A solar filter is also needed for observing all phases of an annular eclipse, because the disk of the Moon does not block the entire face of the Sun. Even if 99% of the Sun is covered, the remaining crescent or ring is dangerously bright. Failure to use a solar filter can result in serious eye damage or permanent blindness. ***Do not look directly at the Sun without proper eye protection!***

During totality, however, when the Sun's disk (photosphere) is fully covered by the Moon, it is completely safe to look at the eclipse without any solar filter. In fact, you *must* remove the solar filter during totality or you will not be able to see or photograph the exquisite solar corona and prominences.

Solar filters for camera lenses and telescopes come in several different kinds of materials. Metal-coated glass filters are the most expensive but they offer excellent resolution and natural-looking color. Handle glass filters with care since they are fragile and break easily when dropped.

Black polymer is composed of carbon particles suspended in a resin matrix. It produces a natural yellow image of the Sun and is found in both cardboard eclipse glasses and telescope/binocular/camera filters. Black polymer has largely replaced aluminized polyester or Mylar[6] as the inexpensive solar filter of choice. Black polymer filters come mounted in a slip-on cell in any number of diameters to fit most camera lenses and telescopes.

For the do-it-yourselfer, Baader Planetarium AstroSolar Safety Film is a metal-coated resin sold in sheets and has excellent optical quality. The company offers instructions on how to make an inexpensive cardboard cell to mount the filter on your lens or telescope.

All these solar filters are designed for use on the front end of the lens or telescope (the end pointed at the Sun). They should never be used anywhere else. Materials that should *not* be used for solar filters include exposed color film, stacked neutral density filters, smoked glass, crossed polarizer filters, floppy disks, and CDs (compact disks). These materials are *not safe* even if the image appears dim and no discomfort is felt while viewing the Sun.

If your telescope has a finder scope, be sure to place a small solar filter over its objective lens to protect your eyes and to keep the finder cross hairs from burning. If you don't have a filter for the finder, keep its upper lens covered with a cap or small piece of cardboard taped in place.

Most telescope manufacturers and dealers offer solar filters. Advertisements for them can be found in all the major astronomy magazines (*Astronomy*, *Astronomy Now*, *Sky & Telescope*, *Sky at Night*, *SkyNews*, etc.). You can also search the web for "solar filter for telescope" (via Google.com, Yahoo.com, etc.).

## Photographing the Partial Eclipse

A solar filter MUST be used with your lens to photograph the partial phases of the eclipse. The filter changes the shutter speed and f-ratio of your camera, so you must determine the proper shutter speed and f-ratio for your equipment with the filter in place well in advance of the eclipse.

If your camera has a built-in spot meter that covers a smaller area than the Sun's image, you can simply meter on the Sun's disk through your solar filter and use that exposure throughout the partial phases.

If your camera does not have a spot meter, you need to perform a simple exposure test. Set your equipment up on a sunny day. Carefully center the Sun in your lens or telescope using a solar filter. For telephoto lenses,

The total solar eclipse of August 21, 2017, is captured in a sequence of seven individual images (later combined in Photoshop). The central image of the corona is a composite of several exposures. [Nikon D810a, Orion ED80 T refractor, fl = 480 mm, Thousand Oaks Solar Filter, partials: 1/2500 sec, total: 1/4000 to 1/10 sec, ISO 200. © 2017 Philippe Jacquot]

open the aperture to its widest setting. Set the camera in manual exposure mode and shoot one exposure with every shutter speed from 1/15 through 1/2000 of a second. Take notes and use the camera's histogram function to help choose the best exposure. Your camera manual has more information on using the histogram function.

For handy reference, write down the best exposure and tape it to your tripod or solar filter. It should include the ISO, f-number, and shutter speed—for example, "Sun: ISO 400, f/8, 1/125." The exposure doesn't change during the partial phases because the surface brightness of the Sun remains the same throughout the eclipse.[7]

Your best exposure should be determined on a clear, sunny day. If eclipse day has haze or clouds, a longer exposure will be needed to compensate. A thin haze may require an exposure one or two shutter speeds slower than normal, while thicker clouds could call for three or more shutter speeds slower. Use your planned exposure and several longer ones. Memory cards are cheap and eclipses don't happen often.

## Photographing the Total Eclipse

The brightness of the solar corona changes tremendously as you move away from the Sun's disk. The inner corona shines as brightly as the full moon but the outer corona is thousands of times fainter. The challenge is to capture both the brightest and faintest parts of the corona. Unfortunately, this variation in brightness is impossible to record in any one exposure because your camera's sensor just doesn't have the dynamic range of human vision. Only your eyes can see the exquisite detail of this

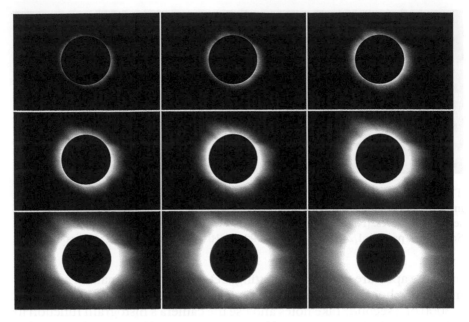

Some photographers shoot a series of exposures to capture the wide range of brightness in the corona. This sequence was made with a DSLR during the March 29, 2006 total solar eclipse from Libya. [Nikon D200 DSLR, Vixen 90 mm fluorite refractor, fl = 810 mm, f/9, 9 exposures: 1/125 to 2 seconds, ISO 400. © 2006 Fred Espenak]

celestial event in all its glory. That's why you should view totality with your eyes and not just with your camera.

The good news is you can photograph some aspect of the corona with almost any exposure you make. There is no one "correct" exposure. Nevertheless, here are some guidelines.

Several factors determine the best shutter speed to use to get a good exposure in your final photograph. The tables accompanying this chapter provide recommended shutter speeds for various eclipse phenomena over a range of ISO speeds and lens f-numbers. Each eclipse phenomenon (partial phases; prominences; inner, middle, and outer corona) has a different brightness value for determining the proper exposure for that aspect of the eclipse. The exposures in the table were determined after photographing more than a dozen solar eclipses, but they are only suggestions. Each eclipse is different and the corona's brightness varies. Weather conditions (haze or clouds) may require longer exposure times. And remember, the partial phases *always* require a solar filter.

During totality, bracket your shots on both sides of the ideal exposure and take several sets of photographs at the same settings to assure success. If you use an ISO speed that is too high, you may discover that your camera does not have a fast enough shutter speed for proper exposure.

## Exposure Guide for Solar Eclipse Photography

These exposure tables are given as guidelines only. The brightness of prominences and the corona can vary. You should bracket your exposures to be safe.

SUN—partial phases and prominences
Use full aperture solar filter for partial phases
No filter for prominences

SUN—total eclipse: middle corona (2° field)
No filter

(Film Speed—ISO)

| f/no. | 100 | 200 | 400 | 800 |
|-------|-----|-----|-----|-----|
| 2.8 | 1/4000 | 1/8000 | — | — |
| 4 | 1/2000 | 1/4000 | 1/8000 | — |
| 5.6 | 1/1000 | 1/2000 | 1/4000 | 1/8000 |
| 8 | 1/500 | 1/1000 | 1/2000 | 1/4000 |
| 11 | 1/250 | 1/500 | 1/1000 | 1/2000 |
| 16 | 1/125 | 1/250 | 1/500 | 1/1000 |
| 22 | 1/60 | 1/125 | 1/250 | 1/500 |

(Film Speed—ISO)

| f/no. | 100 | 200 | 400 | 800 |
|-------|-----|-----|-----|-----|
| 2.8 | 1/60 | 1/125 | 1/120 | 1/500 |
| 4 | 1/30 | 1/60 | 1/125 | 1/250 |
| 5.6 | 1/15 | 1/30 | 1/60 | 1/125 |
| 8 | 1/8 | 1/15 | 1/30 | 1/60 |
| 11 | 1/4 | 1/8 | 1/15 | 1/30 |
| 16 | 1/2 | 1/4 | 1/8 | 1/15 |
| 22 | 1 sec | 1/2 | 1/4 | 1/8 |

SUN—total eclipse: inner corona (1° field)
No filter

SUN—total eclipse: outer corona (4° field)
No filter

(Film Speed—ISO)

| f/no. | 100 | 200 | 400 | 800 |
|-------|-----|-----|-----|-----|
| 2.8 | 1/1000 | 1/2000 | 1/4000 | 1/8000 |
| 4 | 1/500 | 1/1000 | 1/2000 | 1/4000 |
| 5.6 | 1/250 | 1/500 | 1/1000 | 1/2000 |
| 8 | 1/125 | 1/250 | 1/500 | 1/1000 |
| 11 | 1/60 | 1/125 | 1/250 | 1/500 |
| 16 | 1/30 | 1/60 | 1/125 | 1/250 |
| 22 | 1/15 | 1/30 | 1/60 | 1/125 |

(Film Speed—ISO)

| f/no. | 100 | 200 | 400 | 800 |
|-------|-----|-----|-----|-----|
| 2.8 | 1/4 | 1/8 | 1/15 | 1/30 |
| 4 | 1/2 | 1/4 | 1/8 | 1/15 |
| 5.6 | 1 sec | 1/2 | 1/4 | 1/8 |
| 8 | 2 sec | 1 sec | 1/2 | 1/4 |
| 11 | 4 sec | 2 sec | 1 sec | 1/2 |
| 16 | 8 sec | 4 sec | 2 sec | 1 sec |
| 22 | 15 sec | 8 sec | 4 sec | 2 sec |

Determine all your camera settings before the eclipse so you know in advance what ISO works best with your equipment.[8]

Many eclipse photographers plan their exposures with the following strategy. Using the ISO and f-number, determine the shortest shutter speed for the prominences (bright) and longest shutter speed for the outer corona (dim). After totality begins, shoot a sequence of exposures using every shutter speed, starting with the one for prominences and ending with the one for the outer corona. For instance, at ISO 400 and f/11, the recommended exposure for prominences is 1/1000 second and for the outer corona 1 second. The shutter speed sequence would then be: 1/1000, 1/500, 1/250, 1/125, 1/60, 1/30, 1/15, 1/8, 1/4, 1/2, and 1. This is a total

Details in the corona are revealed in this Photoshop composite that was processed to emphasize fine structure. The images were obtained during the March 29, 2006 total eclipse from Jalu, Libya. [Nikon D200 DSLR, Vixen 90 mm fluorite refractor, fl = 820mm, f/9, composite of 22 exposures: 1/1000 second to 1 second, ISO 200. © 2006 Fred Espenak]

of 11 exposures. For additional insurance, repeat the sequence in reverse, ending with 1/1000.

Even if your camera's exposure is completely automatic, or if you simply don't want to keep changing exposure settings, you can still get good pictures of the diamond ring and totality using automatic exposure. Set your camera ISO to 400 and grab some shots on auto-exposure. Although the inner corona and prominences may be overexposed, you will still have some fine souvenirs of the event. Best of all, you can devote most of your time to watching the eclipse rather than fiddling with camera settings. Simplicity is especially recommended if you are a novice photographer or have never seen a total solar eclipse (see vignette on Quick and Easy Eclipse Photography).

## Eclipse Photography at Sea

For eclipse photography at sea, the pitching and rolling of the ship place certain limits on the focal length and shutter speeds that can be used. In most cases, lenses with focal lengths greater than 500 mm can be ruled out because the ship would need to be virtually motionless during totality. You must also contend with vibration from the engines, wind across the deck, and hundreds of people stomping around, so try to find a location that will minimize these problems.

The ISO choice can be determined on eclipse day by viewing the Sun through the camera lens (with a solar filter) and noting the image motion caused by the ship. In most cases, an ISO of 400 or higher will be needed. As the ship rocks, notice the range of motion and try to snap each picture at one of the extremes.

Some manufacturers have introduced image stabilization (or vibration reduction) in their longer zoom and telephoto lenses. It helps steady the image and lets you use shutter speeds two to three stops slower than possible without this feature. Many photographers relied on image-stabilized lenses to shoot the 2005 hybrid eclipse from ships in the South Pacific because no islands were in the path of totality.

## Shooting Video

Total solar eclipses and video are a perfect match. Imagine capturing the excitement of people—their voices and actions—as the Moon's shadow sweeps over the landscape, turning it into an eerie twilight. Virtually all modern point-and-shoot cameras as well as mobile phones are capable of capturing high-definition video.

One of the great things about video is that you can turn it on and let it run. This is especially true if you don't want to be fiddling with a camera during totality. Just start your video recording several minutes before totality begins (no solar filter is needed for wide-angle video). A small tripod is a great asset here because it frees you from holding the camera. Clamps and adapters are available for attaching most mobile phones to a tripod. Just use the same recommendations given previously for landscape eclipse photography.

Video is also the best way of capturing the illusive shadow bands often seen rippling across the ground immediately before and after totality. In this case, the shadow bands show up best against a light-colored surface (snow, sand, a white sheet, even the side of a white car).

If you want close-up video of the eclipse itself, you'll find that super-zoom cameras and newer DSLR models all shoot high-definition video. Check your instruction manual for details. Otherwise, follow the same directions for shooting still images. Just remember that you need a solar filter for the partial phases.

## Final Thoughts

If there is one key to successful eclipse photography, it is *preparation*. Set up and test all your equipment at home to ensure that everything works perfectly. Design a photographic plan or schedule and stick to it. Keep things simple. Don't try to do too much. Practice for the eclipse with a full dress rehearsal. Bring extra memory cards, batteries, and other crucial items.

Finally, don't get so overwhelmed with taking photographs that you deprive yourself of viewing the eclipse with your own eyes.[9]

### Checklist for Solar Eclipse Photography

**Camera Items**

Camera (two, if possible, in case one fails)
Lenses (16 mm to 1000 mm)
Cable release (spare advisable)
Heavy-duty tripod
Extra batteries
Several 32GB or larger memory cards
Sack or zip-lock bags and heavy string (fill with sand or rocks and hang
    from tripod for stability)
Exposure list for partial phases and totality

**Telescope Items**

Telescope (consult owner's manual for photographic equipment)
Photo adapters (T-ring, etc., to attach camera to telescope), lens cleaning
    supplies
Solar filter (full aperture)
Inexpensive solar filter for finder scope

**General Items**

Eclipse glasses (for viewing partial phases)
Penlight flashlight
Digital voice recorder (to record comments during eclipse)
Pocketknife
Tools for minor repairs: screwdrivers (regular, Phillips, and jeweler's),
    needle-nose pliers, tweezers, small adjustable wrench, Allen wrench set, etc.
Plastic trash bags (to protect equipment from dust or rain)
Roll of duct tape or masking tape for emergency repairs
Optional: GPS (global positioning system)

## 2017 Eclipse Stamp

On June 20, 2017, the US Postal Service (USPS) held a First-Day-of-Issue ceremony for a special stamp to commemorate the (then) upcoming total solar eclipse of August 21, 2017 that passed exclusively across the USA. The two images chosen for the stamp were both taken by Fred Espenak. The first is a composite of the 2006 solar eclipse featured in this chapter. The second is an image of the full moon.

This was a first-of-its-kind stamp for the USPS because it changes when you touch it. Using a special thermochromic ink, the "Total Eclipse of the Sun" Forever stamp reveals a second image. By rubbing the eclipse image with a thumb or finger, its warmth causes an underlying image of the full moon to appear. The image reverts back to the eclipse once it cools.

No word yet on whether the US Postal Service will issue a special stamp for the 2024 eclipse. But it is a great way to spread the news about North America's great eclipse to millions of people.

Using a special thermochromic ink, the Total Eclipse of the Sun Forever stamp reveals a second image. By rubbing the eclipse image with a thumb or finger, its warmth causes an underlying image of the full moon to appear. The image reverts back to the eclipse once it cools.

## NOTES AND REFERENCES

1.  Epigraph: Anonymous. Bartlett's *Familiar Quotations* explains that this saying is the creation of Fred R. Barnard, writing in *Printer's Ink*, March 10, 1927. Barnard called it "a Chinese proverb so that people would take it seriously."
2.  JPEG stands for Joint Photographic Experts Group. This committee created the JPEG image format in 1992.

3.   Mirror lenses are made by the major camera manufacturers (Canon, Minolta, Nikon, Pentax, etc.), as well as independent lens makers (Phoenix, Sigma, Tamron, Tokina, Vivitar, etc.).

4.   Manufacturers of apochromatic, fluorite, and extra-low-dispersion refractors include AstroPhysics, Meade, Takahashi, Tele-Vue, Vixen, and William Optics.

5.   Bogen/Manfrotto, Gitzo, and Slik all make heavy-duty tripods.

6.   Mylar is a registered trademark of DuPont, which does *not* manufacture this material for use as a solar filter. Other aluminized polyester products such as "space blankets" should *not* be used for solar filters.

7.   Some photographers like to expose one extra stop during the thin crescent phases because the Sun's limb is a little darker than the disk center.

8.   Weather conditions (haze or thin clouds) may require longer exposure times. Use the recommended exposures as a starting point and then bracket during the eclipse, especially if the weather is a factor.

9.   For tips from Fred Epsenak on still more adventuresome eclipse photography and processing, see Chapter 14: Getting the Most from Your Eclipse Photos in Mark Littmann and Fred Espenak: *Totality: Eclipses of the Sun* (3rd edition or 3rd edition updated) (New York: Oxford University Press, 2008, 2009), pages 201–213.

# A MOMENT OF TOTALITY

## The Sounds of Totality

Michael Rogers, a young writer for *Rolling Stone* magazine, covered the 1973 total eclipse of the Sun in Mauritania—second longest of the 20th century. His award-winning story, "Totality—A Report,"[*] describes his first total eclipse and the moment when the diamond ring flares brilliantly, then fades, leaving the sky a deep purple and the corona spread out around the blackened Sun.

> The courtyard fills instantly with the simultaneous sound of 200 focal-plane shutters firing constantly and rhythmically—ka-chick-ka-chick-ka-chick—a legion of mechanical crickets. At the same moment, outside the courtyard walls, the high steady chanting of a small group of Mauritanians rises in the sudden darkness: "*Allah-tlag es-shems, Allah-tlag es-shems*"—*Allah*, release the Sun—an ancient prayer that has, thus far, never failed to work.

Rogers describes "the frozen flaring light of the corona" and the black disk of the Moon that blocks the Sun: "a perfect gaping ebony cavity in the fabric of the dusk-dark sky." All too soon, totality nears an end.

> Outside the courtyard the *boubou*-clad Mauritanians are prostrate on the dry dust of the mine road; inside, the astronomers, knowing that time is short, launch another frantic volley of photographs: "*Allah tlag*"—ka-chick-ka-chick—"*es-shems*"—ka-chick-ka—"*Allah Allah*"—ka-chick-ka-chick—"*tlag*"—ka-chick—"*es-shems.*"
> The sounds blend musically in the hot darkness and neither chorus appears particularly aware of the other. One, relentlessly Sun-blasted virtually all their days, prays fervently for the Sun's return. The other, knowing return is inevitable, prays inwardly that it be delayed just a moment longer.

[*] Michael Rogers: "Totality—A Report," *Rolling Stone*, October 11, 1973. These passages are slightly condensed.

# 13

—◄◦►—

# Remembering the All-American Eclipse of 2017

*A total solar eclipse is the best show on Earth.*
Julie O'Neil, eclipse veteran (England)[1]

*How did you and you and you experience the eclipse?*
*Whose hands did you clasp at the wonder of the moon*
*blotting out the sun, breaking the day in two?*
Susannah Felts, writer and first-time eclipse observer[2]

In 2017, 24 million American adults saw the Sun in total eclipse. They saw it in person. That was 8.6% of the adults in the United States—better than one out of 12. The Sun in total eclipse.

Those observers tucked themselves into the 70-mile-wide path of totality that cut across the United States from coast to coast.[3] But the 2017 total eclipse of the Sun also provided partial phases that encompassed every state in the Union.

Even if the path of totality didn't come to them or they couldn't go to it, Americans knew the eclipse was going to happen and they were interested. They were interested enough that 216 million American adults stopped what they were doing to view the phenomenon in some way, either as a partial or total eclipse, watching it either in person or electronically on television, computer, or smartphone. That was 88% of American adults.[4] Jon Miller, the researcher who led this survey, "couldn't find evidence of any other event in history witnessed by so many Americans."[5]

And not just Americans. Many thousands of visitors from other countries made a pilgrimage to the United States in 2017 to see the Sun in total eclipse.

Those Americans and world travelers who journeyed into the path of totality—what did they see? And how did they feel about it? Here are a few of their stories.

The August 21, 2017 eclipse track ran diagonally across the United States [Map © 2016 Michael Zeiler, GreatAmericanEclipse.com]

# The View from Salem, Oregon—Ralph Chou

"I owe my entire life's work to eclipses," says Ralph Chou. He saw his first total eclipse of the Sun in 1963 at the age of 13. His mother, a widow, drove him and his younger sister 700 miles (1,100 kilometers) to the wilderness of northern Ontario. "She was adventurous," he says, "She took us where many dared not go."

Ralph developed an abiding interest in science fiction and then in astronomy while in high school in Windsor, Ontario, just across the river from Detroit. He majored in astronomy at the University of Toronto. He got a job as a science demonstrator at the Ontario Science Center in Toronto. But after 1973, jobs in astronomy and the space industry were scarce. Ralph came from a family of physicians and dentists. He earned a doctorate in optometry at the University of Waterloo and spent his career as a professor there.

But his interest in astronomy did not subside. He continued to travel to eclipses—19 total eclipses so far—and to share his knowledge of the eyes with parents and students and groups to help them observe eclipses safely.

An estimated 215 million people saw the 2017 eclipse directly as a partial or total eclipse but, says Dr. Chou, there were only about 100 cases of eye damage reported—less than one in 2 million. Most of the eye damage reported was very minor, with no long-term consequences.[6] There were no reports of anyone going blind from recklessly viewing the partial phases of the eclipse. Dr. Chou's many speeches, articles, and interviews have formidably contributed to that success.

His fame as a champion of eye safety during solar eclipses has often put him in awkward circumstances. There he was in Oregon at the 2017 eclipse. The Moon was carving the Sun into a deeper and deeper crescent. Suddenly, Ralph's cell phone started ringing. Somehow, Canadian radio and television broadcasters had gotten his phone number and started calling him in the midst of observing. One call. Two. Three. As totality neared, Ralph ended the third call: "I'm sorry, I can't talk to you anymore." One broadcaster begged to do a follow-up after the eclipse was over.

Ralph takes photographs of every eclipse. But he makes sure to stop now and then as the eclipse progresses, especially in the midst of totality, to look around—at the appearance of the landscape, at the behavior of animals, at birds roosting, at people "losing it." Case in point: He observed the 1994 total eclipse of the Sun at Iguazú Falls on the border between Brazil and Argentina. As daylight dimmed, he heard shotgun blasts. Farmers were firing at the sky, trying to scare the evil creature that was swallowing the Sun.

Ralph still gets a huge emotional reaction from seeing the corona. No partial eclipse, no annular eclipse can match it, he says. "A total eclipse of the Sun is a life-altering event."

## The View from Madras, Oregon—Dimitry Rotstein

In August 2017, Dimitry Rotstein made his way from his home in Israel to Madras, Oregon for the total eclipse of the Sun. He could have traveled more directly, but he doesn't like long flights. So he flew from Israel to Zaporizhzhia (Ukraine) to Kyiv (Ukraine) to Paris to London to Belfast to New York City to Washington, D.C. to see his brother and sister-in-law who live nearby. They declined Dimitry's invitation to join him for the eclipse because they had two young children. They couldn't fully understand what would possess their brother to travel across 10 time zones for two minutes of darkness. After a pleasant visit and some Washington, D.C. sightseeing, he flew on to Austin then Phoenix then Portland, Oregon and rented a car to drive to Madras.

He positioned himself with Mount Jefferson, a 10,500-foot (3,200-meter) dormant, snow-capped volcano behind him to the west. He came to Oregon because he read that it was beautiful. Indeed, it was. And climatology showed Madras to have the least cloudy weather at that time of year at that time of day along the entire path of totality. The town of 6,600 lived up to its weather promise—except for wildfires around the state that added a dusty haze to the sky.

Dimitry was especially interested in seeing the diamond ring effect that occurs an instant before totality begins and again an instant after totality ends as a single spot of sunlight peaks out from behind the Moon. At his two previous total eclipses, he had missed seeing the diamond ring because as totality was about to begin he had been too busy watching for the black shadow of the Moon to come racing at him from the western horizon. Then, at the end of totality, he had been too busy watching for

One of many improvised recreational vehicle campsites near Madras, Oregon along the path of totality in 2017. The people were in line to see the partial phase of the eclipse through a telescope. [Photo courtesy of Dimitry Rotstein]

Mount Jefferson is one of 61 volcanoes in Oregon. It is considered to be dormant, but two—Mount Hood and Newberry—are active. [Photo courtesy of Wikimedia Commons]

the Moon's shadow to rush off to the east. He may have missed seeing the diamond rings with his eyes, but he caught the event with his cameras.

This time he wanted to be sure to see the diamond ring effect with his own eyes. But how could he not look off to the west, toward Mount Jefferson, to see the eclipse shadow rushing toward him? In that last minute before totality, the sunlight was fading fast. The remaining light was silvery—a color Dimitry has never seen outside a total eclipse. Unreal. Except very real.

One minute to totality. There was Mount Jefferson. The next moment it was gone—lost in the darkness to the west. The volcano had vanished in the shadow of the Moon. He swung around. Halfway up the eastern sky, there was the diamond ring. And a vivid corona, despite the haze. And around the horizon, 360° of twilight color with the dark cone of Mount Jefferson in silhouette.

Moments before the eclipse ended, color returned to Mount Jefferson as it emerged from the Moon's shadow and the edge of the shadow sped eastward toward and past Dimitry and on across the American continent. Above him, the second diamond ring.

It had taken Dimitry 2½ hours to drive from Portland to near Madras, Oregon. On the way back, it took him 9 hours as 100,000 eclipse observers tried to make their way back to the big city on a mostly two-lane highway.

Dimitry celebrated the eclipse by driving up and down the West Coast, including Seattle, San Francisco, Los Angeles, and San Diego. Then he started home. But Dimitry finds that flying long distances eastward gives him a bad case of jet lag. So he flew home by going west, in the shortest legs he could arrange. He flew from Los Angeles to Hawaii (now his favorite place on Earth) to Sydney (Australia) to Perth (Australia) to Singapore to Sri Lanka to Dubai (United Arab Emirates) to Amman (Jordan), to Tel Aviv (Israel)—sightseeing, of course, along the way.

Dimitry Rotstein has read many descriptions of total solar eclipses. "They are never equal to the real thing," he says. "Everyone should see a total eclipse of the Sun."

And when you go, Dimitry says, never go just to see the eclipse. Be sure to visit other places and see other things.

## The View from Prairie City, Oregon—Jeanne Loring and David Barker

Jeanne Loring and David Barker met when he was a young professor of neurobiology and she was a graduate student in neurobiology at the University of Oregon. "But I wasn't his student," Jeanne says. Both are now retired professors, but both continue to be active in their fields. David

Jeanne Loring and David Barker invited their extended family and friends to join them in Prairie City, Oregon for the total solar eclipse of 2017. Most of the party slept in tents. The RV sheltered those who could not sleep on the ground. [Photo courtesy of Jeanne Loring and David Barker]

serves on the boards of three biotech companies. Jeanne founded and still runs a different biotech company. Both are amateur astronomers and have been attending selected total solar eclipses together since 1979.

For the total eclipse of 2017, they wanted to bring their extended family together. They chose Prairie City, Oregon as their observing site and rented two spaces in the city's recreational vehicle park. They flew from their home in Del Mar, California (San Diego area) to Boise, Idaho, picked up their rented RV, and drove to Prairie City. One campsite accommodated the RV; the adjoining campsite was for tents—20 people total. Jeanne and David slept in a tent. They gave the camper to family members and friends who couldn't sleep on the ground. It was cold at night, David says. The elevation of Prairie City is two-thirds of a mile (1 km). The night-time temperatures in August were in the 40s (about 6˚C).

David loves not just the spectacle of total solar eclipses, but also how, year by year, they leap about the globe: Bolivia, Libya, Easter Island. "We see places we never expected to," he says. And while there, David and Jeanne relish excursions to appreciate the landscapes and the cultures of the lands where eclipses have taken them.

While in Prairie City, Oregon to see the 2017 total eclipse of the Sun, Jeanne Loring and David Barker bought this quilt. The Sun's corona is made of men's ties. [Photo courtesy of Jeanne Loring and David Barker]

How do they decide where to view the eclipse along the path of totality? "Easy," says Jeanne. "If Jay Anderson [a Canadian meteorologist famed for his eclipse weather advice] says it's a good place to go, we go there."

Prairie City proved to be a good choice, David says. "It was just far enough east so that traffic wasn't too bad. The town wasn't overrun by eclipse chasers." Artists in Prairie City made artifacts for sale to the eclipse visitors. Jeanne and David bought a quilt showing the Sun in total eclipse, with a stylized corona made of men's ties.

Approaching and during totality, David makes sure to take note of the environment—animals going back to their barns, day birds that stop singing, owls that start hooting. Jeanne is struck by the darkness during totality—"really eerie."

Jeanne and David are both fascinated by the corona. In the days before the eclipse, by looking at the number and position of sunspots, Jeanne tries to predict what the Sun's corona will look like and where prominences will be on the rim of the Sun. "It never looks like what I expect," she says with delight. The corona, the prominences, Baily's beads—every eclipse they're different.

David especially likes to watch for the approach of the Moon's shadow. For that, you need an open vista to the west, he says. He has seen the shadow approach only twice in his 14 eclipse expeditions, but when it is visible, it is really dramatic. "It rushes at you and covers you up."

2017 was "a short eclipse," Jeanne says, but "outstanding. I'll take any amount of totality."

For the 2024 total eclipse, David and Jeanne are going to Mexico. "You plan for eclipses far in advance," Jeanne says. "Eclipses won't wait. They can't be canceled. There are no excuses."

## The View from Jackson Hole, Wyoming—Tony Crocker and Liz O'Mara

On August 21, 2017, Tony Crocker and Liz O'Mara rode the ski lift to the highest point—10,450 feet (3,190 meters)—at Jackson Hole Mountain Resort. From that summit, they could look north across the soaring peaks of Grand Teton National Park. From there they could look east and down on the Snake River meandering south across the valley and through the town of Jackson, Wyoming. It was from there they watched the 2017 total eclipse of the Sun. When the eclipse ended, they got married. It was her ninth total eclipse; his eleventh. The bride wore a knee-length white dress, leather boots, and eclipse earrings. The groom was grateful that the best man remembered to bring a suit coat and tie for him to wear.

Tony and Liz met on-line seven years earlier because their favorite activities were total eclipses and skiing. She was living in New York City;

Tony Crocker and Liz O'Mara getting married on a mountaintop after watching the 2017 total eclipse of the Sun. Conducting the ceremony is retired judge Terry Waters. Offering a reading is friend Tony Richardson. [Photo courtesy of Judy Flayderman © 2017]

he in Glendale, California. They finally met in person in 2011. There were no total eclipses of the Sun that year so they settled for a ski trip—then a second and a third trip that year, including one to Antarctica. Liz and Tony have attended all total solar eclipses together since 2012, including the pandemic years of 2020 and 2021. "Before I met Tony," Liz says, "everyone thought what I wanted to do was crazy."

Liz especially enjoys watching shadow bands at total eclipses. Tony's favorite aspect is solar prominences. Both agree that viewing an eclipse at high altitude is extra rewarding because the thin atmosphere accentuates the details and allows the corona to be seen extending farther from the Sun. In 2017, from their mountain perch, they estimated that the corona stretched 6 diameters on each side of the Sun.

## The View from Casper, Wyoming—Donald Bruns

Donald Bruns observed the 2017 eclipse from Casper, Wyoming. Well, "observed" is an understatement. Don is a retired Ph.D. optical scientist who conducted military and industrial research in optical systems and lasers. He saw his first total eclipse of the Sun in 2006 from a cruise ship in the eastern Mediterranean Sea. It was dramatic, he says. "The Sun turns into a black hole in the sky." He wanted to see another . . .

... and that's when his favorite hobby kicked in. Don likes to repeat momentous scientific experiments from the past to see if he, using modern technology, can get even better results than the original experiment. He thought of Arthur Eddington who used the total solar eclipse of 1919 to measure if rays of distant starlight passing by the eclipsed Sun were bent by the Sun's gravity according to Einstein's new general theory of relativity. They were—and Einstein instantly became the most famous scientist in the world.

Don, with his scientific training, thought through the experiment very carefully. He borrowed an extremely precise refracting telescope and CCD camera. He borrowed the sturdiest of tripods. His tools were far beyond the capabilities of equipment a century earlier. He used star positions plotted by the European Space Agency's *Gaia* space observatory. He practiced with his equipment over and over again for months before the eclipse.

In Casper, four days before the eclipse, he doubled over in pain and was transported to a Casper hospital emergency room. Diagnosis: kidney stones. The first one passed two days before the eclipse, and Don rushed back to his campsite ...

... only to find that, in order to avoid overhead lights coming on automatically as the eclipse neared totality, the campground director was

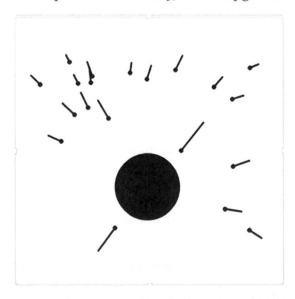

Don Bruns measured the deflection in positions of stars near the Sun during the 2017 total eclipse as the starlight was bent by the Sun's gravity. The circles mark the actual positions of the stars and the lines show the change in positions of the stars observed during the eclipse. The amount of bending was precisely what Einstein predicted in his general theory of relativity—the best results ever measured. [Illustration courtesy of Donald G. Bruns]

Don Bruns and his wife Carol with the 4-inch (101-mm) refracting telescope and CCD camera he used to measure the relativistic deflection of starlight around the eclipsed Sun in 2017. [Photo courtesy of Steve Lang]

planning to shut off electricity to the site for the duration of the eclipse. At Don's request, the director found a 400-foot extension cord and a very tall ladder and climbed a power pole to plug in a connection so that Don would have power for the telescope mount to track the Sun.

Don's measurements topped the accuracy Eddington achieved with two teams of professional observers, one in Brazil, one in Africa—measurements in 1919 that proved Einstein's general theory of relativity was correct, triggering a revolution that transformed science. Donald Bruns published his results in a scientific journal.[7] The American Astronomical Society awarded him their Chambliss Amateur Achievement Award in honor of his precision and ingenuity.

Don is thinking about using the 2024 total to repeat another famous eclipse experiment: the use of the 1878 total eclipse of the Sun to search for Vulcan, a theoretical planet closer to the Sun than Mercury that was thought to be responsible for the odd precession of Mercury's orbit. That search and searches that followed never found substantial evidence of

Vulcan. The precession of Mercury's orbit was completely explained by Einstein in his general theory of relativity. Don doesn't think that Vulcan exists but he thinks it is now possible to use a total eclipse of the Sun to settle the matter once and for all. With modern equipment, he can see a planet 100 times dimmer and 10 times smaller than previous searches.[8] And he still has this hankering to repeat famous experiments from the past with greater accuracy than ever before.

## The View from Agate Fossil Beds, Nebraska—Mark Kidger

Astronomer Mark Kidger is a busy man. He's calibrating and coordinating the instruments that will fly on the European Space Agency's Plato Space Telescope[9] in 2026. Its mission is to examine a quarter of a million stars in search of planets passing in front of them and minutely dimming their light.

Dr. Kidger also serves as Deputy Project Scientist for ESA's Comet Interceptor, a spacecraft to be launched in 2029 and then parked in space. It will go into chase mode when a comet that has never been close to the Sun approaches the inner solar system.[10]

Mark's interest in astronomy dates back to age 5. His family found him in front of their black-and-white television set spellbound by a docking in space. A few months later, when the family moved to Bristol in southwest England, he remembers being entranced by the stars in the sky. At 9 years old, Mark watched Neil Armstrong walking on the Moon. When a sample of lunar soil visited Bristol, he stood in line for two hours to see a few specks of moondust. "Spaceflight," he says, "is the greatest adventure man has ever had."

However, when it comes to seeing solar eclipses up to 2017, Mark has not had the best of luck. "I have been to two totals and two annulars," he says, "and have been clouded out of three of the four." He managed to see the 2005 annular eclipse from south of Madrid, Spain, where he lives. He narrated the event live for BBC Television, even though he didn't know exactly what to expect. The sky darkened a bit. But the landscape colors changed a lot. "All the greens in the park around me became so much darker and more intense," he says. He saw about 20 seconds of Baily's beads twinkling along the rim of the Sun as the annular phase was ending. "It really made me want to see the real thing," he says, "a total eclipse."

For 2017, Mark booked himself, his wife, and his daughter with a tour company because "I was attracted by their plan to see a part of the USA that I had never visited before." They would observe the eclipse from

Nebraska and then travel on to Mount Rushmore in South Dakota and Devils Tower in Wyoming. He was especially delighted that "my family would get to see the real USA, having never traveled out of Europe."

As the bus headed from Denver toward Alliance, Nebraska, the Kidgers soon realized they weren't in Europe anymore. They were passing a ranch. They saw a long fence, then a road that passed through a gate in the fence to meet their highway. Standing at the gate was a sign that announced the name of the ranch and then "23 miles to the house."

The eclipse tourists planned to see the eclipse from near Alliance, close to Carhenge, a full-size replica of Stonehenge, but made from junked automobiles. "This is the sort of monument that you could find only in America," Mark says fondly.

But in Alliance, the weather forecast for eclipse day was threatening. The tour leaders and eclipse seekers got together the night before the eclipse and agonized about running to a more promising site. They woke up at 4 a.m. on eclipse day, after a night of very little sleep, and hit the road for Agate Fossil Beds National Monument, 71 miles (114 km) to the northwest. The bus crept along through a dense fog. Then, suddenly, the fog lifted. From the bus they could see people with telescopes and cameras lining both sides of the road for miles and miles.

They pulled into the campground at Agate Fossil Beds. A good tourist day at the Monument might bring 300 visitors. Eclipse Day brought 7,000. The handful of park rangers at Agate Fossil Beds had mowed the grass in the campground—and posted warnings at the edge of the surrounding tall grass: "Beware of rattlesnakes." They directed traffic as people flooded in. They had plenty of portapotties ready and made sure they were always in good order. They passed out solar filters in cardboard eyeglass frames and offered instruction on eye safety.

Here and at every national monument the Kidgers visited, "the rangers were the friendliest, most helpful people in the world," Mark says. "Americans do not appreciate their park rangers."

Mark set up his camera. His wife Lourdes and daughter Paula were a little bewildered. They didn't know what to expect. Mark didn't know either. He had never succeeded in actually seeing a total eclipse of the Sun.

First contact. The Moon notched the rim of the Sun. The weather was clear. A chill went through Mark. "We're going to see it."

An hour passed. There was less and less Sun. His wife and daughter were struck by the changing tone of the landscape, the changing color of the sky.

And then the Sun vanished. Mark was suddenly staring at the corona. "So diaphanous," he says. "It looks nothing like a photograph." Extending from the corona was a massive streamer at the 7 o'clock position. "If you have never seen a total solar eclipse," Mark says, "nothing can prepare you for the experience."

Mark Kidger, his wife Lourdes, and his daughter Paula at Devils Tower National Monument in Wyoming. [Photo courtesy of Mark Kidger]

A prairie dog at Devils Tower National Monument takes interest in the Kidger family. [Photo courtesy of Mark Kidger]

All too soon, it was as if a switch snapped back on and the Sun began to return. Mark kept photographing until the Moon pulled clear of the Sun. His eclipse group had a picnic lunch to talk about what they had seen.

Everyone had a different impression. Some couldn't get over how ghostly the corona was. "They were haunted by the sight," Mark says. His wife and daughter were fascinated by the very noticeable drop in the temperature. "It was a warm, sunny day," his daughter Paula says. "Then it got really dark—and it got cold."

After the eclipse, the group set off for the Black Hills in South Dakota. The tour group stayed at a hotel run by Native Americans. They learned how the Indians hunted bison. At the hotel, they also found that each bathroom was equipped with a cloth intended for cleaning their guns.

Wherever they went in the American West, they ate buffalo burgers and buffalo stew. "It's healthier than beef," Mark says. "And it's not European cooking."

At Mount Rushmore, they had lunch and ice cream and stared at the faces of four American presidents carved into the mountainside. "I could have died after that and been happy," says Mark.

The tour bus moved on to Devils Tower in Wyoming. The Kidgers remembered *Close Encounters of the Third Kind* but they didn't see any landing site there to welcome the aliens. They did see prairie dogs—bigger and fatter than they had imagined—scampering from hole to hole. They saw 30 wild turkeys running through the woods. They saw Native American offerings placed in the branches of trees. They watched the Sun set behind Devils Tower.

The bus stopped for lunch at a little restaurant near Fort Laramie, Wyoming. It was run by a veteran of the United States military. The vet came over with a pad of paper. The first tourist started to recite his order. "No, no," said the vet. "I get very confused if you tell me the food you want. Just give me the number on the menu." He took the numbers and vanished into the kitchen. A long while later, the veteran reappeared with lunch. It was the best burger Mark had ever eaten.

The bus stopped for malts and sundaes in Chugwater, Wyoming at the oldest soda fountain in the state.

The Kidgers felt they had truly visited America. "For me," says Mark, "2017 was the trip of a lifetime. It was wonderful watching my wife and daughter learn about rural America, about the size of the country, and the fact that their idea of a typical American from films was perhaps not in line with reality."

Mark is looking forward to seeing more eclipses. Beginning in 2026, he won't have far to go. In a period of less than a year, Spain will play host to two total eclipses. For the 2027 eclipse, however, he would like to take Lourdes and Paula to Egypt to see the Sun blink out over the temples at Luxor.

## The View from Marion, Illinois—Alex Barbovschi

Alex Barbovschi grew up in Moldova, a small country wedged between Romania on the west and south and Ukraine on the east and north. He earned his bachelor's and master's degrees there and directed his university's observatory and planetarium. He currently works as a software engineer—when he's not traveling the world to see and photograph astronomical events. He may be the only Moldovan to travel to eclipses.

Alex has been passionate about astronomy since childhood. In 1999 at age 9, he saw his first solar eclipse. It was a 93% partial eclipse in Moldova. He wanted to see totality, but for that he would have had to travel to Romania, a distance of 250 miles (400 km), far out of reach. He filed that dream away. He got a telescope on his 19th birthday.

As 2017 approached, he dusted off his dream of a total solar eclipse, lured by the chance to see it in the United States. He talked a friend into going. He quit his job at the university and took two jobs to raise funds for his trip. At the last moment, his friend decided against going. Alex decided to go on his own. His sister Olga and her husband Iura lived in Chicago. They would drive him into the path of totality in southern Illinois, near Marion. That was a trip of 330 miles (530 km)—longer than the greatest distance across Moldova.

Alex Barbovschi improvises a tent to protect him and his sister and brother-in-law from the scorching heat of his eclipse site in southern Illinois. In front of Alex are his still and video cameras. [Photo courtesy of Alex Barbovschi]

Alex Barbovschi photographed and video recorded the diamond ring effect as totality began in 2017—and accidentally captured a passenger jet passing across the darkened Sun. [Photo courtesy of Alex Barbovschi]

It was the evening before the eclipse. They found a campsite. Alex began setting up his equipment. The weather was ferociously hot. He could get only 4 hours of sleep. Traffic was so heavy there was no chance they would be able to move to a better site if clouds rolled in. They would have to hope for good weather. Clouds during the partial phases of the eclipse reduced the number of photos Alex hoped to capture. The heat was terrible. Alex was drowning in sweat. His sister and brother-in-law played cards during the partial phases of the eclipse. "Yeah. Sure. Whatever." They couldn't understand why Alex would travel so far and spend so much money . . . for this.

And then the Sun vanished. The corona and a huge red prominence appeared at the rim of the now-invisible Sun. It was "crazy beautiful," says Alex. The people around them screamed and shouted and cheered. Alex was trembling. It was his first total eclipse of the Sun. The clouds that had hindered viewing of the crescent Sun were gone for the duration of totality. "Now I see why you came so far and spent so much money," Iura said.

The eclipse ended. It had taken Alex, his sister, and her husband 5 hours to drive from Chicago to Marion for the eclipse. It took them 12 hours to drive home in the 330-mile traffic jam. Alex tried to ignore the creep-and-halt traffic by examining the still photos and video recording he had made of the eclipse. A scream nearly sent them off the road. It was Alex screaming. There it was in a still photo. And there it was in the video recording as well. Just as the diamond ring effect was fading at the beginning of totality, a jet airliner had flown right across the face of the Sun. Alex hadn't noticed it at the time. The transit took less than a second. But there it was in his photography. That picture won an eclipse photography contest.

Alex checked later and found that the aircraft caught in the eclipse was a Boeing 737—United Airlines flight #632 from Washington, D.C. to Los Angeles.

The total solar eclipse of 2017 boosted Alex's passion for astronomy and astrophotography to a new level. He went to Chile for the 2019 eclipse. Covid undid his plans for 2020, and the Antarctic eclipse of 2021 was "way too expensive, sadly," but he has booked travel to Australia for 2023 and he's working on plans to see the 2024 total solar eclipse in Mexico or Texas. Meanwhile, he was back in the United States for the November 2022 lunar eclipse. He flew 2,800 miles (4,500 km) from Moldova to the Azores Islands in the Atlantic Ocean to see if the Tau Herculid meteors would produce a huge storm of shooting stars in May 2022. They didn't. "I regret nothing," Alex says. "I use astronomy as an excuse to see the world."

## The View from Madisonville, Tennessee—Dickson Despommier

Dickson Despommier is a retired Columbia University professor of public and environmental health and a leader in vertical farming. He and his wife, Marlene Bloom, drove to Madisonville, Tennessee for the 2017 eclipse. "We located a high school there," he says, "settled down on a piece of unobstructed grass on the school grounds, and watched as the Moon slowly completely covered the Sun. We were essentially alone. Closest to us was another couple 100 yards away from our chosen site.

"It was truly unearthly," says Dickson. "At the moment of totality, the lighting was spectacularly surreal. The vision of a black Sun surrounded by a bright, irregular halo of fire hovering motionless above us was so foreign to anything I had ever experienced that I am still processing the sensation of those two minutes.

"We had traveled 770 miles from Fort Lee, New Jersey, to Madisonville to see the Sun in total eclipse. We now want to do the same thing again in 2024!"[11]

# The View from Clemson, South Carolina—Eric and Janice Brown

Eric Brown did more than observe the 2017 total eclipse of the Sun. He narrated it for the Clemson University student body and faculty and staff and community members—50,000 people.

He saw his first total eclipse in 1970 at age 17. As of 2021, his count has risen to 19. His observing site in 2024 was easy to choose. His daughter Ilana is getting married then at a ranch outside of Dallas.

Eric has seen two eclipses by plane, the rest on land. "It's best on land," he says. "So much is going on"—the temperature drop, the wind, the shadow bands, the 360° twilight colors, the reaction of the animals, the reaction of the people. His advice: Don't take photos. Yours won't be as good as the ones professionals take. "Just watch the eclipse. And watch the people watching the eclipse."

What does he see when he watches people watching the eclipse? He sees some of them crying, others mesmerized. Most stand, as if in reverence. Some jump up and down. Some can't move. "The eclipsed Sun looks like the eye of God."

"Eclipses are so much more than you read or hear," says Eric's wife Janice. No photograph you can see, no account you can hear can fully do justice to a total eclipse of the Sun—watching the light in the sky change to a color never seen outside an eclipse. Watching a dog that had been running around crawl under a chair during totality. Watching birds fly back to their nests and twitter as if confused. Watching fish jump out of the water.

Does she have a favorite of all the eclipses she has seen? "That's like asking which is my favorite child," Janice says.

A total eclipse of the Sun is the most dramatic of all demonstrations of nature, Eric says. What a gift it is that the Moon is just the right size, just the right distance, and in just the right orbit to blot out the Sun. And it won't always be so. The Moon is drifting farther from Earth, making it smaller in our sky. There will come a time when there will be no more total eclipses of the Sun.

## NOTES AND REFERENCES

1. Epigraph: Julie O'Neil, interview, April 28, 2015.
2. Epigraph: Susannah Felts: "Astonish Me: Anticipating an Eclipse in the Age of Information," *Catapult*, August 21, 2017.
3. The width of the path of totality across the United States ranged from 62 miles (100 km) in Oregon to 71 miles (115 km) from Illinois eastward.

4.  Jon D. Miller: "Americans and the 2017 Eclipse: A Final Report on Public Viewing of the August Total Solar Eclipse," University of Michigan report for NASA, June 12, 2018. Miller, Director of the International Center for the Advancement of Scientific Literacy, conducted this survey as Research Scientist, Institute for Social Research, University of Michigan.

5.  Sarah Kaplan: "2017 solar eclipse was one of most-watched events in American history, survey finds," *Washington Post*, August 13, 2018. Also, Jon D. Miller: Interviewed by Mark Littmann, August 9, 2022.

6.  See also J. Kelly Beattie: "Eye Damage Reported from August's Eclipse," *Sky & Telescope* (online), December 8, 2017.

7.  Donald G. Bruns: "Gravitational Starlight Deflection Measurements During the 21 August 2017 Total Solar Eclipse," *Classical and Quantum Gravity*, Volume 35, Number 7 (6 March 2018).

8.  For the 2024 eclipse, Don Bruns will also be advising Toby Dittrich, Professor of Physics at Portland Community College in Oregon. Toby has received grants so that he and his students can man at least four observing stations to repeat Don's 2017 light-bending measurements.

9.  Plato is an acronym that stands for PLAnetary Transits and Oscillations of stars.

10. Comet Interceptor must be positioned in space ahead of time because once a pristine comet is identified inbound for the Sun, it's too late to launch a spacecraft to fly by it.

11. Dickson Despommier contributed the original version of this story to the website of Sigma Xi, The Scientific Research Honor Society for "Members Share Their Solar Eclipse Stories" by Heather Thorstensen. The article appeared on September 6, 2017.

# A MOMENT OF TOTALITY

## The Eclipse Trip from Hell

Science writer Dawn Levy and three friends set off in a 21-year-old Volkswagen bus from San Francisco to Mexico to see the almost-7-minute-long 1991 total solar eclipse. The van's engine, recently rebuilt, began smoking outside Los Angeles. They replaced a seal and pushed on. A day's drive past the Mexican border, near Guaymas, the engine began spewing smoke. At the VW dealership, the repairman said the engine was irreparable and offered a rebuilt engine.

While they waited for the installation, the eclipse chasers went for a swim in the ocean. Within minutes Dawn had excruciating jellyfish stings on her face, neck, and hands. Then things got worse. Soaked in sweat, her stomach cramped, her hands went numb, and she couldn't breathe—a severe allergic reaction. An ambulance rushed her to the hospital where she received antihistamine injections.

The next day, when she could breathe normally again, Dawn and her friends were informed that the rebuilt engine was bad. The owner abandoned his van and the four travelers waited 8 hours for a train to Mazatlán. But the arriving train was full. Still determined to see the eclipse, they took a bus, arriving just in time for eclipse day.

The morning was overcast and, as the hours passed, the clouds refused to part. The sky turned peach as if it were dusk—then black. "Darkness at noon," Dawn says. People cheered, although they could not see the Moon covering the Sun. A cool breeze replaced the oppressive heat and humidity. Then, like the beam of a giant flashlight, brightness swept back across the sky. After a journey of 1,700 miles (2,700 kilometers), Dawn and her friends returned to their motel and watched a replay of the eclipse on television.[*]

[*] Dawn Levy: "An Eclipsed Vacation," *Los Altos* [California] *Town Crier*, August 7, 1991, pages 22–23. Also, interview, May 23, 2015. Interviewed for Second Try section, September 25, 2022.

## Second Try

Twenty-six years later, Dawn, now a resident of Tennessee, was really looking forward to 2017. The eclipse was coming to her.

Even though the path of totality included her home, Dawn opted to co-host a party with a friend whose house was closer to the central line of the eclipse. They invited a dozen friends, and those friends invited their friends. Total: 60. Refreshments for the party were eclipse-themed: Corona beer, Sunkist orange soda, Starburst candy. "None of us had seen a total eclipse of the Sun," says Dawn. "We were excited and curious, but no one knew what to expect.

"The partial phase of the eclipse proceeded slowly," Dawn says. "Very slowly. And then . . . Words fail me—and I'm a writer. It's utterly outside what you experience in your normal life. The beginning of totality is so unexpected—the diamond ring, the corona. Most shocking to the people at the party was darkness in the middle of the day. At the end of totality, everyone cheered. And I got to share that with people, with friends. It was comforting and joyous. I will definitely see the eclipse in 2024."

# 14

◄◦►

# Coming Back to America—The Total Eclipse of 2024

*Few can imagine how much I longed for another minute, for what I had witnessed seemed very much like a dream.*

Edwin Dunkin (1851)[1]

Party animal. That's the total solar eclipse of April 8, 2024. It visits many big cities—seems to be genuinely fond of people. And it's generous. As a party favor, it hands out more than 4 minutes of totality along its central line throughout Mexico and much of the United States and never less than 3 minutes until the party ends in Newfoundland. And as it sweeps across North America, it spreads the party widely. The swath of totality is as wide as 126 miles (202 kilometers) in Mexico and never less than 101 miles (163 km) through the United States and Canada—a much wider band of totality than the 2017 eclipse. Party, party. More people are invited.

In Mexico, Mazatlán, Durango, and Torreón lie within the zone of totality. In the United States, San Antonio, Austin, Dallas, Fort Worth, Indianapolis, Cincinnati, Dayton, Columbus, Cleveland, Buffalo, Rochester, and Syracuse have the honor of lying wholly or partly within the path where the eclipse is total. In Canada, Hamilton and Montreal. Many middle-size and small cities are equally blessed. Other major cities lie just outside of the path of totality. If people are shy about going to a total eclipse, this one comes to them.

The totality of the 2024 eclipse first touches Earth south of the equator about halfway between Australia and Mexico. Its darkness flies north-eastward across the Pacific, visiting only one island, Socorro, an active volcano that is part of Mexico.

It sweeps ashore in mainland Mexico at Mazaltlán. This is Mazatlán's second total eclipse in 33 years. It played host to the 1991 eclipse. For cities that sit still and wait, a total eclipse usually comes along once in 375 years.

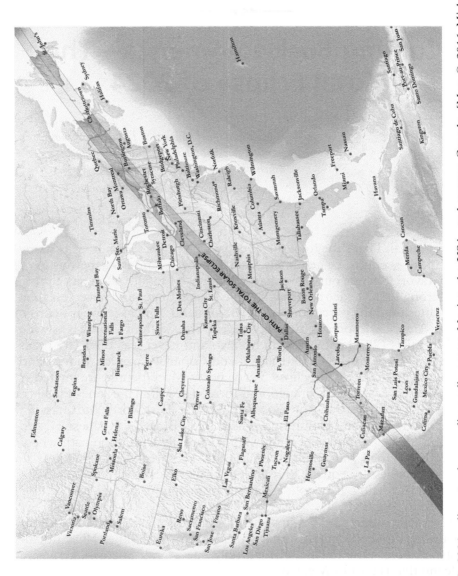

The April 8, 2024 eclipse track runs diagonally across Mexico, the USA, and eastern Canada. [Map © 2016 Michael Zeiler, GreatAmericanEclipse.com]

The path of totality as it crosses Mexico. Points along the central line give the duration of totality in minutes and seconds. [Map © 2016 Michael Zeiler, GreatAmericanEclipse.com]

The Moon and Sun, both fond of bright lights, grant El Faro, the historic lighthouse on a peak at the south end of downtown Mazatlán, 4 minutes 19 seconds of totality. The central line of the eclipse passes 18 miles (30 km) south of downtown, offering an extra 8 seconds in the shadow of the Moon. April is a comparatively dry month in Mazatlán.

The 2024 eclipse heads northeast, up into the Sierra Madre Occidental, a southward continuation of the Rocky Mountains. From Mazatlán to Durango, the eclipse embraces scenic Mexican Highway 40D. In 143 miles (230 km) of sensational canyons, the expressway passes through 61 tunnels and across 115 bridges. Most spectacular is the leap over the Baluarte River gorge on a cable-stayed bridge suspended 1,280 feet (520 meters) above the river. The Baluarte Bridge is the highest roadway bridge in North America.

Next the eclipse arrives in Durango, second largest of the Mexican cities on its itinerary, with a metropolitan population of 660,000. Durango lies 30 miles (50 kilometers) southeast of the central line. In the western part of downtown, amid colonial buildings and plazas, is an aerial tramway that carries passengers to the top of a hill 270 feet (82 meters) high with a fine view of the city—and the approach of the eclipse shadow from the west that will provide a total eclipse of 3 minutes 51 seconds.

Highway 40D continues on northeastward from Durango to Torreón, always within the zone of totality, but well south of the central line. Along the central line, about two-thirds of the way between Durango and Torreón, the 2024 eclipse reaches its point of longest duration—4 minutes 28 seconds.

Torreón, with a metropolitan area population of 1.5 million, is the largest Mexican city in the path of the 2024 total eclipse. Downtown Torreón is 25 miles (40 km) southeast of the central line, providing it with 4 minutes 9 seconds of totality.

Durango and Torreón lie between the western and eastern ranges of the Sierra Madre. April is the second driest month of the year for this region.

The eclipse then bounds over the Sierra Madre Orientale to reach the Rio Grande, which provides a watery boundary between Mexico and the United States for over a thousand meandering river miles. Straddling the border, almost on the central line of the eclipse, are Piedras Negras and much smaller Eagle Pass. On the International Bridge connecting them, totality lasts 4 minutes 25 seconds.

## The Eyes of Texas Are Upon Totality

The 2024 eclipse gives a big Texas hug to four of its five largest cities. It starts by sideswiping San Antonio. The western and northern suburbs of the city fall within the southern limit of the eclipse. The central line crosses Interstate Highway 10 about 75 miles (120 km) northwest of the San Antonio, just west of Kerrville, with 4 minutes 26 seconds of totality.

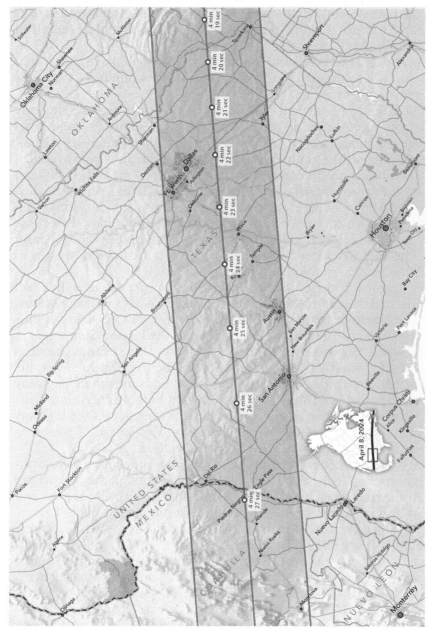

The path of totality as it crosses Texas. Points along the central line give the duration of totality in minutes and seconds. [Map © 2016 Michael Zeiler, GreatAmericanEclipse.com]

Next along the eclipse route is Austin, a second American city with a metropolitan population over 2 million. All of Austin's city limits lie within the path of totality, although at the southern edge of the zone. The Texas State Capitol Building receives a tax-free 1 minute 52 seconds of celestial umbrage.

Flying across Texas, the central line of the eclipse passes just north of Killeen and Waco, bringing times of total eclipse of 4 minutes 24 seconds for those who walk the line.

For 200 miles from Austin to Dallas and Fort Worth, Interstate Highway 35 runs continuously within the path of totality. Within the Dallas-Fort Worth metropolitan area—fourth largest in the United States—live 7 million people who will be completely submerged in the Sun-cast shadow of the Moon. The central line passes about 35 miles (60 km) to the southeast and east of downtown Dallas, bringing 4 minutes 22 seconds of totality.

Interstate Highway 30 escorts the eclipse 320 miles (515 km) from Dallas to Little Rock. Almost halfway between the two cities, the eclipse crosses the meandering Red River where Oklahoma and Arkansas border Texas. The Red River got its name from the soil it carries downstream from its headwaters in the Texas panhandle.

On its diagonal path across the southeastern corner of Oklahoma and the middle of Arkansas, the total eclipse envelops all of Ouachita National Forest as its stretches from the middle of Arkansas into Oklahoma. It is the largest national forest in the eastern half of the United States.

Toward the southern edge of the zone of totality is a second American state capital for the 2024 eclipse: Little Rock, Arkansas. From the capitol building, the total eclipse lasts 2 minutes 31 seconds. Interstate Highway 40 runs northwestward from Little Rock toward Fort Smith. Almost halfway in between, the central line of the eclipse dashes across the highway and totality lingers for 4 minutes 16 seconds. The length of totality is slowly declining, but 2024 is still a long total eclipse.

As the eclipse path leaves Arkansas, its southern edge nips a tiny northwestern corner of Tennessee at a tight horseshoe bend in the Mississippi River. At the top of that bend is New Madrid, Missouri, close to the epicenter of four of the most powerful earthquakes in United States history. These earthquakes, all close to magnitude 8, occurred between December 1811 and February 1812. The last, on February 7, 1812, created ground upheaval and shock waves so violent that the Mississippi River in that region flowed backward for several hours. The seismic jolt damaged many homes in St. Louis, 146 miles (235 km) away, and shook church steeples in New York and Boston enough to set church bells clanging—at a distance of 1,000 miles (1,600 km) and more. The US Geological Survey expects major earthquakes in the future in the New Madrid area.

## *Déjà Vu* All Over Again

As it prepares to cross the junction of the Mississippi and Ohio Rivers, the total solar eclipse of 2024 crosses the path of the total solar eclipse of 2017, bringing totality to a region twice in 6⅔ years, a rare gift. For most of the larger communities within this region of almost 8,840 square miles (22,900 sq km), the eclipse of 2024 brings even longer totality.

Paducah, Kentucky is a twice-blessed city although it lies near the southern edge of totality for 2024. Paducah is home to the National Quilt Museum and late each April hosts an international quilt festival.

Across the Ohio River from Paducah and to the west is Metropolis, Illinois, home of Superman—that is, a 15-foot-tall (4.6-meter-tall) statue of him. It stands in Superman Square. Two blocks away there is a statue of Lois Lane, less than half as tall. Metropolis was founded a century before the comic-strip character appeared on magazine racks. Superman, disguised as a newspaper reporter for the *Daily Planet*, constantly saved his great city of Metropolis. The newspaper of the real city is the *Metropolis Planet*.[2]

Across the Ohio River from Paducah and to the east is Kincaid Mounds State Historic Park.[3] Here, almost a thousand years ago, Native Americans of the Mississippian Culture built a trading and ceremonial center with at least 11 mounds. One mound was 30 feet (9 meters) tall with a flat top larger than a football field. Another, 20 feet tall, had a flat top closer to two football fields in size. At the top

### Sampling of Cities that Share Totality in 2017 and 2024

| | Length of Totality in minutes and seconds | |
| --- | --- | --- |
| | 2017 | 2024 |
| **Missouri** | | |
| Cape Girardeau | 1m 46s | 4m 5s |
| Ste. Genevieve | 2m 40s | 2m 40s |
| **Illinois** | | |
| Chester | 2m 40s | 3m 23s |
| Carbondale | 2m 37s | 4m 8s |
| Marion | 2m 31s | 4m 7s |
| Metropolis | 2m 23s | 2m 37s |
| **Kentucky** | | |
| Paducah | 2m 20s | 1m 41s |

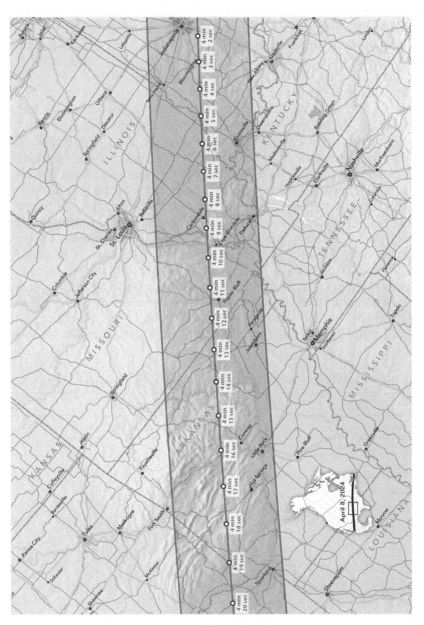

The path of totality as it crosses Arkansas, Missouri, Illinois, Kentucky, and Indiana. Points along the central line give the duration of totality in minutes and seconds. [Map © 2016 Michael Zeiler, GreatAmericanEclipse.com]

of each mound was a building. Some buildings were for ceremonies, some for important families. In the good soil along the river the people grew corn. The Mississippian Indians abandoned their settlement about 1450 CE, perhaps because of drought or a long cold spell or because they had used up the local wildlife and the trees upon which the settlement depended.[4]

As the central line of the 2024 eclipse slides into Illinois, it cuts across the central line of the 2017 eclipse. The crossing occurs south of Carbondale, 5 miles (8 kilometers) northwest of the town of Makanda, near the western shore of Cedar Lake in Giant City State Park.[5] The name Giant City comes from sandstone bluffs that tower like skyscrapers above a trail.

Where the two eclipse paths cross, 2017's totality lasted 2 minutes 40 seconds. Totality in 2024 will last 4 minutes 9 seconds.[6] The central line then passes between Carbondale and Marion—party hub for celebrating two total eclipses so close together.

The eclipse of 2024 does not linger. Its central line cannonballs across the Wabash River into Indiana, hurtles into Vincennes, and steams up 4th Street through the heart of town at 2,000 miles per hour (3,200 km/h). The central line passes a block south of the Red Skelton Museum of American Comedy. The great comedian was born in Vincennes.

Up the Ohio River and along the eclipse route about 130 miles (210 km) from Paducah, Kentucky is Evansville, Indiana. Both are river cities and both are situated near thousand-year-old American Indian ceremonial and trading centers that are National Historic Landmarks. Kincaid Mounds, near Paducah, and Angel Mounds, near Evansville, were created by Native Americans of the Mississippian Culture who built cities throughout the South and the Midwest. Their trade in pottery and other products reached east to the Atlantic coast and west to the Rocky Mountains. Angel Mounds was home to about 1,000 people. Like Kincaid Mounds, it rose about 1050 CE and was abandoned about 1450 CE.

Almost on the central line of the eclipse—less than 1 second of totality to the south—is Bloomington, Indiana and the main campus of Indiana University. Most of the buildings are made of limestone quarried a few miles south of town. This Indiana limestone is also the face of the Empire State Building, the Pentagon, and 35 of the 50 state capitol buildings in the United States.

Indianapolis, capital of Indiana, lies just north of the central line, but close enough that the entire metropolitan area of 1.8 million people lies inside the zone of totality. The Sun and Moon will look down on the Indianapolis Motor Speedway seven weeks before the Indianapolis 500 race is run. The highest speed for one lap in an Indianapolis 500 race was

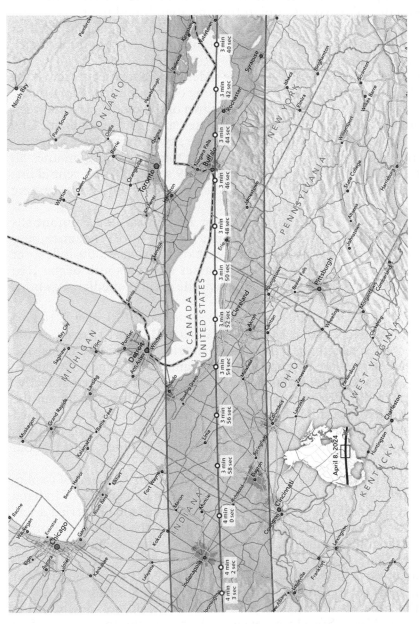

The path of totality as it crosses Indiana, Ohio, Pennsylvania, New York, and Ontario. Points along the central line give the duration of totality in minutes and seconds. [Map © 2016 Michael Zeiler, GreatAmericanEclipse.com]

236 mph (380 km/h). As the eclipse rolls over Indianapolis, its shadow is traveling at 1,990 mph (3,200 km/h). No pit stop.

## Ohio-o-o-h

Just before the central line of the 2024 total eclipse leaves Indiana and enters Ohio, the duration totality falls below 4 minutes. The path of totality skims the northwest suburbs of Cincinnati and Columbus, but it does honor Dayton, birthplace of the airplane, with 2 minutes 44 seconds of totality. Wilbur and Orville Wright would be delighted that millions of people have used their invention to fly around the world to station themselves along eclipse paths. And more. April showers are frequent along much of the 2024 eclipse route though the United States and Canada. The Wright brothers made it possible for people to take to the air when necessary to see total eclipses above the clouds.

The central line of the 2024 eclipse passes very close to downtown Cleveland—just 9 miles (15 km) offshore on Lake Erie. Along the shoreline, standing side by side, are the Cleveland Browns' football stadium, the Great Lakes Science Center, a Second World War submarine, and the Rock and Roll Hall of Fame. Overhead, the music of the spheres performs 3 minutes 50 seconds of totality for Cleveland.

As the eclipse of 2024 approaches Cleveland, it clips the southeastern-most corner of Michigan, but misses Detroit, skips Canadian customs at the border, and enters Ontario, bringing roughly 2 minutes of totality for Canadians along the edge of Lake Erie from Leamington to Hamilton. Almost all of Lake Erie falls under the shadow of the Moon.

Along Lake Erie's southern shore, the path of totality visits northern Pennsylvania. Halfway between Cleveland and Buffalo, Erie, Pennsylvania will thrill to an eerie darkness lasting 3 minutes 42 seconds.

## Over the Falls

The central line of the 2024 eclipse climbs out of Lake Erie and stampedes straight through Buffalo, New York, 2½ miles (4 km) south of downtown. With 3 minutes 45 seconds of totality, downtown is less than 1 second off the central line.

Running north out of town is the very short Niagara River, only 36 miles (58 km) in length. It drains Lake Erie into Lake Ontario. But halfway downstream, it flows around an island and tumbles off two cliffs. One cliff is on the American side; the other, horseshoe in shape, is mostly in Canada. Together they are Niagara Falls. The water drops 170 feet (52 meters) with the highest flow rate of any waterfall in the world. The water

## Eclipse Times[a] for the 2024 Total Solar Eclipse in the United States[b]

| State | City | Partial Begins | Total Begins | Total Ends | Partial Ends | Max. Eclipse | Sun Alt. | Eclipse Mag. | Duration Totality |
|---|---|---|---|---|---|---|---|---|---|
| Arkansas | Hot Springs | 12:32 p.m. | 01:49 p.m. | 01:53 p.m. | 03:10 p.m. | 01:51 p.m. | 62 | 1.055 | 03m36s |
| | Jonesboro | 12:38 p.m. | 01:56 p.m. | 01:58 p.m. | 03:15 p.m. | 01:57 p.m. | 59 | 1.055 | 02m21s |
| | Little Rock | 12:34 p.m. | 01:52 p.m. | 01:54 p.m. | 03:12 p.m. | 01:53 p.m. | 61 | 1.055 | 02m28s |
| California | Los Angeles | 10:06 a.m. | — | — | 12:22 p.m. | 11:12 a.m. | 55 | 0.579 | — |
| | San Francisco | 10:14 a.m. | — | — | 12:16 p.m. | 11:13 a.m. | 50 | 0.448 | — |
| Colorado | Denver | 11:28 a.m. | — | — | 01:54 p.m. | 12:40 p.m. | 58 | 0.714 | — |
| D.C. | Washington | 02:04 p.m. | — | — | 04:33 p.m. | 03:21 p.m. | 47 | 0.890 | — |
| Florida | Miami | 01:48 p.m. | — | — | 04:13 p.m. | 03:02 p.m. | 60 | 0.557 | — |
| Georgia | Atlanta | 01:46 p.m. | — | — | 04:21 p.m. | 03:05 p.m. | 57 | 0.846 | — |
| Illinois | Carbondale | 12:43 p.m. | 01:59 p.m. | 02:03 p.m. | 03:18 p.m. | 02:01 p.m. | 57 | 1.054 | 04m08s |
| | Chicago | 12:51 p.m. | — | — | 03:22 p.m. | 02:08 p.m. | 52 | 0.942 | — |
| Indiana | Bloomington | 01:49 p.m. | 03:05 p.m. | 03:09 p.m. | 04:22 p.m. | 03:07 p.m. | 54 | 1.054 | 04m02s |
| | Evansville | 01:46 p.m. | 03:03 p.m. | 03:06 p.m. | 04:20 p.m. | 03:04 p.m. | 56 | 1.054 | 03m04s |
| | Indianapolis | 01:51 p.m. | 03:06 p.m. | 03:10 p.m. | 04:23 p.m. | 03:08 p.m. | 53 | 1.054 | 03m50s |
| | Marion | 01:52 p.m. | 03:08 p.m. | 03:10 p.m. | 04:24 p.m. | 03:09 p.m. | 52 | 1.054 | 02m18s |
| | Muncie | 01:52 p.m. | 03:08 p.m. | 03:11 p.m. | 04:24 p.m. | 03:09 p.m. | 52 | 1.054 | 03m46s |
| | Terre Haute | 01:48 p.m. | 03:04 p.m. | 03:07 p.m. | 04:21 p.m. | 03:06 p.m. | 54 | 1.054 | 02m54s |
| Kentucky | Paducah | 12:43 p.m. | 02:01 p.m. | 02:02 p.m. | 03:19 p.m. | 02:02 p.m. | 57 | 1.055 | 01m40s |
| Louisiana | New Orleans | 12:30 p.m. | — | — | 03:09 p.m. | 01:50 p.m. | 65 | 0.846 | — |
| Maine | Presque Isle | 02:22 p.m. | 03:32 p.m. | 03:35 p.m. | 04:41 p.m. | 03:34 p.m. | 35 | 1.050 | 02m49s |
| Massachusetts | Boston | 02:16 p.m. | — | — | 04:39 p.m. | 03:30 p.m. | 40 | 0.932 | — |
| Michigan | Detroit | 01:58 p.m. | — | — | 04:28 p.m. | 03:14 p.m. | 49 | 0.991 | — |
| Missouri | Cape Girardeau | 12:42 p.m. | 01:58 p.m. | 02:02 p.m. | 03:17 p.m. | 02:00 p.m. | 57 | 1.055 | 04m07s |
| New York | Buffalo | 02:05 p.m. | 03:18 p.m. | 03:22 p.m. | 04:32 p.m. | 03:20 p.m. | 46 | 1.052 | 03m45s |
| | Jamestown | 02:04 p.m. | 03:18 p.m. | 03:21 p.m. | 04:32 p.m. | 03:19 p.m. | 46 | 1.052 | 02m52s |
| | New York | 02:11 p.m. | — | — | 04:36 p.m. | 03:26 p.m. | 43 | 0.911 | — |
| | Plattsburgh | 02:14 p.m. | 03:26 p.m. | 03:29 p.m. | 04:37 p.m. | 03:27 p.m. | 40 | 1.051 | 03m33s |
| | Rochester | 02:07 p.m. | 03:20 p.m. | 03:24 p.m. | 04:33 p.m. | 03:22 p.m. | 44 | 1.052 | 03m39s |
| | Syracuse | 02:09 p.m. | 03:23 p.m. | 03:24 p.m. | 04:35 p.m. | 03:24 p.m. | 43 | 1.052 | 01m28s |
| Ohio | Akron | 01:59 p.m. | 03:14 p.m. | 03:17 p.m. | 04:29 p.m. | 03:16 p.m. | 49 | 1.053 | 02m49s |
| | Cleveland | 01:59 p.m. | 03:14 p.m. | 03:18 p.m. | 04:29 p.m. | 03:16 p.m. | 49 | 1.053 | 03m49s |
| | Dayton | 01:53 p.m. | 03:09 p.m. | 03:12 p.m. | 04:26 p.m. | 03:11 p.m. | 52 | 1.054 | 02m42s |
| | Lima | 01:55 p.m. | 03:10 p.m. | 03:14 p.m. | 04:26 p.m. | 03:12 p.m. | 51 | 1.053 | 03m51s |
| | Mansfield | 01:57 p.m. | 03:12 p.m. | 03:16 p.m. | 04:28 p.m. | 03:14 p.m. | 50 | 1.053 | 03m15s |
| Ohio | Toledo | 01:57 p.m. | 03:12 p.m. | 03:14 p.m. | 04:27 p.m. | 03:13 p.m. | 50 | 1.053 | 01m48s |
| Pennsylvania | Erie | 02:02 p.m. | 03:16 p.m. | 03:20 p.m. | 04:31 p.m. | 03:18 p.m. | 47 | 1.052 | 03m42s |
| | Philadelphia | 02:08 p.m. | — | — | 04:35 p.m. | 03:24 p.m. | 45 | 0.900 | — |
| Texas | Austin | 12:17 p.m. | 01:36 p.m. | 01:38 p.m. | 02:58 p.m. | 01:37 p.m. | 67 | 1.056 | 01m55s |
| | Dallas | 12:23 p.m. | 01:41 p.m. | 01:44 p.m. | 03:03 p.m. | 01:43 p.m. | 65 | 1.056 | 03m48s |
| | Fort Worth | 12:23 p.m. | 01:40 p.m. | 01:43 p.m. | 03:02 p.m. | 01:42 p.m. | 65 | 1.056 | 02m39s |
| | Houston | 12:20 p.m. | — | — | 03:01 p.m. | 01:40 p.m. | 67 | 0.943 | — |
| | San Antonio | 12:14 p.m. | — | — | 02:56 p.m. | 01:34 p.m. | 68 | 0.998 | — |
| | Texarkana | 12:29 p.m. | 01:47 p.m. | 01:49 p.m. | 03:08 p.m. | 01:48 p.m. | 63 | 1.056 | 02m28s |
| | Waco | 12:20 p.m. | 01:38 p.m. | 01:42 p.m. | 03:01 p.m. | 01:40 p.m. | 66 | 1.056 | 04m12s |
| Utah | Salt Lake City | 11:26 am | — | — | 01:41 p.m. | 12:32 p.m. | 55 | 0.577 | — |
| Vermont | Burlington | 02:14 p.m. | 03:26 p.m. | 03:29 p.m. | 04:37 p.m. | 03:28 p.m. | 40 | 1.051 | 03m15s |
| | Montpelier | 02:15 p.m. | 03:28 p.m. | 03:29 p.m. | 04:38 p.m. | 03:28 p.m. | 40 | 1.051 | 01m41s |
| Washington | Seattle | 10:39 am | — | — | 12:21 p.m. | 11:29 am | 45 | 0.310 | — |

[a] All Times are in Local Time (including Daylight Saving Time)

[b] Eclipse times calculations by Fred Espenak, <http://www.eclipsewise.com>.

in Lake Superior, Lake Michigan, Lake Huron, and Lake Erie—one-fifth of all the fresh water in the world—takes the plunge over Niagara Falls. Weather permitting, the Sun in eclipse over Niagara Falls for 3 minutes 31 seconds might be a prize photo and an indelible memory.

Like it did for Lake Erie, the eclipse gives almost all of the Lake Ontario a bath of totality, but gives Toronto, largest city in Canada, not even a splash.

Between Buffalo and Rochester, New York, the central line of the eclipse closely traces the westernmost 80 miles (130 km) of the Erie Canal. The canal, started in 1817 and completed in 1825, connected Albany to Buffalo (363 miles [584 km]), allowing west-bound passengers and freight to travel by water from New York City up the Hudson River, along the canal, and into the Great Lakes, an engineering triumph of its age that did much to settle and unify the western territories of the young United States.

Rushing on beyond Buffalo, the central line of the eclipse passes just north of Rochester, New York, bringing 3 minutes 39 seconds of totality, and promptly dives into Lake Ontario. To the north of the central line, at the northeastern end of Lake Ontario, is Kingston, Ontario. Meanwhile, to the south and away from the Lake Ontario, the eclipse clips the nails of the Finger Lakes in upstate New York and takes in Syracuse, near the southern limit of totality. Syracuse has the dubious distinction of being the snowiest city in the United States, far snowier than Alaskan and Rocky Mountain cities. Syracuse even edges out Buffalo and Rochester. The reason for so much snow is the lake effect. The Great Lakes, especially Lake Ontario because of its depth, store their summer warmth. Cold Canadian air moving southeast over the slightly warmer waters of Lakes Erie and Ontario gathers moisture from the lakes and dumps it on the cooler land east and southeast of the lakes. This lake-effect snow usually ends by March. Besides, would the elements really be so cruel as to snow out a total solar eclipse in April?

The central line of the eclipse emerges from the east end of Lake Ontario and passes just southeast of Watertown, New York, heading toward Lake Champlain. To the south of the central line are the Adirondack Mountains of northern New York State, headwaters of the Hudson River which flows south to New York City and the sea.

To the north of the central line, the path of totality moves into the province of Quebec and across the southern half of Montreal, including downtown. In the midst of Montreal is a hilltop with three peaks—Mount Royal—from which the city takes its name. There, totality lasts 1 minute 17 seconds. The metropolitan areas of Toronto and Montreal together account for almost 30% of Canada's population.

While the northern edge of totality is enveloping Montreal, the central line, still offering 3 minutes 34 seconds of totality, is passing just northwest of downtown Plattsburgh, New York, on the western shore of Lake Champlain. The lake forms the border between New York and Vermont, with the northernmost tip of the lake resting in Canada.

## Saros 139—Portrait of a Young Superstar

Getting strong now. If the 2024 total eclipse of the Sun could sing, it and its eclipse family—its saros series—might be chanting the song from *Rocky*. The 2024 eclipse belongs to up-and-coming Saros series 139.

Saros 139 gave birth to its first eclipse—a weak partial (as saros series always do)—on May 17, 1501 near the north pole. The odd number of this saros series indicates that the Moon is at its ascending node as it blocks the Sun to cause each of these eclipses. Subsequent eclipses migrate southward. Six partial eclipses followed the first in 1501, each covering more of the Sun's face.

In 1627, a strange eclipse occurred. It started as an annular, became total for just an instant, then returned to being annular. The spherical shape of Earth had brought its surface close enough to the Moon so that the Moon appeared large enough to hide the entire disk of the Sun from view for just a moment. An eclipse that is annular in some places and total in others is called a hybrid—the rarest type of eclipse. Only 4.8% of solar eclipses are hybrids. Saros 139 produced 11 more hybrids. With each eclipse, there was less annularity and more totality.

Then, in 1843, Saros 139 generated its first eclipse that was total throughout its path. At its peak, totality lasted 1 minute 43 seconds. Thereafter, each eclipse created by Saros 139 was total, and the length of totality was increasing.

In 1970, a Saros 139 total eclipse lasting 3 minutes 28 seconds visited the east coast of the United States, an event vividly remembered by some eclipse veterans who tell their stories in this book. Saros 139 presented eclipse chasers with totals in 1988 in Indonesia and 2006 in Africa. With the dawn of the 21st century, the duration of totality passed the 4-minute mark.

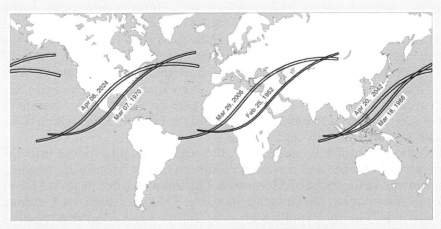

The paths of six central eclipses of Saros 139. The tracks shift west with each succeeding eclipse and include total eclipses in 1952, 1970, 1988, 2006, 2024, and 2042. Note how the path of the 2024 eclipse through the USA resembles the one in 1970. [Map and eclipse calculations by Fred Espenak]

And now comes the total eclipse of April 8, 2024, with totality lasting 4 minutes 28 seconds. It is the 30th eclipse of the 71 that Saros 139 will produce. And with each return—2042, 2060, 2078 (across Mexico and the southeastern United States)—the length of totality keeps on growing. Saros 139 is in the prime of its life.

On the 39th eclipse in the series, on July 16, 2186, a total eclipse of Saros 139 will last 7 minutes 29 seconds, the longest totality on record in the 8,000 years from 3000 BCE to 5000 CE. That duration is only 3 seconds off the theoretical maximum of 7 minutes 32 seconds. The eclipse peaks over the Atlantic Ocean, but Colombia and Venezuela will experience more than 7 minutes of totality. Total eclipses last their longest in the tropics.

After 2186, as eclipses of Saros 139 drift farther south, the duration of totality will dwindle. In 2294, a Saros 139 total eclipse will slip below the 5-minute mark. Yet that's still a very long eclipse. Most saroses never generate an eclipse that long.

In 2601, deep in the southern hemisphere, there will be a total eclipse lasting 35 seconds. It will be the last total for Saros 139. Saros 139, in old age, will produce 9 more partials near the south pole. Saros 139 ends it career in partial eclipse on July 3, 2763.

In a span of 1,262 years, Saros 139 will have created 71 eclipses, starting with 7 partials, then 12 hybrids, then 43 totals, and concluding with 9 partials. Among those 43 total eclipses is one of epic duration and five with totality lasting more than 7 minutes.*

What a career!

* For additional information about Saros 139, see <http://eclipsewise.com/solar/SEsaros/SEsaros139.tml>.


To the south of the central line, on the eastern shore of Lake Champlain, lies Burlington, Vermont's largest city (metropolitan area: 215,000). Forty miles (64 km) to the southeast but still within the path of totality is Montpelier, capital of Vermont. Its population, 8,000, makes it the smallest of all the American state capitals. Burlington gets 3 minutes 16 seconds of totality; Montpelier gets 1 minute 40 seconds.


## The Line Crosses

For nearly 2,000 miles the central line of the 2024 eclipse has been traveling through the United States. Since it approached Cleveland, the path of totality has been skimming along the border with Canada, dispensing totality to some Canadian cities and towns north of the central line along Lakes Erie and Ontario.

Now, for the first time, the central line of the eclipse enters Canada, just east of the tiny village of East Richford, Vermont. There, a two-lane bridge crosses the Missisquoi River. On opposite sides of the border, Canadian and American officials man inspection stations. Meanwhile, East Richford Slide Road, the other street in town, bends eastward to

### Eclipse Times[a] for the 2024 Total Solar Eclipse in Canada and Mexico[b]

| Country/ Provence | City | Partial Begins | Total Begins | Total Ends | Partial Ends | Max. Eclipse | Sun Alt. | Eclipse Mag. | Duration Totality |
|---|---|---|---|---|---|---|---|---|---|
| **CANADA** | | | | | | | | | |
| Ontario | Hamilton | 02:04 p.m. | 03:18 p.m. | 03:20 p.m. | 04:31 p.m. | 03:19 p.m. | 46 | 1.052 | 01m52s |
| | Belleville | 02:08 p.m. | 03:22 p.m. | 03:24 p.m. | 04:34 p.m. | 03:23 p.m. | 44 | 1.052 | 02m00s |
| | Kingston | 02:09 p.m. | 03:22 p.m. | 03:25 p.m. | 04:34 p.m. | 03:24 p.m. | 43 | 1.052 | 03m02s |
| | Toronto | 02:05 p.m. | — | — | 04:32 p.m. | 03:20 p.m. | 45 | 0.997 | – |
| | Brockville | 02:11 p.m. | 03:23 p.m. | 03:26 p.m. | 04:35 p.m. | 03:25 p.m. | 42 | 1.051 | 02m47s |
| | Cornwall | 02:13 p.m. | 03:25 p.m. | 03:27 p.m. | 04:36 p.m. | 03:26 p.m. | 41 | 1.051 | 02m10s |
| Quebec | Drummon-dville | 02:16 p.m. | 03:29 p.m. | 03:29 p.m. | 04:38 p.m. | 03:29 p.m. | 39 | 1.051 | 00m32s |
| | Montreal | 02:14 p.m. | 03:27 p.m. | 03:28 p.m. | 04:37 p.m. | 03:28 p.m. | 40 | 1.051 | 01m16s |
| | Sherbrooke | 02:17 p.m. | 03:28 p.m. | 03:31 p.m. | 04:38 p.m. | 03:29 p.m. | 39 | 1.050 | 03m25s |
| New Brunswick | Fredericton | 03:24 p.m. | 04:34 p.m. | 04:36 p.m. | 05:42 p.m. | 04:35 p.m. | 35 | 1.049 | 02m18s |
| Newfound land | Channel Port Aux Basques | 04:03 p.m. | 05:10 p.m. | 05:13 p.m. | 06:15 p.m. | 05:11 p.m. | 28 | 1.048 | 02m45s |
| | Gander | 04:07 p.m. | 05:13 p.m. | 05:15 p.m. | 06:16 p.m. | 05:14 p.m. | 25 | 1.047 | 02m14s |
| **MEXICO** | | | | | | | | | |
| | Durango | 11:55 am | 01:12 p.m. | 01:16 p.m. | 02:37 p.m. | 01:14 p.m. | 70 | 1.057 | 03m47s |
| | Mazatlan | 09:51 am | 11:07 am | 11:12 am | 12:32 p.m. | 11:10 am | 69 | 1.057 | 04m18s |
| | Torreon | 12:00 am | 01:17 p.m. | 01:21 p.m. | 02:41 p.m. | 01:19 p.m. | 70 | 1.057 | 04m12s |

[a] All times are in local time (including Daylight Saving Time)

[b] Eclipse times calculations by Fred Espenak, <http://www.eclipsewise.com>.

the East Richford Cemetery. The cemetery itself is half on the American side, half on the Canadian, with graves straddling the border. The central line goes right through the middle of the tiny cemetery, while, a few yards away, East Richford Slide Road strays across the Canadian border, stays in Canada for 200 yards (200 meters), and then nonchalantly slips back into the United States with no inspections.

The central line of the eclipse, however, remains in Canada for 120 miles (194 km)—passing south of Granby and Sherbrooke, Quebec. Sherbrooke receives 3 minutes 25 seconds of totality, just 5 seconds less than the central line is now providing.

South of the central line, the northernmost 60 miles (100 km) of New Hampshire enjoys some totality. Pittsburg, population 869, the northernmost town in New Hampshire, proudly proclaims that it has more moose than people. Both get 3 minutes 15 seconds of total eclipse.

Then the central line's brief foray into Canada ends as it reenters the United States for a farewell visit that crosses north-central Maine. After trekking across the United States from Texas to Maine, the central line of the 2024 eclipse passes 7 miles (12 km) north of Mt. Katahdin, tallest mountain in Maine. Mt. Katahdin rises nearly a mile above sea level. It is the northern end of the 2,200-mile (3,500-km) Appalachian Trail, the longest hiking-only trail in the world. The Appalachian Trail traverses 14 states. The trail of 2024 totality traverses 14 states. Both are beloved by their followers. On its way out of Maine, the eclipse enshrouds Presque Isle with 2 minutes 41 seconds of totality. The town is located on a peninsula where Presque Isle Stream angles into the Aroostock River, making the city "almost an island"—Presque Isle.

## Going from America

Meanwhile, the eclipse's central line passes 9 miles (15 km) north of Houlton, Maine. Houlton marks the northern end of Interstate Highway 95, America's east coast expressway. It runs 1,920 miles (3,090 km), the longest north–south *Interstate* highway in the country. At Houlton, I-95 crosses US Route 1. Route 1 was the principal east coast highway in the United States until I-95 replaced it. But U.S. 1 starts farther south—in Key West—and continues north past Houlton. It is 23% longer than I-95 and holds the distinction of being the longest north-south *road* in the United States.

Almost side by side, I-95 and the central line of the 2024 eclipse— carrying with it 3 minutes 22 seconds of totality—cross from Maine into Canada, ending the eclipse's journey in the United States. The central line of the 2024 eclipse travels a total of 2,140 miles (3,444 km) within the United States, more than Mexico (535 miles [861 km]) and Canada (705 miles [1,135 km]) combined.

On its way across the Canadian province of New Brunswick, the eclipse visits its capital city, Fredericton, with so many church spires that its 106,000 citizens have nicknamed their town the Celestial City.[7] Two celestial bodies wink back for 2 minutes 17 seconds.

The southern limit of the eclipse passes through the northern outskirts of Moncton, largest city in New Brunswick. Moncton is just up the Petitcodiac River from the eastern end of the Bay of Fundy, famous for its extreme tides—yet another collaboration of the Moon and Sun. At full moon and new moon, when the Moon and Sun combine their gravitational effects on ocean waters, Bay of Fundy high tides rise 50 feet (15 meters) above low tides.

The central line of the eclipse then leaps into the Gulf of St. Lawrence, part of the Atlantic Ocean, and barely misses the northern tip of Prince Edward Island, although the southern portion of totality laps over 40% of the island. Prince Edward Island, smallest of the 13 Canadian provinces and territories, is about the size of Delaware, the second smallest

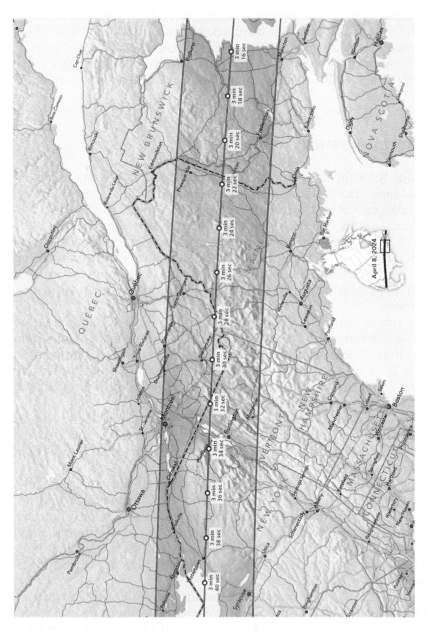

The path of totality as it crosses New York, Vermont, New Hampshire, Maine, Quebec, and New Brunswick. Points along the central line give the duration of totality in minutes and seconds. [Map © 2016 Michael Zeiler, GreatAmericanEclipse.com]

## Total Eclipses for America in the 21st Century

If the Moon was stingy about using the Sun to cast its shadow on the continental United States from 1979 to 2017 (38 years), the Sun and Moon will be more generous through the remainder of the 21st century—especially between 2017 and 2052.

In 2017 and 2024, less than 7 years apart, total eclipses cross the country diagonally to one another. Nine years later, on March 30, 2033, Alaska will have a chilly eclipse. Then will follow a burst of three total eclipses for the contiguous United States over a period of 7½ years, with two eclipses less than a year apart:

| | | |
|---|---|---|
| August 23, 2044 | Montana and North Dakota | 2 minutes 4 seconds |
| August 12, 2045 | California through Florida | 6 minutes 6 seconds |
| March 30, 2052 | Louisiana through South Carolina | 4 minutes 8 seconds |

Three more total eclipses will visit the United States before the century is out:

| | | |
|---|---|---|
| May 11, 2078 | Texas through North Carolina | 5 minutes 40 seconds |
| May 1, 2079 | Maryland through Maine | 2 minutes 55 seconds |
| September 14, 2099 | Montana through North Carolina | 5 minutes 18 seconds |

American state. Near the northern tip of the island, where totality lasts longest—3 minutes 14 seconds—is a tiny settlement called Seacow Pond. This name does not refer to the manatee (or sea cow) of Florida, the threatened marine mammal that prefers warm waters. Instead, Seacow Pond honors the Atlantic walrus, sometimes called a sea cow because it makes mooing sounds.[8] Walruses, also threatened with extinction, have tusks 3 feet (1 meter) long which they use to pull themselves up onto ice floes. That talent is noted in their scientific name: *Odobenus rosmarus*—tooth-walking sea-horse.

Midway between Prince Edward Island and Newfoundland, the eclipse of 2024 covers all of the Magdalen Islands, part of Quebec, a prime site where ecotourists go in late February to see newborn, white-coated harp seals.

The southern limit of totality then catches the northern tip of Cape Breton Island, part of Nova Scotia.

The final landfall of the 2024 eclipse is Newfoundland. As it moves across the island, the duration of totality subsides from 3 minutes 5 seconds to 2 minutes 54 seconds. Most of Newfoundland's cities and towns lie on or near the coast—primarily the east coast—and that's where the roads are too.

Gander, Newfoundland played a major role in the development of aviation. Lying at the northeast corner of North America, closest to Europe,

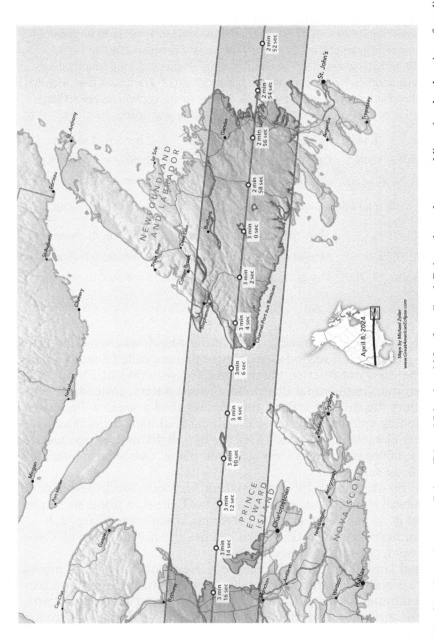

The path of totality as it crosses Prince Edward Island and Newfoundland. Points along the central line give the duration of totality in minutes and seconds. [Map © 2016 Michael Zeiler, GreatAmericanEclipse.com]

Shadow bands are usually seen on the ground but in rare cases like this one, they appear as a series of fast moving, nearly horizontal streaks on thin cirrus clouds during the diamond ring effect. Total solar eclipse of Dec. 14, 2020 - Fortin Nogueira, Argentina [Canon EOS 5D, Canon 600mm F4L IS III, f/8, 1/1250 s, © 2020 Larry Stevens]

A statue of Tyrannosaurus Rex in front of the Tate Geological Museum, Casper College, WY) serves as a compelling foreground for totality during the 2017 eclipse [Nikon D90, Nikkor 18-70 mm (at 18 mm), 1 sec, f/11, ISO 400, Casper, WY, © 2017 Fred Espenak]

This spectacular image of the Sun's corona was made from 161 exposures with two identical telescopes. Perfect seeing made it possible to create an image of extraordinary sharpness which is limited only by the movement of coronal features during the total solar eclipse. Shot from Mitchell, Oregon on Aug. 21, 2017. [Nikon D810, TS Photo Line 115mm Refractor, Focal Length 800mm, f/7, 1/1000 sec - 4 sec, © 2017 Miloslav Druckmüller, Peter Aniol, Shadia Habbal]

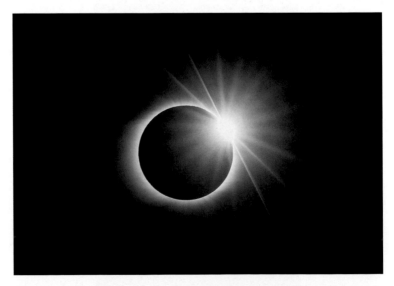

The diamond ring effect is captured seconds before totality begins during the July 2, 2019 total solar eclipse. [Nikon D7200, Sigma 170-500mm APO Lens (at 500mm), 1/125 second, f/11, ISO 400, Mamalluca Observatory, Vicuña, Chile, © 2019 Fred Espenak]

Witnesses to the total solar eclipse of July 3, 2019 appear in silhouette during this time-lapse sequence that required careful planning [Nikon D800, 24mm, f/8, 1/250 and solar filter for partial phases every 3 minutes, totality, no filter, 1/10 sec. © 2015 Thanskrit Santikunaporn]

On July 3, 2019, a total solar eclipse was visible over ESO's La Silla Observatory, located in the Chilean Atacama Desert. ESO invited over 25 scientists, communicators, and educators to observe and document the phenomenon. [Credit: ESO/M. Zamani]

A total eclipse "selfie" captures two photographers with their cameras during the 2019 eclipse in Chile. [GoPro Hero 5, auto-exposure, Mamalluca Observatory, Vicuña, Chile, © 2019 Fred Espenak]

A few minutes after annularity, an oddly shaped crescent Sun sets behind giant wind turbines of a wind farm near Elida, New Mexico on May 20, 2012. [Canon EOS Rebel XSi, Canon EF 400mm lens at f/7.1, 1/400 sec, ISO 200, © 2012 Evan Zucker]

This 2019 eclipse composition of three separate images features an HDR (High Dynamic Range) composite of the corona (72 exposures 1/1000 to 2.5 seconds) flanked by two shots of the diamond ring before and after totality (1/1000 second). [Nikon D850, Vixen 90mm Fluorite Refractor (90mm, f/9, fl=810mm), f/9, ISO 250, Mamalluca Observatory, Vicuña, Chile, © 2019 Fred Espenak]

Wonderful reflections appear in Lake Juin Cuesta del Viento, Argentina during a time-lapse sequence of the total solar eclipse of July 3, 2019. [Sony Alpha 7R III, Sigma Art 40mm f/1.4 lens, partial phases every 5 minutes, totality: f/2.8 1/2 second. © 2019 Thierry Legault]

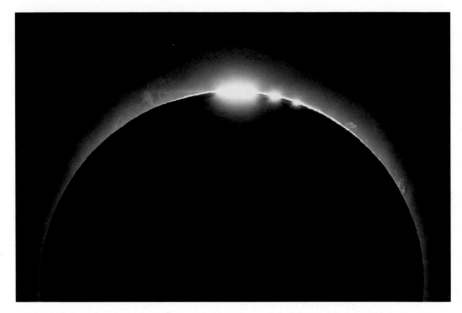

Baily's beads appear along the Moon's limb as totality ends during the July 3, 2019 total solar eclipse. Several red prominences are also visible. [Nikon D7200, Vixen 90mm Fluorite Refractor (90mm, f/9, fl=810mm), 1/1000 second, f/9, ISO 250, Mamalluca Observatory, Vicuña, Chile, © 2019 Fred Espenak].

To view the total solar eclipse of Dec. 4, 2021 two separate flights of B787-8 Dreamliners (the second one is visible near the tip of the wing) flew into the path of totality near the Falklands. [Canon 6D, Samyang 24mm, f3.5, 1/50s, ISO 640, © 2021 Petr Horálek/Institute of Physics in Opava]

Andreas Möller was one of a small band of hearty eclipse chasers who traveled to Antarctica for the total solar eclipse of Dec. 4, 2021 [still frame from video shot with a GoPro HERO 7, Union Glacier, Antarctica, © 2021 Andreas Möller, Miloslav Druckmüller]

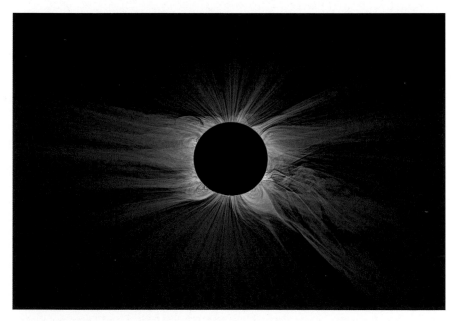

This unique HDR (High Dynamic Range) eclipse image shows a huge CME (Coronal Mass Ejection) during the total solar eclipse of Dec. 4, 2021 from Antarctica [Nikon Z6, Nikkor 80-400mm, f/5.6, exposures: 1/400 to 2 seconds, ISO 100, Union Glacier, Antarctica, © 2021 Miloslav Druckmüller, Andreas Möller]

This composition captures the entire total solar eclipse of August 21, 2017 in a sequence of partial phases, diamond rings, and the corona during totality. The partial phases were shot with a Nikon D810, Vixen ED100SF refractor, and Thousand Oaks solar filter. Same camera/telescope for diamond ring shots but with filter removed. The image of the corona was produced with an Astro-Physics 105EDT refractor and a Nikon D800 DSLR. The corona is a composite of 57 separate exposures (1/2000 to 2 seconds) combined and processed using Photoshop CC. [Casper College, Casper, WY, © 2017 Fred Espenak]

Gander was a crucial refueling stop for the first flights from the United States and Canada to England and France and back. During the Second World War, by refueling in Gander, fighters and bombers could ferry themselves from North America to England, ready for missions against Nazi forces. Thus Gander became known as the Crossroads of the World. On April 8, 2024, the Moon and Sun cross paths there, providing 2 minutes 15 seconds of totality (at the airport). Transatlantic flights no longer need to stop for fuel in Gander, but the city still provides runways and hospitality for intercontinental flight emergencies. So taken with aviation are Ganterites that they have named their streets Charles Lindbergh, Amelia Earhart, Chuck Yeager, pilots who paved the way for man to roam the Moon—and for eclipse chasers to roam the Earth.

The Sun and Moon in total eclipse leave Newfoundland with a nod to two villages on the southeast coast—Sunnyside and Come By Chance. Then the central line of the eclipse leaves Newfoundland—and land altogether—at Maberly, an east coast village so small it doesn't make the list of provincial towns. The eclipse will spend its shrinking path and time of totality racing ever faster eastward across the Atlantic, to fall off the edge of the Earth at sundown before it can reach Europe.

The party is over for the April 8, 2024 total eclipse of the Sun. Seldom are eclipses so generous, entertaining so many people for so long.

But the celebration of eclipses is never really over. The party just moves on to another site. And you are invited.

## NOTES AND REFERENCES

1.  Edwin Dunkin: *Autobiography*, unpublished (compiled by Peter Hingley, Royal Astronomical Society). Dunkin is referring to the total solar eclipse of July 28, 1851 as seen from within the northern limit of the path of totality in Scandinavia.
2.  The *Metropolis News* changed its name in 1972 to the *Metropolis Planet*.
3.  The mounds are only 6 miles from Paducah, but the road takes one north and east around the mounds, increasing the distance to 20 miles (32 km).
4.  John E. Schwegman: "A Prehistoric Cultural and Religious Center in Southern Illinois." <http://www.southernmostillinoishistory.net/kincaid-mounds.html>.
5.  Latitude 37° 38' 42" north; longitude 89° 16' 37" west.
6.  The 2017 eclipse reaches its longest duration (2 minutes 40.1 seconds) at a point 6 miles (10 km) southeast of the location in Giant City State Park where the central lines of the 2017 and 2024 eclipses cross. The 2024 reaches longest duration in Mexico.

7.  <http://www.fredericton.ca/en/communityculture/churches_placesofwor ship.asp>.

8.  Todd Dupuis, Assistant Deputy Minister of Environment, Prince Edward Island, personal correspondence, December 2, 2015. Jeremy Berlin: "The Unexpected Walrus," *National Geographic*, December 2013 (online).

—— ◯◑●◐ ——————

# A MOMENT OF TOTALITY

## Vacation Planning

"If you see one total solar eclipse, you know where you'll be going for vacations for the rest of your life," says Sheridan Williams, retired British rocket scientist.

"In 2028, I'll be in Australia, looking at the eclipse with the Sydney Harbor Bridge in the foreground. For the eclipse of 2037, I'll be on the North Island of New Zealand. And I think I'll stay there a year and a half to see the eclipse of 2038 that crosses the path of 2037. Total eclipses in the same place in consecutive years. I'll be 90 years old at the time."*

* Interview, May 23, 2015.

## Vacation Planning

# 15

◄o►

# The Weather Outlook

*If you've got a weather forecast and a day or two to travel, there's no reason to miss the eclipse of 2024.*

Jay Anderson, Canadian meteorologist and
eclipse weather climatologist[1]

Total solar eclipses may be wondrous but they are not always easy to reach. The 2010 eclipse swept across the southern Pacific Ocean, missing almost every inhabited plot of land except Easter Island, and ended in the mountains near the southern tip of South America. The 2015 eclipse passed over a few very cold, frequently overcast islands in the Arctic Ocean. The 2021 eclipse visited only Antarctica.

The three rules for seeing a solar eclipse are—just like buying real estate—location, location, location. First, of course, make sure you choose a site located inside the zone of totality. You can use the maps in this book (and others) or the Google interactive eclipse map by Xavier Jubier[2] to position yourself perfectly. Second, check climate conditions to find an observing location with a good probability of clear skies. Third, make sure you have a means of transportation and study the location of roads leading from your site to other good observing sites in case you need to move to a more promising location a day or two in advance because of incoming clouds.

For total eclipses of the Sun, the saying actually should be "location, location, location, location." Many people travel a substantial distance to see a total solar eclipse. Why not take advantage of the travel to enjoy the location—to see some cultural and natural wonders as part of your trip?

The 2024 North American eclipse offers plenty of land to observe from. And it's easy to get to the eclipse path. The central line of the path of totality crosses Mexico from Mazatlán to Piedras Negras—535 miles (861 km). It crosses the United States from Eagle Pass, Texas to Houlton, Maine—2,140 miles (3,444 km). It crosses Canada from Mansonville, Québec to Bonavista, Newfoundland—705 miles (1,135 km).[3] Plenty of places to watch the eclipse—and, in most places, plenty of good roads in case you need to change your location.

How many people live within a few hundred miles of the 2024 path of totality? This infographic illustrates the number of people living within 100 miles (75.3 million) to 1000 miles (298.5 million) off the path. [Map © 2022 Michael Zeiler, GreatAmericanEclipse.com]

For the 2024 total eclipse, 4 million people in Mexico, 75 million people in the United States, and 15 million people in Canada live within the path of totality or within a picnic's drive (100 miles [160 km]) of the path.

The total solar eclipse of 2024 could be an even bigger affair than the treasured eclipse of 2017. The duration of totality is longer—4 minutes 28 seconds in 2024 versus 2 minutes 40 seconds in 2017. The path of totality is wider—124 miles (200 km) in 2024 versus 71 miles (115 km) at its widest in 2017. The length of the path of totality across the United States in 2024 is not as long as the coast-to-coast eclipse of 2017, but 2024 brings midday darkness to all or parts of many more large cities. In 2017, at least some portions of three cities with metropolitan populations over 1 million experienced totality: Kansas City, St. Louis, and Nashville. In 2024, there are a dozen cities with populations over one million that will have a total eclipse within the city limits: San Antonio, Austin, Dallas, Fort Worth, Indianapolis, Cincinnati, Columbus, Cleveland, Buffalo, and Rochester in the United States. Add to that Torreón in Mexico and Montreal in Canada.

If the weather cooperates, the total eclipse of the Sun on April 8, 2024 could mark the biggest outdoor spectator event in American history—a 2,140-mile-long (3,444-km-long) tailgate party to watch the heavenly performance of the Moon and Sun.

## Be Your Own Weatherman

You've picked an observing site with reasonably good chances of sunny weather. You have a means of moving to another site if bad weather threatens as eclipse day approaches. Now here's how you can follow weather forecasts and, if necessary, make the best possible decision a day or two in advance about moving to a site with better weather.

The National Weather Service is directly or indirectly the source for almost all the forecasts Americans hear on television and radio and read in newspapers and on the internet. It's government information, so it's all free. You can get forecasts directly from the National Weather Service website. Commercial weather services provide their own forecasts using National Weather Service information and their own tools—also available free on the internet. All offer radar maps showing clouds in motion. Among the easiest to use are the Weather Channel, AccuWeather, and Weather Underground.

The National Weather Service and the commercial providers offer weather predictions for 8 to 14 days into the future. These long-range forecasts are useful as alerts about what weather systems are developing, but the NWS urges people not to put much faith in forecasts for a particular spot more than a week in the future. Clouds could arrive earlier or later than expected.

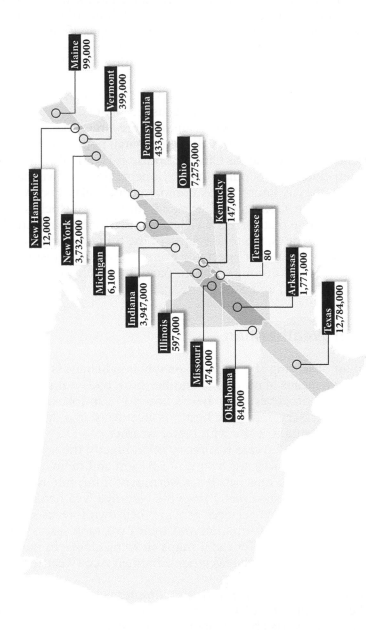

How many people live inside the 120-mile-wide (190-km-wide) path of totality across the USA? Approximately 31,600,000 people according to this infographic which breaks the population down by individual states. [Map © 2022 Michael Zeiler, GreatAmericanEclipse.com]

Forecasts for a week in advance are much more accurate. Five days in advance, more accurate still. Keep track of predictions for your chosen site and alternate sites. Once forecasts give the same prediction for eclipse day several days in a row, you can begin to trust it, says Jay Anderson, who spent his career as a meteorologist with Environment Canada. He is known around the world for the weather information he offers for every total and annular eclipse and for his many observations of eclipses everywhere on Earth.[4]

By two days before eclipse day, commercial weather forecasts are quite dependable. A day or two before the eclipse is when you should make your decision to move if your site is likely to be cloudy and the forecast for another site is quite favorable. Allow more time than usual for driving. Other folks may be altering their eclipse-observing plans too.

## Fearsome Clouds

Clouds are the enemy of eclipse observing. No place along the eclipse path is immune to the threat of clouds. Some clouds pose a great threat to eclipse watching. Other clouds pose almost no threat at all; they will vanish as the eclipse approaches.

Clouded out. Glum observers at the August 9, 1896 eclipse in Lapland see only the twilight glow on the horizon in this painting by Lord Hampton. [Annie S. D. Maunder and E. Walter Maunder: *The Heavens and Their Story*]

Eclipse chasers aboard the MS Paul Gauguin experienced a real cliff-hanger. Hidden behind thick clouds, the Sun emerged into view just as the diamond ring formed at second contact. The total solar eclipse of April 9, 2005 was observed from the South Pacific. [Canon EOS 20D DSLR, 10–20 mm lens at 10 mm, f/4, ¼ second, ISO 100. © 2005 Alan Dyer]

The clouds you need to escape are frontal systems. Cold fronts and warm fronts bring overcast skies for many hours or a day or more. Even if you catch a glimpse of the Sun through broken clouds, the clouds may quickly cover the Sun again. Weather systems in North America push across the continent generally from west to east. Cold fronts move northwest to southeast. Warm fronts move southwest to northeast.

A weather forecast two days in advance of the eclipse that calls for a cold front or warm front to cover your viewing site on eclipse day begs you to move to better location. Remember, clouds build up even before the front officially arrives. Check the professional weather forecasts for your alternative sites.

Even though the 2024 eclipse occurs in April, there is still a threat of thunderstorms, particularly in the southern United States. Reducing that threat is the early afternoon time of the eclipse in the American South. Thunderstorms most often happen later in the afternoon.

"Pop-up" thunderstorms develop because the Sun steadily warms the Earth's surface and heat and moisture rise from the ground. The ground warms unevenly—more heat rises from dark forests and asphalt roads than from lighter-colored meadows and farm fields, Anderson notes.[5] So clouds billow up here and there—separate from a frontal system. Winds arise as well, created by the same uneven heating of the ground. The winds carry the clouds generally eastward. The probability of thunderstorms is part of every forecast. That probability differs from county to county. Even more, it differs by terrain.

The cumulonimbus clouds of a thunderstorm in America may tower to a height of 10 miles (52,000 feet; 16 km). Thunderclouds in the tropics can soar even higher—14 miles (75,000 feet; 22 km). An average thunderstorm is about 15 miles (24 km) in diameter, says Anderson. An isolated thunderstorm moving about 30 miles per hour (50 km/h) will drench you and move on in about 30 minutes. The trouble is, not all thunderstorms are isolated. There may be another heading your way.

You know from driving into a thunderstorm that you usually come out the other side in a matter of minutes. If you drive down a highway going the same direction as a thunderstorm, you can outrun it. But you have to get well ahead of it or it will catch up with you and eclipse the eclipse. It's usually better to go westward through a thunderstorm or go north or south of it to find clearer skies. If you need to dodge a thunderstorm, just be sure that the sunny spot you find is in the path of totality.

## The Eclipse as Cloud Killer

Your chance of seeing totality is greatly increased by the eclipse itself. Clouds are caused by transparent water vapor rising from the ground to high in the sky. The higher the water vapor goes, the lower the air pressure, and the more the gas expands—which cools it. As the water gas cools, it changes to tiny droplets of water (and ice) suspended in the air—an all-too-visible, non-transparent cloud.

Clouds are always changing, either growing or dissipating. More water vapor may be arriving from below and condensing into droplets. Water droplets elsewhere in the cloud are constantly vaporizing—changing back to transparent water gas. The pretty little cumulus clouds of a summer day will last only 5 to 10 minutes if the updrafts of heat and moisture are shut off. Unless thunderstorms are constantly fed a new supply of hot air and water vapor from below, they will die out in 20 to 30 minutes.

So if there are scattered small cumulus clouds as the partial phase of the total eclipse begins, don't worry, says Anderson. They will disappear. As the Moon blocks more and more of the Sun's light from hitting the land around you, the heating of the ground slows. Less heat and moisture rise into the sky. The clouds stop building. But the water in the cloud continues to change from tiny liquid droplets back to transparent water vapor. The cloud steadily shrinks and vanishes.

A solar eclipse eats clouds. The partial phase of an eclipse can't destroy a thunderstorm, especially if it has begun to rain. And it can't erase a frontal system. But as the Sun becomes an ever thinner crescent and the atmosphere and ground cool, the vanishing Sun helps enormously to clear the sky of smaller cumulus clouds for the show that's about to happen.

## Clouds and Mountains

You can also use the terrain to influence the weather you experience. Winds generally carry weather systems in North America to the east. When air encounters a mountain, it flows up the slope, expands, cools, and condenses into clouds. So clouds tend to build up on the windward (usually west-facing) sides of mountains.

If the weather is clear, mountaintops are great places to watch a total eclipse. It's easier to see the Moon's shadow approaching from the west and rushing toward you. After totality, you can see it rushing away to the east. But mountaintops can be risky observation sites: they are cloudier than the valleys below them, particularly the valleys to the east.

Anderson explains: As air flows up a mountainside, it cools. After crossing the mountain peak, the cool air sinks. As it descends to a lower altitude where the air pressure is greater, the sinking air warms by compression. The tiny water droplets of the cloud vaporize back into transparent water gas as the air warms and dries, clearing or partially clearing the skies on the downwind (leeward) side of mountains, usually the east. So the valleys east of mountains tend to have clearer skies than mountaintops and the western slopes of mountains.[6]

This cloud-eating "rain shadow effect" is most pronounced if the mountains are tall, but even modest mountains produce some cloud clearing on their downwind sides.

## Weather and the 2024 Eclipse

In choosing a site to view the 2024 eclipse of the Sun, there is one overriding climate circumstance to be considered. The eclipse occurs on April 8. That means that the eclipse occurs as winter is giving way to spring. But, alas, not all at once, not neatly, not dependably. One day it may be wintery and the next day springlike. And a day or two later, it's winter again.

In Mexico, far to the south, April marks the last month of the dry season. Temperatures are comfortable. Skies are usually clear. The Mexican west coast at Mazatlán is quite promising. Then the eclipse path climbs the Sierra Madre Occidental, a western branch of the Rocky Mountains extending through Mexico. Upper air currents flowing east carry moist air up the mountains. As moist air rises, it expands and cools, clouds form, and weather prospects decline.

When the cool air descends on the eastern side of the mountains, the air contracts and becomes warmer. The water droplets that form clouds change back into water vapor, which is transparent. Clouds dissipate. So Mexico's Central Plateau between Durango and Torreón is also a region with high probability of a clear eclipse day.

Then the eclipse track ascends the Sierra Madre Oriental, an eastern branch of the Rockies, and again clouds are more likely to form on the western slopes. On the downwind side of these mountains, though, the clearing is not as dependable because the air is now mixing with wet, warm air from the Gulf of Mexico.

In northern Mexico and southern Texas, the major weather threat comes from frontal systems that sweep through the region. In this region at the beginning of spring, there is also an occasional thunderstorm, even a tornado.

At the beginning of April, the farther northeast the eclipse path goes in the United States, the less spring it encounters and the more winter. From the Mexican border throughout its pilgrimage through the United States and Canada, the farther north the eclipse track ventures, the cloudier the weather. But there are occasional exceptions.

## Getting Specific

For the eclipse's arrival at the Mexican west coast at Mazatlán and then for its trek across the Central Plateau of Mexico from Durango through Torreón, the skies for April in the early afternoon—eclipse time—are cloudy only about 25% of the time. For that time of year and that time of day, that's the sunniest weather along the entire eclipse route. The mountains—the Sierra Madre Occidental and the Sierra Madre Oriental—are scenic but less sunny, especially on the western slopes.

As the eclipse speeds through northern Mexico and across the Rio Grande into Texas, the Pacific Ocean air flow mixes with warm, wet air from the Gulf of Mexico and the percentage of cloudiness rises. In April, the farther northeast the eclipse route goes, the more frequently it encounters frontal systems bringing cloudy weather. Along the eclipse path through most of Texas, the April cloudiness at eclipse time hovers around 50%. In Arkansas, about 55% cloudiness. In Illinois, about 65% cloudiness. In eastern New York State, about 75%. In Newfoundland, expect clouds 80% to 85% of the time.

"Don't go to Newfoundland in April for an eclipse experience," says Jay Anderson. "Go for the Newfoundland experience."

For a long eclipse along a wide route through three large countries, the weather outlook is not as promising as one could hope. However, this

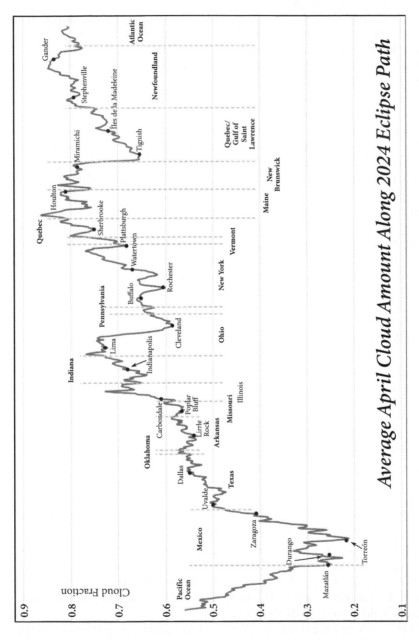

*Average April Cloud Amount Along 2024 Eclipse Path*

The average cloud cover along the central line of the 2024 eclipse path has been derived from 18 years of satellite observations by the Aqua satellite during its 1:30 p.m. passage. The locations of cities and towns are indicated. Cloud fraction may be treated as percentage of cloudiness. [Graph courtesy of *Eclipse Bulletin: Total Solar Eclipse of 2024 April 08* by Fred Espenak and Jay Anderson]

perspective is based on one kind of measurement: a *cloudiness index*, how often clouds would block a view of totality at eclipse time along the path of the eclipse in April. There is also a *sunniness index*, the average amount of clear skies during the daytime at various cities along the eclipse track. They don't quite agree. The sunniness index offers a sunnier outlook.

For instance, Dallas shows cloudy at eclipse time 55% of the time. That would suggest a 45% likeliness of sunny skies. But the sunniness index shows the percentage of sunshine during April days in Dallas to be 61%.

Rochester, 60% cloudy. Does that mean 40% sunny? The sunniness index says 51%. Montreal, 71% cloudy—but 44% sunny.

Many factors contribute to this disparity. One is different vantage points. Is the observation made from a satellite or from the ground—and is it made by instruments or by human estimates?

If instruments are making the measurements, a second factor is that different instruments are sensitive to different reflected wavelengths from clouds and from the ground, so that different instruments reach slightly different conclusions about cloudiness.

A third factor is different angles of view. On a partly cloudy day, if you look straight down from orbit or straight up from the ground, there are abundant gaps between the clouds. But if you look toward the horizon, you see the sides of clouds, making the sky look cloudier.

Do high, thin, cirrus clouds look to a satellite like cloud cover although those clouds would not block a view of the Sun with an eclipse in progress?

So, says Anderson, don't think that the cloudy-day percentage is precise. The cloudiness index is more valuable for comparing one location on the eclipse path with another. He thinks the cloudiness index is a bit more gloomy than is warranted. The sunshine index may be a better estimate of your chances of seeing the eclipse at your chosen site. "At the very least," says Anderson, "the sunshine data should reassure the eclipse observer that things are not as bad as they seem from the cloudiness index."

Even where climate statistics are not so encouraging, there are always places along the eclipse route that have a better chance of clear weather than the spots around them. Here are some of those favored places and the curious reasons why they experience less cloudiness.

## Lake Effect Cloud Clearing

The first surprise is along or near the south shore of Lake Erie stretching from Sandusky, Ohio through Cleveland to Erie, Pennsylvania and Dunkirk, New York. If a gentle, not-too-cold breeze from the north is flowing across Lake Erie, it will be cooled by the lingering winter temperature of the lake water, which will suppress cloud formation until the air has moved inland about 20 miles (30 km) and heat rising from the land can form new clouds. Away from the lake, the cloudiness exceeds 70%. In Cleveland and east and west along the south shore of Lake Erie, cloudiness lessens to about 60% or a bit clearer.

The second surprise is in New York State from Niagara Falls through Rochester to Oswego. Farther east in New York, cloud cover is between 70% and 75%. But here, along and near the south shore of Lake Ontario, the cloudiness is between 60% and 65%. Different lake but same reason: the cold lake water cools the air above it, impedes cloud formation, and pushes existing clouds inland, away from the shore.

These Lake Erie and Lake Ontario anomalies of slightly better weather suggest to Jay Anderson an eclipse-chasing strategy for those regions. Frontal systems and their clouds move west to east. The south shore of the lake will clear faster than places inland. If a day or two before the eclipse the weather forecast for your observing site shows it blanketed by clouds, try moving west along the south shore of Lake Ontario or Lake Erie. Stop when clear skies appear over the lake to the north.

A third climate surprise is in the Canadian Maritime Provinces, along the shore of the Northumberland Strait that separates New Brunswick from Prince Edward Island. For most of New Brunswick, midafternoon in April brings cloudiness about 80% of the time. But here, along the coast, water again makes the difference. The land warms more quickly than the ocean. The warm air rises, forming clouds and drawing in the denser, cooler, clearer air over the ocean. The cloudless ocean air flows inland as a sea breeze, pushing coastal clouds a few miles inland, leaving the shore clearer than expected. Particularly favored are points of land that project out into the ocean, such as Point Escuminac and Richibucto Head (Cap-Lumière) in New Brunswick and at Tignish on the northern tip of Prince Edward Island. Here, instead of about 80% cloudiness, weather records show cloud cover about 65% of the time. Not optimal, Anderson notes, but these are the least cloudy areas east of Lake Ontario.

## Closing Thoughts

Remember, says meteorologist Jay Anderson, the keys to optimizing your chances of seeing totality are:

- Weather forecasts more than a week ahead of the eclipse are not a reliable prediction for eclipse day.

- Weather forecasts a week in advance are useful and rapidly become more accurate as the eclipse approaches.

- When forecasts are essentially the same from day to day, they are quite reliable.

- On the night before or the day of the eclipse, the local weather forecast provides the information you need.

- If you have friends at observing sites within traveling distance, check with them about weather conditions there.

- Above all, prepare carefully and stay as mobile as you can.

## Sunniest Spots by Regions Along the Eclipse Route

The 2024 total eclipse of the Sun pursues a northeasterly course through Mexico, the United States, and Canada. Region by region, the farther north and east this eclipse goes, the greater the likelihood of clouds. But within most regions there are some places where the chance of clear skies improves enough to merit attention—at least compared to the regions around them.

Here are the places most likely to be sunny within certain regions along the path of totality, based Jay Anderson's research on April weather reports through the years.*

- In Mexico: Mazatlán and from Durango to Torreón
- In Texas: Eagle Pass and from Del Rio to Brady (although far from the central line)
- In Arkansas: From Hot Springs to Little Rock
- In Kentucky and Illinois: From Paducah to Carbondale
- In Ohio and Pennsylvania: South shore of Lake Erie from Cleveland west to Sandusky and from Cleveland east to Erie, Pennsylvania and Dunkirk, New York
- In New York: South shore of Lake Ontario from Niagara Falls through Rochester to Oswego
- In Maine: Greenville and Millinocket
- In New Brunswick, Canada: Point Escuminac & Richibucto Head (Cap Lumière)
- On Prince Edward Island, Canada: Tignish and North Cape

* Based on climatology research by Jay Anderson. See Fred Espenak and Jay Anderson: *Eclipse Bulletin: Total Solar Eclipse of 2024 April 8* and Jay Anderson's Eclipsophile website: <https://eclipsophile.com/2024tse/>.

## Additional Eclipse-Planning Resources

Fred Espenak and Jay Anderson's *Eclipse Bulletin: Total Solar Eclipse of 2024 April 08* provides additional tables, charts, maps, weather data, and eclipse circumstances for hundreds cities in the USA, Canada, Mexico, and elsewhere. Go to <http://astropixels.com/pubs/EB2024.html> for more information.

Espenak has also published *Road Atlas for the Total Solar Eclipse of 2024* with detailed road maps covering the entire path from Mexico to Newfoundland. The duration of totality is plotted in 30-second steps, making it easy to estimate the length of the total eclipse from any location in the eclipse path. Visit <http://astropixels.com/pubs/Atlas2024.html> for details.

## NOTES AND REFERENCES

1.  Epigraph: Jay Anderson, interview, September 14, 2022.
2.  Also available at EclipseWise.com. The path of totality shown on maps in this book and on the Google interactive maps use eclipse calculations by Fred Espenak.
3.  This distance of 705 miles (1,135 km) is measured along the central line of the path of totality. The central line runs very close to the Canadian border for an equal distance. The path of totality in this region is about 110 miles (178 km) wide, so parts of Ontario and Québec, from Kingsville to Montreal, are actually in the zone of totality even though the central line has not yet crossed into Canada. If the part of Canada that experiences totality but not the presence of the central line is included in the mileage across Canada, the distance increases to about 1,521 miles (2,450 km).
4.  Jay Anderson has his own eclipse weather website: <http://www.eclipsophile. com>. Eclipsophile provides additional climate information for everywhere along the path of the 2024 total solar eclipse and for other eclipses and celestial events. Much of that information for 2024 also appears in the weather chapter of *Eclipse Bulletin: Total Solar Eclipse of 2024 April 8* by Fred Espenak and Jay Anderson.
5.  White or silver surfaces (concrete, mirrors) *reflect* more visible light than dark surfaces. Dark surfaces (forests, asphalt) *absorb* more visible sunlight, warm up more quickly than white surfaces, and re-emit the energy they absorbed as infrared radiation—heat—rather than visible light. Heat rising from forests is a major source of storms that provide fresh water for the land, although they challenge eclipse seekers.
6.  In most cases, valleys east of mountains have clearer skies than mountaintops and the western slopes of mountains. But over the Appalachian Mountain chain, storms from the Atlantic Ocean can approach from the east, so winds can blow from either direction depending on the daily weather pattern. Thus, Anderson cautions, eclipse seekers in the Appalachians should pick a leeside (downwind) location based on the eclipse-day forecast.

# A MOMENT OF TOTALITY

## Paris to Zambia for 38 Hours to See an Eclipse

*by Luca Quaglia*

I traveled from Italy, where I grew up, to France to see the great total eclipse of 1999— and was clouded out. Even though I didn't see the Sun in eclipse, I was stunned by the passage of the Moon's shadow. I promised myself: Wherever the next total eclipse is, I will go!

I did not know the 2001 eclipse would be visible only in southern Africa. At the time I was a student with very little money. Every travel option I found was well out of my price range. But I did not give up. One day I saw an ad for a flight from Paris to Zambia and back in less than 60 hours just to see the eclipse. It was still quite expensive but it only took me 5 minutes to decide. I withdrew almost half the money I had in my bank account and bought that ticket. It was the best money I ever spent.

On June 19, 2001 we left Paris on a chartered overnight flight to Lusaka, capital of Zambia. We arrived on June 20 and headed for a specially built campsite in the midst of the bush north of Lusaka. Night fell and gave me my first view of southern hemisphere stars.

June 21 dawned: eclipse day. It was the middle of the dry season, so the sky was cloudless—perfect. We were told not to leave the campsite but I wanted a different kind of eclipse experience. I left our group of more than 300 people and walked down a path into the bush.

I reached a small humble village with a dozen tiny mud huts with thatched roofs. It was almost deserted. Only a young boy came my way. I gave him eclipse glasses and we stayed together to watch the eclipse. We found a leafy tree and observed beneath it the images of a hundred crescent Suns. About 15 minutes before totality, as the light was dimming, some cows laid down and started bellowing. A hen and a row of small chicks quietly walked past us, heading for the box where they would roost at night.

Some women with a couple of young children joined us. I gave them eclipse glasses and we all looked at the now very slim crescent Sun.

One minute to go. I did not know what to expect. I had read books and articles. I had stood in the lunar shadow once. But I had never seen the corona.

All of a sudden moving bands of shimmering light wavered across the ground and the walls of the huts—a striking display of shadow bands. The sunlight was fading fast. The lunar shadow on the horizon was rushing toward us. A last ray of sunlight gleamed brightly for one last moment on the limb of the Moon.

We all screamed—the kids, the women, and me. The black disk of the Moon was surrounded by the pearly corona with beautiful tendrils. The pinkish red prominences of the Sun shone above the dark edge of the Moon. Our screaming ended. We fell into rapt silence.

Frogs started croaking in the bush around the village. The boy who had joined me first was the best observer among us. "Stars!" he said suddenly. And there they were—easy to see. The sky around the eclipse was a deep blue with tinges of indigo. The horizon in every direction was bathed in orange.

As abruptly as it started, totality ended. I was so overwhelmed that I cried.

In the evening we flew back to Paris. We had been in Africa only 38 hours.

That was my first time to see totality—an experience etched forever in my mind. Now, many years later, I still think back to that day in that humble African village. I was very lucky to have shared that experience with those kids and those women. It taught me a great lesson: Go see a total eclipse of the Sun. Who would have thought that I would end up in Zambia? Be open to whatever experiences come your way. I've travelled to five more total eclipses since 2001, each of them awesome.

Luca Quaglia has a Ph.D. in physics, conducts research at total solar eclipses, and is completing his master of financial engineering degree at Baruch College, City University of New York.

# 16

————◄◦►————

# When Is the Next One?
# Total Eclipses: 2025–2033

*Now eclipses are elusive and provoking things . . . visiting
the same locality only once in centuries. Consequently, it
will not do to sit down quietly at home and wait for one to
come, but a person must be up and doing and on the chase.*

Rebecca R. Joslin (1929)[1]

What a decade lies ahead. There are no total eclipses of the Sun in 2025—
and no annular eclipses either. It's as if the Sun and Moon were pausing
to prepare a special treat for Earth. It begins in 2026. A total eclipse of
the Sun will pay a visit to Spain in 2026—then a second total eclipse in
2027, less than a year later.

Then it's the Southern Hemisphere's turn. In 2028 and 2030, Australia
will get two total solar eclipses in less than 2½ years. Australia and the
United States are almost the same size, yet the mainland USA waited from
1979 to 2017—38 years—for a return. Seems like the Sun and Moon can't
get enough eclipsing over Australia. In one decade, from 2028 to 2038,
Australia will treated to four total eclipses.[2]

Two other total eclipses occur before the end of 2033 but they are hard
to reach. In 2031, the Sun and Moon will perform their magic over the
Pacific Ocean. In 2033, a total eclipse will return to the United States;
well, one United State—Alaska, to its remote western and northern coast
just as spring is beginning.

But every total solar eclipse has its uniqueness and its special fascination. Let's take a look.

## August 12, 2026—Iceland and Spain

There won't be many people watching when the total eclipse of August 12,
2026 first touches Earth. Its landing site is on the northernmost peninsula of
Siberia. Totality lasts 1 minute 35 seconds. The shadow wants out of Russia

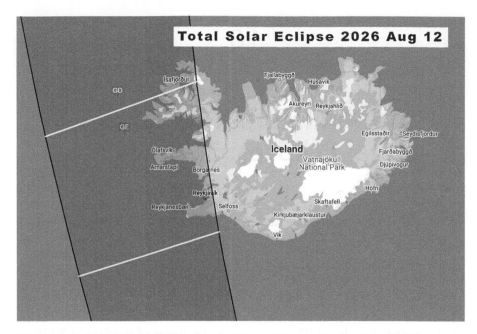

The August 12, 2026 total eclipse track runs across the north pole and clips western Iceland. [Map by Fred Espenak, EclipseWise.com]

fast. From the moment the front edge of the eclipse shadow arrives in Russia to the moment it defects by leaping into the Arctic Ocean is 6 seconds.

The eclipse slides northward across the polar ice and water, sweeps close to the north pole, and heads south. It comes ashore in Greenland, the largest island on Earth, in Northeast Greenland National Park, the largest national park in the world. At the north end of the park is Station Nord, a Danish military and scientific outpost only 580 miles (934 km) from the pole. The camp lies one-third of a mile inside the eastern limit of the total eclipse path, allowing it 10 seconds of totality. Station Nord cannot be reached by boat because of ice. Six soldiers live there year round.[3] And those six constitute the total population residing in the park.

## Chilling Out in Iceland

The eclipse skids off the glacier-lined coast of southeastern Greenland into the waters of the Atlantic Ocean, teeming with flat-topped icebergs spawned by those glaciers. Then, land ahead. The eclipse reaches the northwest coast of Iceland at the point where it is closest to Greenland, only 175 miles (280 km) apart. Offshore, the duration of totality for the

2026 eclipse reaches its maximum—2 minutes 18 seconds. The central line of the eclipse misses Iceland by 25 miles (40 km), but the path of totality is now 182 miles (293 km) wide, so darkness falls on Iceland's west coast. The westernmost point of land enjoys 2 minutes 10 seconds of totality.

Down the coast to the southeast lies Reykjavik, Iceland's capital and its largest city. Almost two-thirds of Iceland's 372,000 people live in the Reykjavik metropolitan area. The city's position near the eastern limit of totality allows it 1 minute under the shadow of the Moon.

The eclipse passes over Reykjavik. The eclipse passes over a crack under Reykjavik. That crack is the Mid-Atlantic Ridge, the longest mountain range on Earth. Most of it lies under water. The Mid-Atlantic Ridge is a vast fracture in the Earth's crust where scorching hot magma from the Earth's mantle breaks through the crust in volcanic eruptions, causing the seafloor to spread and push the North American tectonic plate away from the Eurasian plate at a rate of about 1 inch (2.5 centimeters) per year. The volcanic activity has built up so much solidified lava that it pokes out of the ocean—as Iceland.

Volcanic eruptions remain quite frequent in Iceland, building still more island, and the residual underground warmth gives Iceland its hot springs. Just east of Reykjavik is a gorge where with one step, you can leave Eurasia and arrive in North America—you step from one tectonic plate to the other. The fault line runs from southwest to northeast across all of Iceland.

The eclipse shadow now leaves Iceland, speeding southeast across the Atlantic Ocean. The shadow covers 1,500 miles (2,400 km) in 39 minutes, but finds no land until it sails across the northwest coast of Spain.


## Retiring in Spain

It is the first time since 1905 that Spain has hosted a total eclipse of the Sun—121 years. It is also the first total eclipse for mainland Europe since 1999. This eclipse will repay the wait. Approximately 45% of the land area of Spain will experience a total eclipse. And almost all of Spain has nearly ideal weather for eclipse watching in August.

By the time the eclipse reaches Spain, it is late in the day and the Sun is close to setting in the west. Eclipse observers in Spain must be sure to have a clear view of the western horizon.

Spain welcomes the eclipse with a landscape of cliff-lined shores and mountains not far inland. The central line of the eclipse is now producing 1 minute 49 seconds of totality. Close to the central line are the cities of Gijon, on the coast, and Oviedo, 15 miles (24 km) inland.

The path of eclipse then crosses a historic pathway on the ground—Camino de Santiago, the Footpath of St. James. Medieval stories say that the Apostle walked this route across Spain to its northwestern corner, spreading his new religion. He was martyred when he returned to Jerusalem. In the early 9th century, a hermit claimed to have found St. James' body buried in northwestern Spain. A city and cathedral grew up around his tomb and for 1,200 years Santiago de Compostela has been the destination of a major Christian pilgrimage. The path of this pilgrimage, Camino de Santiago, is also a popular hiking and bike trail.

Camino de Santiago cuts across Spain from Pamplona through Burgos and León, and ends at the cathedral in Santiago de Compostela. Pamplona is where the bulls run—but the path of this eclipse does not. Pamplona is just beyond the northern edge where the eclipse is total. It is, however, only a very short pilgrimage to reach totality.

The same is true at the other end of the trail, in Santiago de Compostela and its cathedral with the tomb of St. James. They lie just beyond the southern limit of totality.

Burgos and León, near the central line of the eclipse, are luckier. They are blessed with about 1 minute 44 seconds of Moon shadow.

The eclipse shadow continues southeastward and its central line crosses highway A-2, the major route between Spain's two largest cities. But both Madrid (population 6.7 million) and Barcelona (5.7 million) find themselves just outside of the path of totality. Between the two cities, though, where the central line crosses highway A-2, the eclipse is total and lasts 1 minute 43 seconds.

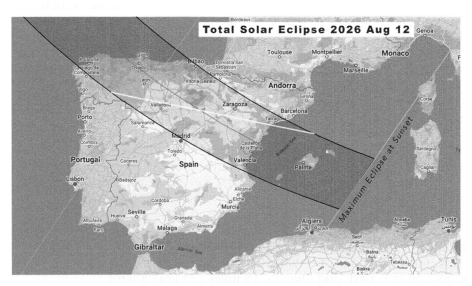

The track of the August 12, 2026 total eclipse crosses northern Spain and the island of Majorca (Mallorca) and ends at sunset in the Mediterranean Sea. [Map by Fred Espenak, EclipseWise.com]

When the eclipse reaches Valencia, it leaves mainland Spain and retires to the Spanish island of Majorca (Mallorca), its last land stop before leaving Earth. Majorca's largest city, Palma (500,000), lies almost exactly on the central line and receives 1 minute 36 seconds of totality.

Especially on Majorca, a clear view of the western horizon is crucial. As the eclipse sweeps southeastward, the Sun sets earlier and earlier within the phases of the eclipse cycle. By the time the eclipse reaches Majorca, there is just time for the Sun to emerge from totality before the Sun sets.

The total solar eclipse of 2026 then heads out into the Mediterranean Sea. About 125 miles (200 km) off the coast of Majorca, the Sun sets before totality starts. The Moon's shadow has left the Earth. The total solar eclipse of 2026 is over.

## August 2, 2027—Spain (again), Egypt, Saudi Arabia

Here it comes again. Less than a year after the total solar eclipse of 2026 flashed its diamond ring and flared its corona for 45% of Spain, a new total eclipse books a return passage to Spain. This time it will visit other parts

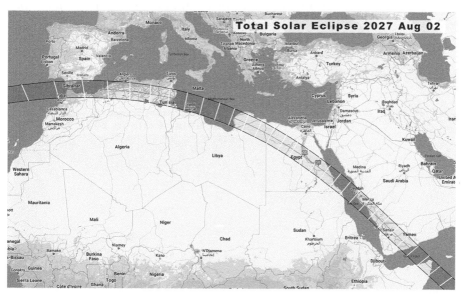

The path of the total solar eclipse of August 2, 2027 crosses southern Spain, Morocco, Algeria, Tunisia, Libya, Egypt, Sudan, Saudi Arabia, Yemen, and Somalia. The duration of totality is an extraordinarily long 6.4 minutes in Egypt. [Map by Fred Espenak, www. EclipseWise.com]

of the country, giving other Spaniards a turn. And many other people as well. The itinerary of the 2027 eclipse includes North Africa, Egypt, Saudi Arabia, Yemen, and Somalia.[4]

And quite a show it will be. The eclipse of 2027 is from the best of saroses—a highly distinguished family of eclipses. It has provided and continues to provide record-setting periods of totality.

The 2027 eclipse begins in the middle of the Atlantic Ocean, about halfway between Miami and Gibraltar, cruises across the waters, and sails through the Strait of Gibraltar into the Mediterranean Sea.

Gibraltar is a dagger of land poking southward out of Spain into the Mediterranean Sea. It is a self-governing Overseas Territory of the United Kingdom. There isn't much territory: only 2.6 square miles (6.7 sq km). It is dominated by its most famous feature, the Rock of Gibraltar, 1,398 feet (426 meters) tall. Historically, whoever controlled the Rock of Gibraltar was in a good position to control the Strait of Gibraltar, the 36-mile-long strip of water where the coast of Africa and the coast of Europe stand as little as 8 miles (13 km) apart. The Strait of Gibraltar is the only entrance from the Atlantic Ocean into the Mediterranean Sea.

The eclipse arrives on the Rock of Gibraltar to ensure its 34,000 citizens the thrill of a whole lifetime. They are beneficiaries of 4 minutes 26 seconds of totality. Watching with them in curiosity and perhaps alarm will be the other citizens of Gibraltar, about 300 tailless macaques that Gibraltarians call Barbary apes. They are the only monkeys living wild in Europe.

In early August, rain is rare in Gibraltar and the skies are mostly clear about 90% of the time.

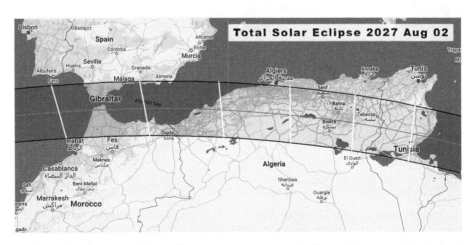

Details of the August 2, 2027 eclipse path across southern Spain, Morocco, Algeria, and Tunisia. [Map by Fred Espenak, www.EclipseWise.com]

## Spain

At the same time as the central line of the 2027 eclipse is passing just south of the Strait of Gibraltar, its wide path of totality (147 miles; 237 km) envelops the south coast of Spain and the north coast of Africa.

The path of totality runs through southernmost Spain, a region called Andalucía. Totality reaches a little north of the city of Cádiz on the peninsula's west coast and a little north of Málaga on the peninsula's east coast. Cádiz gets about 3 minutes of totality; Málaga gets about 2 minutes.

## North Africa

*Before traveling to North Africa, Saudi Arabia,*
*Yemen, or Somalia, American citizens should*
*check the US Department of State Travel Advisory*

On the African continent, Tangiers, Morocco stands on the Atlantic coast at beginning of the Strait of Gibraltar right where the central line of the 2027 eclipse passes with a gift of 4 minutes 51 seconds of totality. The city of 1 million took its name from an orange-colored citrus fruit that originated in southeast Asia and was introduced to Europe from Africa through its port. This fruit from Tangiers became known as a tangerine.

As it passes through North Africa, the eclipse takes in primarily the populated areas on the fertile strip of land that stretches from the Mediterranean Sea inland 50 to 100 miles (100–150 km) before giving way to the Sahara Desert.

The eclipse leaves Morocco and enters Algeria. Algeria is the largest nation in Africa—the size of Alaska and Texas put together. More than 80% of Algeria is desert.

The central line of the 2027 eclipse passes through the harbor of Oran, unloading 5 minutes 8 seconds of totality for its 950,000 citizens. Next it's Algiers' turn—almost. Algiers is the capital and largest city (2.9 million) of Algeria. But Algiers lies just beyond the northern limit of the eclipse. The southern outskirts of the city get a few seconds of totality.

Out of Algeria into Tunisia. Again, the capital and largest city—Tunis—is bypassed. But the eclipse delivers 5 minutes 40 seconds to Tunisia's second largest city, Sfax.

The eclipse crosses the border into Libya and the southern limit of the path of totality reaches to the shores of Tripoli, the capital and largest city. In the 17th and 18th centuries, the Barbary pirates operated out of Tripoli, Tunis, and Algiers, raiding throughout the Mediterranean Sea and in the Atlantic Ocean as far as England and Iceland. In 1801 and again in 1815, US presidents sent the Marines to halt the piracy. Only the northeast corner of Tripoli gets a few seconds to a minute totality.

The eclipse then stays off the Libyan shores until it reaches Benghazi where the central line slices through the center of Libya's second largest city (807,000). The duration of totality has reached rare length: 6 minutes 9 seconds. It is not yet at its maximum, but it will stay above 6 minutes throughout its visit to Egypt and Saudi Arabia.


## Into the Desert

Past Benghazi, heading toward the border between Libya and Egypt, the eclipse plunges inland, across the Sahara Desert. Within the Sahara Desert, it sets off into the Great Sand Sea with its waves of sand dunes. Somewhere in that journey of 400 miles (700 km), it crosses the border from Libya into Egypt.

Where the Great Sand Sea ends, the Qattara Depression begins. The Qattara Depression is the world's largest natural sinkhole, 50 miles (80 km) long and 75 miles (120 km) wide. It is 436 feet (133 meters) below sea level.

Within the Depression, in this orange sand wilderness, there are occasional oases, with houses, schools, stores, orchards, and farm fields grouped tightly around the water source. Olive and dates dominate the diet. The most famous of these oases is Siwa, with a population of 33,000—and, in 2027, 5 minutes 28 seconds of totality. Chances are good that Siwans will see the eclipse. Weather averages at Siwa show 0 days of rain in July, 0 days of rain in August, 0 days of rain in September . . .

Sixty miles (105 km) away is another oasis, Qara. It has a population of 363. From Siwa to Qara is more than a two-day camel ride.

Ancient Egyptians called the desert "the red land" in contrast to "the black land," the wet, fertile soil along the Nile River. Ninety-five percent of Egyptians live in the black land, within 3 to 7 miles (5–12 km) of the river.[5] In Ancient Egyptian, the word for the red land was *deshret*. It is a curious coincidence that the English word *desert* should sound so similar.

The path of the eclipse runs far south of Cairo, Egypt's capital and largest city. It runs south of Giza, just across the Nile from Cairo, with the biggest pyramids and the Sphinx.

But the central line runs almost directly through Luxor, up the Nile River 320 miles (515 km) to the south. In Ancient Egypt, 4,000 years ago, this was the location of Thebes, the capital. That's when the Egyptians began building the Temple of Amun in Karnak, now part of Luxor. Two thousand years later, when they finished the temple complex, they had created the largest religious building in the world. It still is.[6]

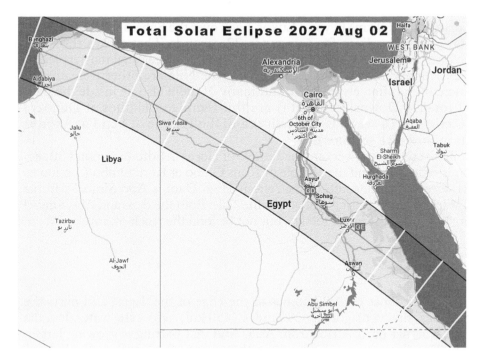

Details of the August 2, 2027 eclipse path across Libya and Egypt. Totality lasts up to 6.4 minutes in Egypt. [Map by Fred Espenak, www.EclipseWise.com]

Just south of the Karnak Temple, also right in the heart of the modern city of Luxor, is another huge temple complex referred to as the Luxor Temple. The Egyptians started building it 600 years after construction began on the Karnak Temple. They finished the Luxor Temple in less than 200 years.

Many Egyptian temples have lines of columns that, from the inside, focus your view on a special spot on the horizon. Archeoastronomers are still calculating whether these temples were likely used in ceremonies that involved the rising of certain stars and the Sun. The pyramids certainly had astronomical orientations.[7] What would the ancient Egyptians have thought of a total solar eclipse happening at the top of their sky—especially one at their capital, Luxor, that lasted 6 minutes 21 seconds?

The wonders of Luxor continue on the west bank of the Nile with the Valley of the Kings where, in the hillsides along the course of a dry river, pharaohs were buried. Tutankhamun is there. A mile and a half (2 km) away is the Valley of the Queens, where the wives of the pharaohs were also buried in splendor. Nefertari is there.

Just as Luxor displays the astonishing monuments of Ancient Egyptian civilization, so now the total eclipse of 2027 reaches its peak of achievement.

Forty miles (60 km) beyond Luxor, back over the Sahara Desert, the Sun stays in total eclipse for 6 minutes 23 seconds. That makes the 2027 eclipse the second longest of the 21st century—a celestial masterpiece.

North and south from Luxor along the Nile are other treasures of ancient Egyptian civilization, and many of them also lie inside the path of totality. Downriver (north) are the Temple of Hathor at Dendera (6 minutes 14 seconds of totality) and the Temple of Seti I at Abydos (6 minutes 21 seconds).

To the south (upriver) are the Temple of Horus at Edfu (5 minutes 30 seconds of totality) and the Temple of Kom Ombo at Kom Ombo (3 minutes 21 seconds). Well south of the Aswan High Dam is Abu Simbel with its towering statues staring east across the Nile. Both the dam and Abu Simbel are beyond the reach of totality, but not beyond the reach of travelers.

## Saudi Arabia

The total eclipse of 2027 comes to the edge of the desert and parts the Red Sea, slicing diagonally 280 miles (450 km) across the water. It is the Red Sea that parts Africa from Asia. And that parting is growing larger. The Red Sea covers a great split in the crust of the Earth. At the bottom of the Red Sea, the sea floor is spreading, widening the gap between Africa and Asia. Marked by earthquakes and volcanic eruptions, the expanding rift is shoving the tectonic plate carrying Egypt northward, while the plate that bears Saudi Arabia and the other Middle Eastern countries is being rammed northeastward. Over millions of years, the Red Sea will become an ocean.

Now in Asia, now in Saudi Arabia, the total eclipse sweeps over Jeddah on the coast and Mecca, 45 miles (70 km) to the east. Only Muslims are permitted in Mecca (population 2.2 million). Jeddah (5 million) is not restricted to Muslims and provides 6 minutes of totality. On average, the August 2 weather in Jeddah offers a 63% chance of mostly clear weather with a 2% chance of rain. The average high temperature for August 2 is 102 °F (39 °C). Because of humidity, Weather Spark lists Jeddah weather on August 2 as "muggy," "oppressive," or "miserable" 75% of the time—overall characterizing it as "sweltering."

Saudi Arabia, like Algeria, is the size of Alaska and Texas put together. Ninety-five percent of its land is desert. As the 2027 eclipse approaches Saudi Arabia's border with Yemen, it encounters the fiercest of those deserts, the Empty Quarter. It fills the southern third of Saudi Arabia and is so vast that it spills into Yemen, Oman, and the United Arab Emirates.

Only 25% of the Sahara Desert is covered with sand. Most of the rest is gravel. The Empty Quarter is almost entirely sand. Nowhere on Earth is there a bigger continuous sea of sand than the Empty Quarter.

Just inside the border of Yemen, the duration of totality drops to less than 6 minutes—still an eclipse of majestic length.

Yemen is 57% desert and only 3% of the land is farmable. Water is in very short supply, including in Sana'a, the capital and largest city (population 3.2 million). Sana'a lies in the path of totality, but toward the southern limit. The center of town gets 2 minutes 15 seconds of totality.

Sana'a is nestled at the base of mountains at an altitude of 7,500 feet (2,300 meters) above sea level. The altitude keeps the average early August high to 82 °F (28 °C) and the humidity low. In early August the chance of rain is only 4%, but the skies are mostly cloudy or overcast 58% of the time. The central line lies over the mountains 75 miles (120 km) away, at the edge of the Empty Quarter.

The central line of the 2027 eclipse surfs over the sand sea, glides over the coastal mountains, leaves Yemen, flies 250 miles (400 km) across the Gulf of Aden, and sets down its shadow on a peninsula that looks like a rhinoceros horn—the Horn of Africa. The country that wraps around that horn is Somalia. The eclipse path clips the tip of that horn.

In the years before the eclipse, Somalia was gripped by a long drought that threatened an already impoverished population. Meanwhile, Somalia has suffered a long civil war with the breakaway region at the top of the Horn that has declared itself independent under the name Somaliland.

In 585 BCE, the Lydians and the Medes were at war for control of the Middle East. According to the story, in the midst of a battle, a total eclipse of the Sun occurred. It scared the soldiers and their leaders so badly that they stopped fighting and signed a truce. Somalia is in desperate need of a new eclipse to halt its civil war. The solar eclipse of 2027 offers 5 minutes 30 seconds of awe as its peace offering.

The total eclipse of 2027 is nearing exhaustion. It flees from Somalia, from Africa, and from all further travel on land to spend the rest of its life over the Indian Ocean. Far to the south of India, long before it can reach Australia, it quits the Earth.

## Progenitor of the 2027 Eclipse: The Amazing Saros 136

Every eclipse belongs to a family, a saros series, that begins, evolves, and ends. Each saros series has distinguishing features and a chance to participate in human history. Here is the story of one remarkable family of eclipses.

The eclipse family known as Saros 136 was born on June 14, 1360 as a very slight partial eclipse over Antarctica and the southern Indian Ocean. There was no one there to see.

The firstborn of every saros family is always a slight partial eclipse that brushes the Earth near one of the poles and gives no visual evidence of the splendor that will come.

## Celestial Clockwork

Time passed. The years rolled by. Forty-one other solar eclipses touched the Earth, but they came from other saros families. Then, in 1378, Saros 136 produced a second partial eclipse in which the Moon cut off a slightly larger portion of the Sun's light.

Every 18 years 11⅓ days brought a new solar eclipse from Saros family 136, each a little farther north on the whole; each a partial eclipse, but each time covering a little more of the Sun. In 1486, the Moon covered 98.6% of the Sun's diameter, leaving only a thin crescent of the Sun.

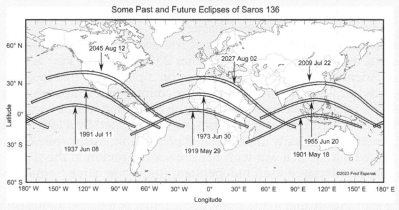

Paths of totality for seven past and two future eclipses of Saros 136. Successive eclipses shift westward and northward. For odd-numbered saroses, eclipses shift westward and southward. [Map and eclipse calculations by Fred Espenak]

The next eclipse in the cycle, in 1504, was different from those that preceded it. The Moon passed directly across the disk of the Sun as seen from near the Antarctic coast. It would have been a total eclipse but the Moon was farther from the Earth than average, so it appeared too small to cover the Sun completely. The Sun's surface still shone around the circumference of the Moon as a bright ring of light—an annular eclipse—lasting 31 seconds.

Throughout the 16th century, each of the six eclipses of Saros 136 was an annular eclipse, and the eclipse paths were gradually migrating northward. At each eclipse, the Moon was a little closer to Earth, so that the Moon's apparent size grew larger. In 1594, the disk of the Moon was almost large enough to completely hide the Sun.

At the next eclipse, it did—for just *1 second*. On November 22, 1612, in the southeastern Pacific Ocean and Antarctica, birds, fish, and whales saw an eclipse that was annular all along the central path until the eclipse reached its midpoint, where the surface of the Earth was closest to the Moon and hence the Moon appeared largest. At that location, the Moon's disk for just an instant completely covered the Sun and the eclipse (technically, at least) became total. The dark shadow of the Moon caused by Saros 136 actually touched the Earth for the first time. But the Earth's surface then curved away from the extended shadow, just enough to make the Moon once again appear too small to completely cover the Sun. The eclipse was annular once more.

The eclipse of 1612 and all five eclipses of Saros 136 in the 17th century were hybrids of this kind: beginning as annular, becoming total, then returning to annular again. But each time, the duration of totality was a little longer.

## The Big Leagues

On the twentieth eclipse of Saros 136, on January 17, 1703, the duration of totality had grown to a maximum of 50 seconds and the eclipse was total along its entire path except for the last 55 miles when it became annular.

Each subsequent eclipse of Saros 136 was total, and the duration of totality was rapidly increasing. The eclipse path continued shifting northward. As the American Civil War ended, Saros 136 crossed the 5-minute plateau into the realm of rare and great eclipses as it stretched from South America to southern Africa. The totality of April 25, 1865 lasted 5 minutes 23 seconds.

Now almost everything was conspiring to make Saros 136 one of the greatest in history. At each eclipse, the Sun was farther from the Earth and thus smaller in apparent size, enabling total eclipses to last longer. At each eclipse, the Moon was closer to Earth, so the Moon appeared larger, allowing total eclipses to last still longer. Every eclipse of Saros 136 in the 20th century would last longer than 6 minutes.

The eclipse of May 18, 1901 brought astronomers from all over the world to Indonesia to continue their analysis of the solar atmosphere. They had at most 6 minutes 29 seconds within which to work.

The next eclipse in the cycle lasted even longer—6 minutes 51 seconds at its peak, a totality long enough to make it memorable in and of itself. But that was not why eclipse number 32 in Saros 136 became the most famous in history. Astronomers used that eclipse to measure star positions around the darkened Sun and concluded that starlight passing by the Sun had been bent, just as Einstein had predicted in his recently completed general theory of relativity. Thus the total eclipse of May 29, 1919 marked a major turning point in the history of science.

Saros 136, however, had its own dramatic schedule to fulfill. It returned on June 8, 1937 with a 7-minute 4-second eclipse—the first to last over 7 minutes since 1098.

Saros 136 reached its climax on June 20, 1955 with an eclipse that lasted 7 minutes 8 seconds, the longest since June 20, 1080 (7 minutes 18 seconds), and the longest until June 25, 2150 (7 minutes 14 seconds). Never again would Saros 136 equal that duration of totality. Yet few saroses even come close.

June 30, 1973 brought an eclipse, visible across Africa, that lasted as long as 7 minutes 4 seconds.*

The last offering of Saros 136 in the 20th century happened on July 11, 1991, bringing up to 6 minutes 54 seconds of totality and tracing a path from Hawaii through Mexico, Central America, Colombia, and Brazil.

The six longest eclipses of the 20th century were in 1901, 1919, 1937, 1955, 1973, and 1991—every one of them a member of the same eclipse family. Saros 136 is one of the greatest eclipse-generation cycles in recorded history.

## Closing Out a Career

Now, gradually, Saros 136 is declining. At each eclipse the Moon will appear a little smaller in the sky while the Sun appears larger. Eclipses will become steadily shorter, but they will still be total and will still be comparatively long for quite some time. The glory will fade slowly.

Saros 136 returned on July 22, 2009 with a 6-minute-39-second display and a path across India, China, and the western Pacific. The total solar eclipse of 2009 was the longest of the 21st century.

On August 2, 2027, a Saros 136 eclipse will race over the North African coast through Egypt bringing darkness for up to 6 minutes 23 seconds. It will pay its first visit to the continental United States on August 12, 2045, streaking from northern California through Florida, the longest totality for the continental United States in the calculated history of eclipses. Over the Caribbean, it will last for 6 minutes 6 seconds. But that will be the last eclipse of Saros 136 to surpass the 6-minute mark.

On September 14, 2099, it will return to North America, slicing across southwestern Canada and plunging southeastward across the United States to the mouth of the Chesapeake Bay and off into the Atlantic. At maximum, the eclipse will last 5 minutes 18 seconds.

Saros 136 is aging but its eclipses are still total and more than long enough to lure people by the thousands into its shadows. By May 13, 2496, however, a total eclipse of Saros 136 will have dwindled to 1 minute 2 seconds for hardy travelers in the Arctic. It is the last of 44 total eclipses in this remarkable saros family.

At the next visit of this saros, May 25, 2514, the dark shadow of the Moon will miss the Earth, passing above the north polar region. The remaining six eclipses in the sequence will all be partial as well, steadily declining in the Moon's coverage of the face of the Sun. On July 30, 2622, there will be one final partial eclipse, virtually unnoticeable near the north pole. Saros 136 will have died.

Saros number 136 will have performed on Earth for 1,252 years and created 71 solar eclipses. Not an exceptionally long career, but what a record! Of its 71 eclipses, 15 will have been partial, 6 annular, 6 a combination annular and total, and *44 total*. The typical saros offers only 19 or 20 total eclipses.

Saros 136 brought the people of the 20th century 3 eclipses with totality exceeding 7 minutes and all 6 of the longest eclipses in that century. Its three eclipses in the first half of 21st century will all exceed 6 minutes, the longest eclipses to be seen in that span of time. If this saros were an athlete, its shadow-black jersey with its corona-white number 136 would be retired to hang in glory in the Eclipse Hall of Fame.

* Michael Rogers wrote about the eclipse of June 30, 1973 in "Totality—A Report," published in *Rolling Stone* magazine, October 11, 1973. Some of the scientific information is wrong, but Rogers does a remarkable job of capturing the excitement of a total solar eclipse. For this article, he won the AAAS-Westinghouse Science-Writing Award.

## Saroses of Total Eclipses 2025–2033

To more fully appreciate how remarkable Saros 136 is, compare it to the saroses of the other total eclipses between 2026 and 2033.

| | 2026 | 2027 | 2028 | 2030 | 2031 | 2033 |
|---|---|---|---|---|---|---|
| Type of eclipse | Saros 126 | Saros 136 | Saros 146 | Saros 133 | Saros 143 | Saros 120 |
| Partial Eclipses | 31 | 15 | 35 | 19 | 30 | 16 |
| Annular Eclipses | 28 | 6 | 24 | 6 | 26 | 25 |
| Hybrid Eclipses | 3 | 6 | 4 | 1 | 4 | 4 |
| **Total Eclipses** | **10** | **44** | **13** | **46** | **12** | **26** |
| Number of eclipses in saros | 72 | 71 | 76 | 72 | 72 | 71 |
| Longest total eclipse in saros | 2m38s 07/10/1972 | 7m8s 06/20/1955 | 5m21s 06/30/1992 | 6m50s 08/07/1850 | 3m50s 08/19/1887 | 2m50s 03/09/1997 |
| Position of this total eclipse in saros | 2026 is 2nd to last total | 2027 is #18 of 44 totals | 2028 is #6 of 14 totals | 2030 is#27 of 46 totals | 2031 is #2 of 4 hybrids; no more totals | 2033 is last total |

# July 22, 2028—Australia and New Zealand

The total solar eclipse of 2028 makes first contact with the Earth in the middle of the Indian Ocean and glides east toward the Australian continent, 3,300 miles (5,300 km) away. But it reaches Australian territory 2,000 miles (3,000 km) sooner. This first landfall is the Territory of the Cocos (Keeling) Islands—a somewhat awkward name because of its awkward history.[8] Some called the island group Cocos because of the coconut trees there. Others used the name of their British discoverer, William Keeling (1609). For more than a century, the islands were part of the British Empire but privately owned by a family. In 1955, the British transferred the islands to Australia. Australia combined the islands' historical names and bought the land from the family. Six hundred people live on two atolls where totality will last 3 minutes 27 seconds.

The eclipse then sails 600 miles (1,000 km) to reach the nearest land, Christmas Island. Christmas Island looks nothing like the Cocos (Keeling) Islands. Christmas Island has a volcano in the center; the Cocos (Keeling) Islands have a lagoon. But really, they are different stages of one another. Over millions of years, Christmas Island may become an atoll.

Christmas Island is a 14,000-foot (4,500-meter) volcano that rose 60 million years ago from the ocean floor until its peak poked out of the water. Coral built a reef around the island. Over time, erosion and tectonic plate movement may cause the volcano to sink below the surface of the water. But the coral that encircles the island may continue to grow, leaving behind a ring of coral islands with a lagoon in the center—an atoll like the Cocos (Keeling) Islands.

The British Empire transferred Christmas Island to Australia in 1958. In 2028, the 2,000 residents of Christmas Island will be gifted with totality lasting 3 minutes 56 seconds.

The Australian continent welcomes the 2028 eclipse with two national parks side by side that extend to the water's edge. As the eclipse speeds diagonally southeast across the country, there is seldom a minute in which the front edge of its 142-mile-wide (228-km-wide) path is not visiting one national park or another. Australia has 726 national parks, far more than any other country.

Just before the eclipse reaches Lake Argyle in Western Australia, the 2028 eclipse reaches its maximum, with totality lasting 5 minutes 3 seconds. At that very second, the eclipse crosses the Great Northern Highway. The road is not remarkable to look at. There are no cloverleaves or high-speed exit ramps. It is a two-lane highway. It starts in Perth, on the southwestern coast of Australia, and runs north all the way across the country to Wyndham on the northern coast—a distance of 2,000 miles (3,200 km). That makes the Great Northern Highway the longest remote paved road in the world.

The tractor-trailers that travel this road are long too. Many trucks pull three trailers so that the whole unit stretches out 175 feet (53 meters). Aussies call them road trains. Imagine a seven-car subway train off its tracks and going down the road at 60 mph (100 km/hr).

The eclipse is flying by places on the map with wonderful-sounding names: Purnululu, Gurindji, Gungalman, and Wingadee. On a typical map, such markings would indicate a town. But there are no buildings there. Instead, these are places where dirt roads meet. They are "localities." Perhaps in the vastness of the Australian Outback, intersections with names help travelers find their way.

From the Wingadee intersection, just up Wingadee Road about 18 miles (30 km), is Come By Chance, a junction where three roads meet. No houses. But a mile and a half south on Come By Chance Road is a sheep station (ranch) with a few buildings and the Come By Chance Cemetery.

The 2028 total eclipse does not come by there. Come By Chance is just outside the path of totality.

Sheep stations are scattered around the Outback, usually 5 or 10 miles from one another. Occasionally there are places like Elong Elong, a road junction where there actually is a tiny village with a church, a social hall,

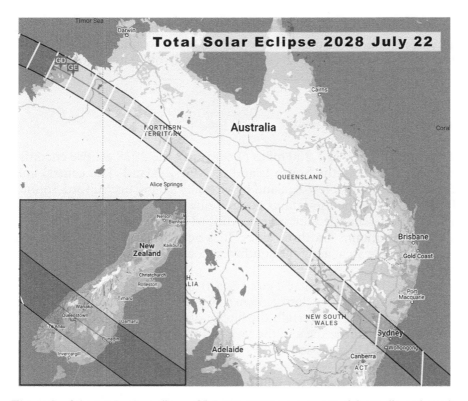

The path of the total solar eclipse of July 22, 2028 crosses central Australia and southern New Zealand. Sydney gets 3.8 minutes of totality. [Map by Fred Espenak, www. EclipseWise.com]

and about a dozen houses. There are a couple more houses in the suburbs. Totality in Elong Elong lasts 3 minutes 55 seconds.

The central line of the eclipse zips between Dubbo and Mudgee. Dubbo is the largest city the 2028 eclipse has visited so far. It has a population of 36,000, complete with a zoo and a university branch. Nearby Mudgee is much smaller (12,000) but has a golf course.

The total eclipse of 2028 has now almost completed its transcontinental flight of 2,100 miles (3,400 km). Ahead lies Sydney, Australia's second-largest city, with a population of 5.4 million. Sydney welcomes the approach of the eclipse shadow from the northwest with three national parks.

The central line of the eclipse goes straight through the center of Sydney, just 2 miles (3 km) from the Sydney Harbour Bridge and the Sydney Opera House. It brings 3 minutes 48 seconds of totality.

At the water's edge, three national parks bid farewell to the eclipse as it leaves Australia.

Then onward goes the eclipse, a wandering worker, come from afar, carrying its bundle of surprises, moving on to stir wonders in other lands. Or, as an Australian might say: a jolly swagman, toting his matilda, waltzing on across the Pacific billabong.

## Sightseeing in New Zealand

After a journey across 1,200 miles (1,900 km), the 2028 eclipse encounters its last landfall, the South Island of New Zealand. The central line of the eclipse enters New Zealand by virtually sailing up Milford Sound. Here, glaciers in the mountains gouged out stream valleys. The glaciers melted. The sea flooded the valleys. Milford Sound is ocean surrounded by steep mountains on three sides. It is actually a fjord, like those in Norway. In this area along the southwest coast of the South Island, there are 14 fjords.

In Milford Sound, Mitre Peak rises from the water to a height of more than a mile (more than 1.6 km). From the sheer cliffs of the fjord pour two permanent waterfalls the height of 50-story buildings. When it rains—about half the days in the year—dozens more waterfalls plunge into the fjord. The eclipse bestows 2 minutes 56 seconds of totality on Milford Sound.

No sooner does the eclipse climb out of New Zealand's fjords than the central line soars over even taller mountains, almost 1½ miles (2.3 km) high. Their name reflects their appearance: The Remarkables. Beneath The Remarkables, along the shore of a lake, sits the city of Queenstown (population 16,000), a winter destination for skiing, a summer destination for hiking, bungee jumping, and whitewater river rafting.

By the time the eclipse gets to Milford Sound and Queenstown, the Sun is preparing to set, only 10° above the western horizon. Eclipse observers must make sure that mountains do not block a clear view of the horizon. Totality lasts 2 minutes 54 seconds in Queenstown.

The central line of the 2028 eclipse departs Queenstown by flying over the biggest ski area in The Remarkables, then 106 miles (171 km) across the South Island to Dunedin on New Zealand's east coast. Dunedin *(Dun-EE-din)* is Gaelic for Edinburgh. It calls itself the Edinburgh of the South. It's the height of winter in New Zealand but Dunedin receives only one-tenth of an inch of snow per year. This city of 130,000 has an oversized railway station with a striking exterior made of dark gray basalt trimmed in white limestone. Residents call it the Gingerbread House.

From Dunedin, the Sun, in the midst of 2 minutes 51 seconds of totality, is now only 8° above the horizon. The last partial phase of the eclipse is just ending when the Sun sets.

The eclipse then dashes on southeastward into the South Pacific Ocean, its darkness skimming across the ocean until the shadow no longer points to Earth—and the total solar eclipse of 2028 is now a memory.

## 2029— No Total Eclipses of the Sun

## November 25, 2030—Southern Africa and Australia (again)

Late in November 2030, the Sun and Moon get together and send their shadow off on a desert tour in the southern hemisphere. The eclipse will darken a 100-mile (160-km) swath of land across two continents in less than a day.

The shadow drops into the Atlantic Ocean 900 miles (1,400 km) off the west coast of southern Africa and skims the water toward dry land in Namibia.

### The Namib Desert

And dry the land is. A cold ocean current flowing northward about 100 miles (160 km) off the southwestern coast of Africa keeps Namibia surprisingly dry, just as a cold current off the California coast provides San Diego and Los Angeles with near desert conditions.

Even at the shoreline, Namibia is a desert. That desert, the Namib, gives Namibia its name. To the south, just outside the path of totality, Namib-Naukluft National Park begins. The park features some of the largest sand dunes in the world. They are shades of red and orange and have names like Big Daddy and Dead End. Dune 7, the tallest, is 1,273 feet (388 meters) high, the height of the Empire State Building.

Namib Naukluft National Park stretches down the coast 370 miles (600 km). It is the largest national park in Africa.

But the boundaries of the national park do not mark the end of the Namib Desert. It extends north from Namibia along most of the coastline of Angola and south from Namibia almost to Cape Town in South Africa. That's 1,200 miles (2,000 km), the distance between Boston and Miami. The name Namib means "vast place."

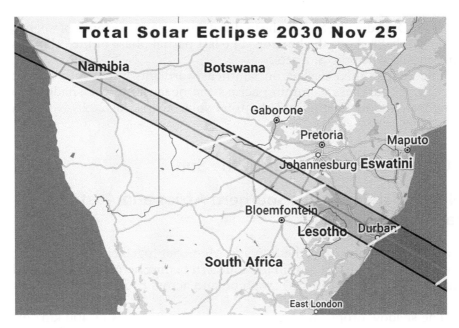

The path of the total solar eclipse of November 25, 2030 crosses Namibia, Botswana, South Africa, and Lesotho. Totality lasts 2.5 minutes in Durban. [Map by Fred Espenak, <http://www.EclipseWise.com>]

This land has been desert for more than 55 million years. For comparison, the Sahara Desert is about 5 million years old. The Namib is the oldest desert on Earth.

A hallmark of the total solar eclipse of 2030 is deserts. There are more to come. A second hallmark of this eclipse is mines. There are several right ahead, and more to come. Fifty miles (80 km) inland, near the southern limit of the path of totality, is the Rössing Mine, an open pit uranium mine that measures almost 2 miles long and three-quarters of a mile wide (3 × 1.2 km). Terrace after terrace descend to a depth of 1,280 (390 meters), as deep as Dune 7 is tall. The Namibian economy is buoyed principally by mining, especially uranium and diamonds.

In the city of Windhoek on November 25, 2030, the Sun rises at 6:00. Twenty-five minutes later, the Moon begins to nibble at the edge of the Sun and an eclipse is underway. At 7:20 a.m., a total solar eclipse is at its maximum, yielding 1 minute 53 seconds of totality. Windhoek is only 6 seconds off the central line. It's a memorable way to start a working week.

Windhoek, right in the center of the country, is the capital of Namibia. It is also the most populous city (325,000). The residents enjoy more than 300 sunny days a year. But Namibia is so dry that less than 1% of its land is farmable, compared to 40% of the United States. Namibia is so dry that there are no rivers or lakes that have water all year round.

Where occasional rivers create occasional shallow lakes, the water quickly evaporates, leaving behind salt deposits that bake in the Sun to a glaring white. As the eclipse leaves Namibia, it sweeps across hundreds of these salt pans (salt flats), some more than a mile across. On the edge of one of these salt pans, a flying club operates an airport that uses the usual hardness and smoothness of the salt to serve as two runways.

In the north of Namibia but not on the eclipse path is a salt pan measuring 81 by 31 miles (130 × 50 km) in size. It is a national park in itself—Etosha, meaning "the great white place."

## The Kalahari Desert

The many dozens of salt pans the eclipse shadow encounters in southeastern Namibia mark the passage of the path of totality into a second desert, the Kalahari. Strictly speaking, parts of the Kalahari receive a little too much rain to qualify as a desert but, as the salt pans testify, the Kalahari is quite inhospitable. The name says as much. Kalahari means "great thirst."

The 90-mile-wide (150-km-wide) front of the eclipse path slides into Botswana without casting shadow upon a town. The front of the whole path speeds across the sand for 150 miles (240 km) before it encounters a road, and still, up and down that north–south road, there is no town within the eclipse path. The front of the eclipse hurries on another 50 miles (80 km) before it finally finds a scattering of towns—along the border of Botswana with South Africa. The eclipse of 2030 has traversed the southwestern part of Botswana—200 miles (325 km) across the Kalahari Desert. Eighty percent of Botswana is Kalahari Desert. The border of Botswana with South Africa is marked by the Molopo River, which is usually dry.

## Straddling Countries

Now in South Africa, the climate is less arid and there are more towns and bigger towns. Johannesburg, population 6.1 million, lies 70 miles (115 km) off the eclipse path to the northeast. Johannesburg is the largest city in the world that is not on a river, lake, or sea coast.

Within the path of totality are modest-sized cities like Welkom (64,130). Some have unusual features. Surrounding Welkom, at the edge of the city and within it, are white shapes often a mile or two (1–3 km) across. But they are not salt pans. They are geometric in shape. This is gold- and, more recently, uranium-mining country. The white shapes are tailings from gold mines that have been impounded by dams on all sides. Welkom is a Dutch word meaning "welcome."

To the southwest of the eclipse path is Kimberly. It was a farm when diamonds were discovered there in 1871. Prospectors swarmed in and started digging furiously with picks and shovels. A big hole developed and a town grew up around it. Over the next 43 years, the hole grew until it was 1,500 feet (460 meters) wide and almost 800 feet (240 meters) deep. The mine closed in 1914. The Big Hole remains in the middle of Kimberly. Three tons of diamonds came out of that hole.

The geological formation in which diamonds are found is called a kimberlite pipe. Other kimberlite pipes were found near Kimberly.

The central line of the 2030 eclipse continues on through South Africa. But the southern portion of the eclipse's path of totality catches the northeastern quarter of Lesotho. The nation of Lesotho is about the size and shape of Vermont and New Hampshire put together—and like the two of them together, Lesotho is two-thirds mountains. Here, in the mountains of northern Lesotho, they mine diamonds. The Sun and Moon look down on open-pit mines that dwarf the Big Hole in Kimberly.

In its travels across Africa, the 2030 eclipse has visited four countries—and all four rank among the world's top eight diamond-mining countries: Botswana (#2), South Africa (#5), Namibia (#7), and Lesotho (#8). The eclipse is now about to move on to #9: Australia.

The eclipse of 2030 has traversed Africa on a safari of 1,250 miles (2,000 km). It has crossed two deserts and a mountain range. It has been a lonely journey. Until it is within 50 miles (80 km) of the coast, the eclipse has visited fewer than a million people. Now, at the end of its sojourn in Africa, just before it sets out to sea, the eclipse casts its spell on the largest city it will enchant, Durban, South Africa, population 3.7 million. After endless miles in deserts, the eclipse happens upon the greenest city in the world (based on how much healthy vegetation it has).[9]

As it crossed Africa, the duration of 2030 total eclipse has risen from 1 minute 53 seconds to 2 minutes 35 seconds. Durban is not far from the central line and receives 2 minutes 25 seconds of totality. In late November, Durban is mostly cloudy or overcast 40% of the time. Greenness comes with some compromises.

## Water, Water Everywhere

The total eclipse of 2030 then flies on to Australia. The eclipse shadow leaves South Africa at breakfast time, travels almost 6,000 miles (almost 10,000 km), and reaches Australia at dinnertime. It's a quick 2-hour 45-minute flight—at an average speed for this eclipse shadow of 2,100 mph (3,400 km/hr).

As it speeds across the southern Indian Ocean, the eclipse shadow has an average width of 100 miles (160 km). Yet in its 6,000-mile journey, when this eclipse is at its best, its 100-mile-wide shadow will darken not one speck of land. The 2030 eclipse celebrates its maximum duration of 3 minutes 44 seconds out in the middle of the ocean by briefly confusing fish.

## Australia

Less than 2½ years after their 2028 show over Australia and New Zealand, the Sun and Moon are back performing over Australia. It's a traveling production so they are playing different venues Down Under.

The eclipse, bringing 2 minutes of totality in its pouch, hops onshore near Streaky Bay (population 1,400) and leaps inland. More than one-third of Australia is desert and another third is semi-arid. With one exception, the 15 largest cities in Australia are located on the coasts.

In fact, more than 85% of Australia's population lives within 30 miles (50 km) of the coast.

If Australia were divided up evenly among its 26 million citizens, each person would have 73 acres of their own (9 people per square mile; 3.4/sq km). By comparison, the United States has a population density of 94 people per square mile (36/sq km).

The eclipse of 2030 visits none of Australia's largest cities. Inland from those cities is the Outback. The eclipse of 2030 visits lots of that.

The population density of the Outback is 0.36 people per square mile (0.14/sq km).

The eclipse bounds into the Outback. It sweeps northeastward across localities named Pinkawillinie and Buckleboo. This is dry country, desert. Ahead lies Lake Gairdner National Park. "Lake" Gairdner is more a hope than a reality. It is a lake when it rains occasionally. A downpour in the area will fill the lake to a depth of 2 feet (half a meter). But most of the time Lake Gairdner is a dry lake bed—a salt flat that is 100 miles long and 30 miles wide (160 × 50 km). Here Australian drivers try to set world land speed records.

Ahead are three more large lake beds, now mostly salt flats too. They and Lake Gairdner are the remnants of a vast inland sea that once filled almost the entire Australian Outback.

The eclipse speeds on to the northeast, out of the state of South Australia and into the northwest corner of the state of New South Wales, past Packsaddle, a hot desert township of 87 people.

The eclipse departs New South Wales and enters the state of Queensland. Right at the border, the central line of the eclipse goes straight through the town of Hungerford, population 38, and no 38 people have ever been more deserving, based on their affection for the stars. Hungerford has 11 streets. As the central line passes through, it crosses Achernar Street,

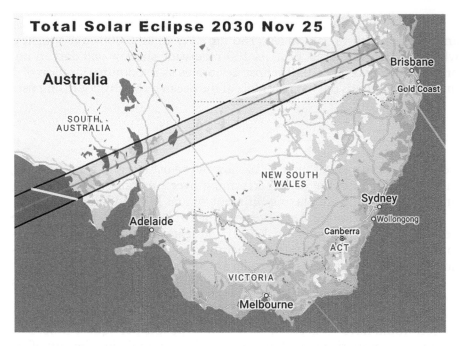

**Total Solar Eclipse 2030 Nov 25**

After crossing southern Africa and the Indian Ocean, the path of the November 25, 2030 total eclipse crosses southern Australia where it ends at sunset. [Map by Fred Espenak, <http://www.EclipseWise.com>]

briefly runs parallel to Canopus Street, then crosses Arcturus Street and Aldebaran Street, but misses Centauri Street on the east end of town. The reward for such thoughtfulness: 1 minute 23 seconds of totality.

And the town needs a gift like that. The covid-19 epidemic of 2020 and 2021 hit Hungerford hard. The Queensland Police blockaded Achernar Street at the Queensland–New South Wales border lest citizens of New South Wales try to cross into Queensland.

North of Hungerford, across the Paroo River, which occasionally has flowing water, is Currawinya National Park. It is a sanctuary for bilbies, an endangered marsupial. The greater bilby is the size of a rabbit, has long ears like a rabbit, but has a longer snout and quite a long tail. The lesser bilby was driven to extinction in the 1950s. Greater bilbies don't hop like rabbits but they do leap 3 to 4 feet (1 meter) into the air. Bilbies serve as the Australian Easter Bunny.

Currawinya National Park has two sizable lakes only 2 miles (3 km) apart. One is a salt lake and the other has fresh water. These lakes are a refuge for waterbirds and migratory shorebirds.

In 2030, the Hungerford-Currawinya area of Australia will be hosting its second total eclipse of the Sun in 2½ years. The eclipse of 2028 swept through here. Now the eclipse of 2030 crosses its path.

But for the 2030 eclipse, its time on Earth is ending. Past Bindebango and Ballaroo the eclipse goes. Ahead lie Fairyland and Burra Burri. It is here, at the end of its journey across the Outback from Sneaky Bay to Chinchilla, that the 2030 path of totality finally encounters the largest Australian city along its walkabout of 1,100 miles (1,750 km). Chinchilla has a population of 6,612.

For Chinchilla, the Sun is setting just as the total phase of the eclipse is ending. The corona vanishes, the final diamond ring flashes, and the Sun dips below the horizon. The eclipse of 2030 vanishes from the face of the Earth.

## What a Decade!

In the 10 years between 2028 and 2038, Australia will host four total eclipses of the Sun.

| Year | Path | Maximum Duration of Totality |
| --- | --- | --- |
| 2028 | Northwest to southeast across continent | 5 minutes 10 seconds |
| 2030 | South to northeast across continent | 3 minutes 44 seconds |
| 2037 | West to east across the center of the continent | 3 minutes 58 seconds |
| 2038 | Northwest to southeast across continent | 2 minutes 18 seconds |

Add to that the eclipse of 2023 that clipped the westernmost tip of Australia and Australians will have enjoyed five total solar eclipses in 15 years.

For comparison, the United States is a little larger than Australia, yet while Australia is welcoming four total eclipses in the 10-year period from 2028 to 2038, the USA will receive 1, visible only from northwestern Alaska.

# November 14, 2031—Pacific Ocean

Spin your Earth globe until the middle of the Pacific Ocean is facing you. Now stand back, as if you are an astronaut on your way to the Moon, so that you see half of Planet Earth. And what you see is almost entirely ocean. North America, Asia, and Australia are barely visible at the limb of the Earth. The Pacific Ocean occupies almost your entire view. As it should. The Pacific Ocean covers about one-third of our planet. It is bigger than all the land masses on Earth put together.

No wonder then that the Pacific Ocean swallows up the entire path of totality of an occasional solar eclipse, with very little land touched, or none at all.

That's the story of the November 14, 2031 solar eclipse. It spends almost all its time on Earth paddling across the Pacific Ocean.

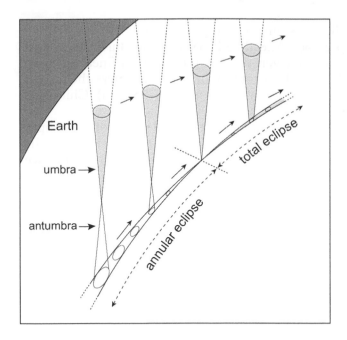

Earth

umbra →

antumbra →

total eclipse

annular eclipse

An occasional solar eclipse may start off annular but become total as the roundness of the Earth reaches up to intercept the shadow. The eclipse then returns to annular as the curvature of the Earth causes its surface to fall away from the shadow. These unusual eclipses are called hybrid or annular-total eclipses.

The 2031 eclipse is different in another way too. It's a hybrid eclipse. It starts off annular for about one-eighth of its journey, transforms into a total eclipse for three-quarters its flight, then reverts to annular for the final one-eighth of its travels.

This circumstance is rare—where the spherical shape of the Earth is enough to convert an eclipse from annular to total or total to annular while the eclipse is in progress. Only 4.8% of eclipses are hybrid—less than one in 20.

Because the spherical shape of the Earth brings a portion of the surface only a bit closer to the Moon, the Moon's apparent size changes only a little, so the total phase of a hybrid eclipse is never very long. In the case of the 2031 eclipse, maximum duration of totality is only 1 minute 8 seconds.

Because the Moon's shadow just barely touches Earth in a hybrid eclipse, the path of totality is never very wide, in this case only 24 miles (38 km). The average width of the path of totality in a total eclipse is about 100 miles (160 km).

The eclipse begins in the western Pacific Ocean in the middle of nowhere. The closest solid ground are islands like Wake, Guam, and Midway—none within 450 miles (750 km) of where the eclipse touches down. The eclipse continues its standoffish behavior along its entire course of 7,600 miles (12,200 km), avoiding every island and atoll as it speeds across the Pacific. Shy? To eclipse chasers, it might seem stingy.

It is only after the eclipse becomes annular again that it briefly touches land in Panama at the end of its visit to Earth. The 2031 eclipse spends its shore time on Panama's lightly populated Azuero Peninsula and, even then, avoids the region's two largest cities, with populations of 9,000 each.

With the 2031 hybrid eclipse, it's catch me if you can.

# 2032—No Total Eclipses of the Sun

# March 30, 2033—Alaska

The center of the shadow of the Moon collides with Earth in the Bering Sea between Alaska and Siberia, creating the total solar eclipse of March 30, 2033. The shadow falls far from the equator, so it strikes the Earth at a shallow angle and the shadow is elongated—greatly widening the path of totality.

When the shadow of a total eclipse falls close to the equator, the shadow on the ground is nearly round in shape and averages about 100 miles (160 km) wide. Near the poles, the stretched-out shadow of a total eclipse can be hundreds of miles across.

Alaska is the largest state in the USA. It is larger than Texas, California, and Montana, the next three largest states, put together. For Alaska in 2033, the zone of totality will average about 500 miles (800 km) wide, so wide that almost half of Alaska will experience totality.

In 2033, the path of totality is so wide that it includes some of Siberia, including Anadyr, the easternmost town in Russia, and then Uelen, the easternmost settlement in Russia.

### Nome

But the central line of the 2033 eclipse takes direct aim at Alaska, at the Seward Peninsula in northwestern Alaska. In the most recent Ice Age, when sea levels were lower, there was a wide strip of dry land connecting this peninsula with Asia. From Asia, across this Bering Land Bridge, came the first people to enter the Americas and settle there. That was 20,000 years ago or more.

On March 30, 2033, the central line of the eclipse comes ashore in Nome, Alaska, plowing right through the middle of downtown. The area around Nome has been inhabited by the Iñupiat People (formerly called Eskimos) for at least 5,000 years. These Native Americans are still more than half the population of Nome.

Nome was a small outpost until gold was discovered in the area in 1898, setting off the Nome Gold Rush that brought 20,000 people to town. The population has settled back to the current 3,700.

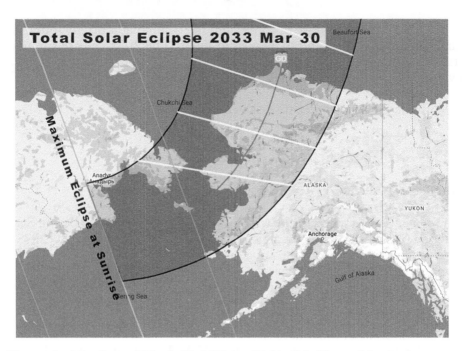

The total solar eclipse of March 30, 2033 is a frigid affair. The path begins in eastern Siberia, sweeps over the northwestern third of Alaska, and ends near the north pole. [Map by Fred Espenak, <http://www.EclipseWise.com>]

There was an outbreak of diphtheria in Nome in the winter of 1925. Five children died, 19 were sick, and an epidemic was imminent. The disease could be treated by antitoxin but how could the medicine get to them? Sea lanes were clogged by ice. Airplanes of the era could not cope with the cold. There were no roads. The only hope was dog sled. A relay of dog sled teams made the trip from near Fairbanks to Nome—674 miles (1,085 km) in 5⅓ days—and the sick children were saved.

In 1973, an annual dog sled race from Anchorage to Nome—938 miles (1,510 km)—was established, in part to honor the 1925 "Serum Run." The race is called the Iditarod. Iditarod means "distant place."

In 2033, Nome receives 2 minutes 30 seconds of totality with the Sun standing only 8° above the horizon. In late March, Nome has overcast or mostly cloudy days 62% of the time, with an average high temperature of 22 °F (−6 °C).

These days, you can reach Nome by plane or by ship. There are roads in Nome, but they do not connect to any other city. You can't get there by car. As the saying goes: No roads lead to Nome.

## Kotzebue

The 2033 eclipse ignores roads and speeds northward. As the central line crosses the Arctic Circle, the town of Kotzebue lies just 35 miles (60 km) to the northwest. Two-thirds of Kotzebue's 3,100 residents are Native Americans.

Kotzebue lies on a fishhook of land extending from the coast into the Arctic Ocean.[10] The town has roads but they don't reach any other towns. As with Nome, you can't get to Kotzebue by car. Instead, almost all travelers arrive in Kotzebue by plane. And what a landing it is. The one runway is short by airline standards for jets carrying 100 passengers or more—6,300 feet (1,920 meters) instead of the usual 10,000 feet (3,000 meters). The Kotzebue runway has water on all four sides, a lagoon on three sides and the Arctic Ocean on the fourth. It is not uncommon for a runway to begin and end at the water's edge. What's unusual about Kotzebue is the sides of the runway—there is almost no apron. The water laps up very close to the sides of the runway. If, in the Arctic weather, a plane skids off the side of the runway . . . Pilots flying into Kotzebue are very skillful.

A curious feature of Alaska is that even the tiniest of towns has an airport with a runway more than a mile in length. That's essential in a state like Alaska where there are very few long-distance roads and the only way to reach these towns is by air.

Another curious feature of Kotzebue is that the airport runway, as short as it is, is longer than the town is wide. That's true of Nome as well and all the airports along the 2033 eclipse path of totality.

The 2033 eclipse brings 2 minutes 31 seconds of totality to Kotzebue, with the Sun 9° above the horizon. At the end of March, the high temperature averages 15 °F (−9 °C) and the skies are cloudy or mostly cloudy 59% of the time.

In Kotzebue, each good weather day in the brief summer season sends a swarm of bush pilots into the air, carrying visitors into the vast backcountry for fishing, hunting, boating, or hiking. Some aircraft head northeast, carrying visitors to see some hard-to-reach national parks.

## Amenities

A hundred miles (160 km) from Kotzebue, the central line flies over the Kobuk River as it flows leisurely across a plain. A plain with sand dunes. In the Arctic. Some dunes are 100 feet (30 meters) high. The Great Kobuk Sand Dunes cover 25 square miles (65 sq. km.). Glaciers pulverized rock to sand and left it there when they melted. This is Kobuk Valley National Park.

The eclipse of 2033 brings 2 minutes 35 seconds of totality to the valley.

As a national park, Kobuk Valley poses special challenges to visitors. There are no roads to the Park. There are no roads inside the Park. And there are no accommodations. And no trails. Some charter pilots will land you on the dunes.

One hundred sixty miles (260 km) northeast of the dunes, yet still within the path of totality, is America's northernmost national park, Gates of the Arctic, with soaring mountains of the Brooks Range that create the headwaters of six wild rivers. Those peaks, stretching east to west, create a continental divide. Rivers on the south side drain into the Pacific Ocean. Rivers on the north side flow across the North Slope, past America's largest oil field, and empty into the Arctic Ocean.

Gates of the Arctic National Park and Preserve has no roads, no trails. It is a wilderness park. Hike or fly in. March 30, 2033 brings it 2 minutes 10 seconds of total eclipse.

## Total Solar Eclipses in the USA Through the 21st Century

The 21st century brings eight total solar eclipses to parts of the contiguous United States. The eclipses occur in 2017, 2024, 2044, 2045, 2052, 2078, 2079, and 2099. There are no total eclipses visible from Hawaii during the 21st century, but Alaska gets two: in 2033 and in 2097.

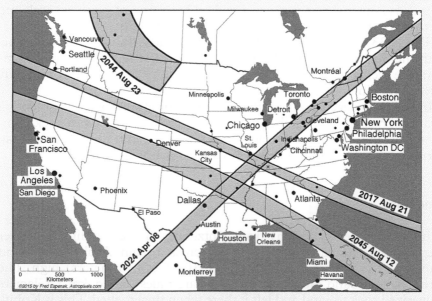

The path of every total solar eclipse through the continental USA from 2001 through 2050 is shown. They include the years 2017, 2024, 2044, and 2045. [Map and eclipse predictions by Fred Espenak, EclipseWise.com]

**The Top of the World**

The 2033 eclipse's final port of call before it departs from Alaska is Barrow. That is, it used to be called Barrow. In 2016, its citizens gave it the name Utqiaġvik (*oot-key-AHG-vick*), reflecting the Iñupiat heritage of most of them. The name means "place to gather wild roots."

It's difficult for things to grow in most of Alaska's soil. In addition to cold temperatures and the short growing season, the ground is permafrost in most of Alaska. The ground under Utqiaġvik is permanently frozen to a depth of 1,300 feet (400 meters)—deeper than the Empire State Building is tall. The result is that the landscape of Nome, Kotzebue, and Utqiaġvik is almost barren of trees. Permafrost keeps tree roots from burrowing deep enough underground to anchor the trees against the wind. Only specialized mosses, lichens, and ground-hugging bushes and flowers can tolerate the cold. This vegetation is called tundra. It has a beauty all its own.

The two maps that follow show the paths of these eclipses. Although the United States mainland experienced a drought of total eclipses from 1979 to 2017, the period from 2017 through 2099 is much richer.

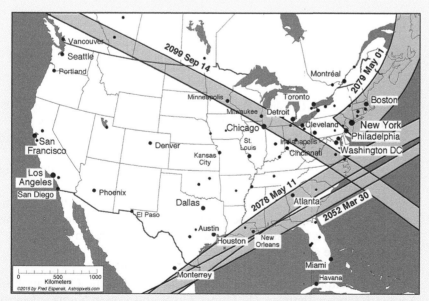

The path of every total solar eclipse through the continental USA from 2051 through 2100 is shown. They include the years 2052, 2078, 2079, and 2099. [Map and eclipse predictions by Fred Espenak, EclipseWise.com]

Utqiaġvik, at a latitude of 71° north, is the northernmost town in the United States. There you will find a Top of the World Hotel and a Top of the World Hardware Store. Utqiaġvik is also the cultural center of the Iñupiat People.

The Sun sets in Utqiaġvik on November 18 or 19 and doesn't rise again for 65 days—usually on January 22. In 2033, just two months after the Sun reappears in the sky for the Utqiaġvikians, they will experience the daytime disappearance of the Sun again, for 2 minutes 36 seconds. It is noontime in Utqiaġvik, so the Sun is its highest in the sky for the day: 11°.

Alas, in late March, Utqiaġvik is overcast or mostly cloudy 92% of the time. The average high temperature on March 30 is a toasty 0 °F (−18 °C).

A favorite activity of visitors to Utqiaġvik is swimming in the Arctic Ocean. Briefly. Once.

As the 2033 eclipse is about to leave Alaska, it reaches a maximum duration of totality, 2 minutes 37 seconds. The eclipse has run half its course on Earth. The remainder of its journey will be spent over polar waters close to the north pole.

The United States will wait 11 years—until August 29, 2044—for its next total eclipse of the Sun.

For the eclipse cycle known as saros 120, the year 2033 marks the 26th total eclipse of the Sun it has created. It is the last.

## NOTES AND REFERENCES

1.  Epigraph: Rebecca R. Joslin: *Chasing Eclipses: The Total Solar Eclipses of 1905, 1914, 1925* (Boston: Walton Advertising and Printing, 1929), pages 1–2.
2.  Counting the year 2023, Australia will have five total solar eclipses in 15 years.
3.  Jennifer Kingsley: "At This Arctic Science Base, Life Is Anything But Lonely," *National Geographic*, September 2019.
4.  The southern side of the 2027 eclipse path passes through the extreme northeastern corner of Sudan, but the central line of the eclipse does not. This portion of Sudan is almost uninhabited and it is not covered in this section.
5.  The same is true along the Suez Canal.
6.  Angkor Wat in Cambodia also claims that distinction.
7.  Mosalam Shaltout and Juan Antonio Belmonte: "On the Orientation of Ancient Egyptian Temples: (1) Upper Egypt and Lower Nubia," *Journal for the History of Astronomy*, volume 36, 2005, pages 273–298.
8.  Also to differentiate it from Cocos Island off Costa Rica in the Pacific Ocean.
9.  According to Husqvarna Urban Green Space Index.
10. This part of the Arctic Ocean is called the Chukchi Sea.

# A MOMENT OF TOTALITY

## The Extinction of Total Solar Eclipses

In 1695, Edmond Halley discovered that eclipses recorded in ancient history did not agree with calculations for the times or places of those eclipses. Starting with records of eclipses in his day and the observed motion of the Moon and Sun, he used Isaac Newton's new theory of universal gravitation (1687) to calculate when and where ancient eclipses should have occurred and then compared them with eclipses actually observed more than 2,000 years earlier. They did not match. Halley had great confidence in the theory of gravitation and resisted the temptation to conclude that the force of gravity was changing as time passed. Instead, he proposed that the length of a day on Earth must be slowly increasing. The Earth's rotation must be slowing down.

If the Earth's rotation had slowed down slightly, the Moon must have gained angular momentum to conserve the total angular momentum of the Earth-Moon system. This boost in angular momentum for the Moon would have caused it to spiral slowly outward from the Earth to a more distant orbit where it travels more slowly. If, 2,000 years earlier, the Earth had been spinning a little faster and the Moon had been a little closer and orbiting a little faster, then eclipse theory and observation would match. Scientists soon realized that Halley was right.

But what would cause the Earth's spin to slow? Tides. The gravitational attraction of the Moon is the principal cause of the ocean tides on Earth. As the shallow continental shelves (primarily in the Bering Sea) collide with high tides, the Earth's rotation is retarded. The slower spin of the Earth causes the Moon to edge farther from our planet.

From 1969 to 1972, the Apollo astronauts left a series of laser reflectors on the Moon's surface. Since then, scientists on Earth have been bouncing powerful lasers off these reflectors. By timing the round trip of each laser pulse, the Moon's distance can be measured to an accuracy of several inches. The Moon is receding from the Earth at the rate of about 1.5 inches (3.8 centimeters) a year. As the Moon recedes from Earth, its apparent disk becomes smaller. Total eclipses become rarer; annular eclipses more frequent. Total eclipses are moving toward extinction. When the Moon's mean distance from the Earth has increased by 14,550 miles (23,410 km), the Moon's apparent disk will be too small to cover the entire Sun, even when the Moon's elliptical orbit carries it closest to Earth. Total eclipses will no longer be possible.

How long will that take? With the Moon receding at 1.5 inches a year, the last total solar eclipse visible from the surface of the Earth will take place 620 million years from now. There is still time to catch one of these majestic events.

# 17

◄◦►

# Epilogue:
# Eclipses—Cosmic Perspective,
# Human Perspective

*No one should pass through life without seeing a
total solar eclipse.*

Leif Robinson (1999)[1]

The Moon's shadow has darkened space for 4½ billion years, since the Sun formed and began to shine and the planets and their moons formed and could not shine. From the newly formed Sun, light streamed outward in all directions. Here and there it illuminated a body of rock or gas or ice. The Sun's dominant light identified that world as one of its own children.

A little over 8 minutes of light-travel outbound from the Sun, a portion of the light encountered two small dark bodies and bathed the sunward half of each in brightness. The surrounding flood of unimpeded light sped on, leaving a long cone of darkness—a shadow—behind the planet and its one large moon.

Not long before—measured on a cosmic timescale—Earth and Moon had been a single body, but they were separate and utterly different now.

## A Cosmic Birth

The Sun, planets, moons, asteroids, and comets had all begun within a cloud of gas and dust, a cloud so large that there was material enough to make dozens or hundreds of solar systems. Where the density was great enough, fragments of the cloud began to condense by gravity. At the heart of each fragment, a star was coalescing. Near those stars-to-be, other bodies, too low in mass ever to reach starhood, also began to form. At first they grew by gentle collisions and adhesions, gathering up a grain of ice, a fleck of dust along their paths around the Sun until icy or rocky planetesimals had taken shape—the beginnings of planets. And still they

A composite of eight NASA GOES-7 weather satellite images captures the path of the Moon's shadow during the total solar eclipse of July 11, 1991. [Photo courtesy NASA, Dr. William Emery, and Timothy D. Kelley, University of Colorado]

gathered dust and small debris until they were so massive that gravity became their prime means of growth, gathering to them still more materials—and other planetesimals. The number of small bodies declined. The size of a few large bodies increased. The planets had been formed by convergence.

Convergence brought near disaster as well. Another planet-size body (the size of Mars today) wandered across the path of a body that would become the Earth. No living thing witnessed the collision. No living thing could have survived the collision that obliterated the smaller world and nearly shattered the Earth. The Earth recoiled from the impact by spewing molten fragments of the intruder and its own crust and mantle outward. Some escaped from Earth; others rained down from the skies, pelting the surface in a rock storm of unimaginable proportions. But many of the fragments, caught in the Earth's gravity, stayed aloft, orbiting the Earth as the Earth orbited the Sun. Quickly, in a century or less, the fragments joined together by collisions and accretion to form a new world circling the first. That new world was the Moon.[2]

From convergence now came divergence. The Earth was 81 times more massive than the Moon. That mass allowed the Earth to hold an atmosphere by gravity, while the Moon could not.

The eons passed. Life arose on Earth and covered the planet. Plants and animals responded to the tides raised by the Moon. The lunar tides slowed the Earth and caused the Moon to spiral slowly outward, diverging ever farther from the Earth in distance and ever further from the Earth in environment as well.

The lifeless Moon withdrew until today its shadow can just barely reach the Earth. As shadows in the universe go, this one is of no great size: a cone of darkness extending at most only 236,000 miles (379,900 km) in length before dwindling to a point. It is long enough to touch the Earth only occasionally and very briefly and with a single narrow stroke.

The black insubstantial cone reaches out, but for most of the time there is nothing to touch. The shadow sweeps on through space unseen, unnoticed.

## Contact

Yet now ahead lies the Earth. It is a special day. Suddenly the Moon's shadow becomes visible as it collides with a world of rock and water and air. The shadow swoops in from the heavens, silently darkening the sky where it alights upon the Earth and begins its ceaseless rush across the planet's surface.

Awesome totality—August 1, 2008 from China. [Canon G9, auto-exposure, ISO 200. © 2008 Fred Espenak]

On these occasions, the people of Earth have gathered and still gather scientific knowledge from the Sun. And more. For long before we drew information from total eclipses, we stared in wonder at them. And long after all knowledge from eclipses has been gleaned, people will still travel to the ends of the Earth to treasure their majesty and beauty.

So compelling is a solar eclipse that when at last the Moon has drifted too far away to touch the Earth with darkness, the beings of that era may use their technology to gently, gradually halt the Moon's retreat—and then reverse it, bringing it closer once again—so that they too will see the sight that their books and visual records can only hint at: a total eclipse of the Sun.

## NOTES AND REFERENCES

1. Epigraph: Leif Robinson: Foreword to *Totality: Eclipses of the Sun*, 2nd edition, by Mark Littmann, Fred Espenak, and Ken Willcox (New York: Oxford University Press, 1999).
2. The leading theory of the Moon's formation involves a collision between a large planetesimal and the proto-Earth.

# Appendix A

# Maps for Every Solar Eclipse 2024–2045

*"I doubt if the effect of witnessing a total eclipse ever quite passes away. The impression is singularly vivid and quieting for days, and can never be wholly lost."[1]*

Mabel Loomis Todd (1900)

Between the years 2024 and 2045, a 22-year period, the Moon will eclipse the Sun 48 times. This interval samples at least one eclipse from every saros series currently producing eclipses. The eclipses during this period fall into the following categories:

| Eclipse Type | Number of Eclipses 2024–2045 | Number of Eclipses 2000 BCE–3000 CE |
|---|---|---|
| Total | 16 = 33.3% | 3173 = 26.7 % |
| Annular | 17 = 34.4% | 3956 = 33.2 % |
| Hybrid | 1 = 2.1% | 569 = 4.8 % |
| Partial | 14 = 29.2% | 4200 = 35.3 % |

The following pages offer 48 global maps, one for each eclipse.[2] The odd saddle-shaped zone in each map shows the region where the partial eclipse is visible. The magnitude of each eclipse (maximum fraction of the Sun's diameter covered) is shown in increments of 25%, 50%, and 75%. This allows you to quickly estimate the magnitude for any location within the eclipse path. For central eclipses, the path of either totality or annularity is plotted.

The Moon's penumbral shadow typically produces a zone of partial eclipse (during both partial and central eclipses) that covers 25% to 50% of the daylight hemisphere of the Earth. In comparison, the Moon's

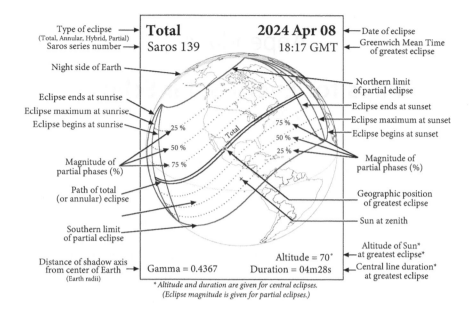

Type of eclipse → **Total**                **2024 Apr 08** ← Date of eclipse
(Total, Annular, Hybrid, Partial)                                    Greenwich Mean Time
Saros series number → Saros 139            18:17 GMT  ← of greatest eclipse

Night side of Earth

Northern limit
of partial eclipse

Eclipse ends at sunrise
Eclipse maximum at sunrise                 Eclipse ends at sunset
Eclipse begins at sunrise                   Eclipse maximum at sunset
25 %                75 %                    Eclipse begins at sunset
                    50 %
                    25 %
50 %
Magnitude of                               Magnitude of
partial phases (%)    75 %                 partial phases (%)

Path of total
(or annular) eclipse                        Geographic position
of greatest eclipse

Sun at zenith

Southern limit
of partial eclipse
                                            Altitude of Sun*
Distance of shadow axis                     at greatest eclipse*
from center of Earth → Gamma = 0.4367    Altitude = 70°   Central line duration*
(Earth radii)                     Duration = 04m28s  at greatest eclipse

*Altitude and duration are given for central eclipses.*
*(Eclipse magnitude is given for partial eclipses.)*

umbral shadow (total eclipse) or antumbral shadow (annular eclipse) is much smaller: its path covers less than 1% of the Earth's surface.

Additional information on each map can be identified using the key above.

On pages 315–317 are three maps: North and South America, Europe and Africa, and Asia and Australia. These detailed maps show the paths of every central solar eclipse (total, annular, and hybrid) from 2024 through 2045. The maximum duration of totality or annularity as well as a list of all countries within each central eclipse path can be found in Appendix B, which covers eclipses over the same period.

Use *Totality: The Great North American Eclipse of 2024* to plan your own voyage into the Moon's shadow. Remember, words and pictures can never fully convey the wonder of a total eclipse of the Sun: to stand in the path of totality, in the light of the corona. You must see one for yourself—or two, or . . .

Hope to meet you there.

| **Total** | **2024 Apr 08** |
|---|---|
| Saros 139 | 18:17 GMT |

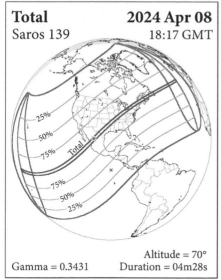

| Gamma = 0.3431 | Altitude = 70° |
|---|---|
| | Duration = 04m28s |

| **Annular** | **2024 Oct 02** |
|---|---|
| Saros 144 | 18:45 GMT |

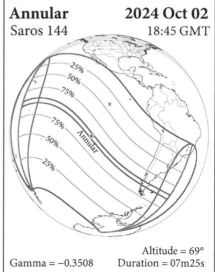

| Gamma = −0.3508 | Altitude = 69° |
|---|---|
| | Duration = 07m25s |

| **Partial** | **2025 Mar 29** |
|---|---|
| Saros 149 | 10:47 GMT |

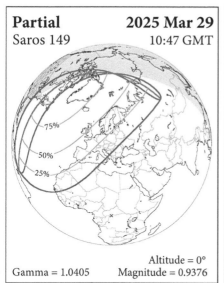

| Gamma = 1.0405 | Altitude = 0° |
|---|---|
| | Magnitude = 0.9376 |

| **Partial** | **2025 Sep 21** |
|---|---|
| Saros 154 | 19:42 GMT |

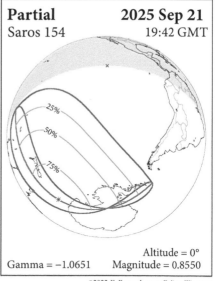

| Gamma = −1.0651 | Altitude = 0° |
|---|---|
| | Magnitude = 0.8550 |

| Annular | 2026 Feb 17 |
|---|---|
| Saros 121 | 12:12 GMT |

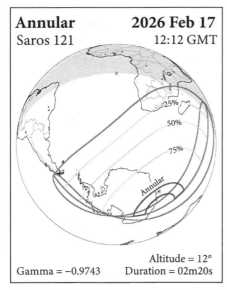

| Gamma = −0.9743 | Altitude = 12°<br>Duration = 02m20s |

| Total | 2026 Aug 12 |
|---|---|
| Saros 126 | 17:46 GMT |

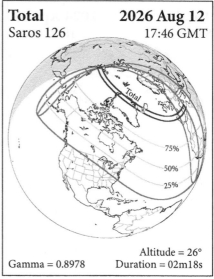

| Gamma = 0.8978 | Altitude = 26°<br>Duration = 02m18s |

| Annular | 2027 Feb 06 |
|---|---|
| Saros 131 | 16:00 GMT |

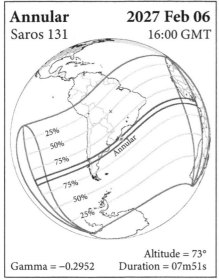

| Gamma = −0.2952 | Altitude = 73°<br>Duration = 07m51s |

| Total | 2027 Aug 02 |
|---|---|
| Saros 136 | 10:07 GMT |

| Gamma = 0.1421 | Altitude = 82°<br>Duration = 06m23s |

| | |
|---|---|
| **Annular** **2028 Jan 26** | **Total** **2028 Jan 22** |
| Saros 141 15:08 GMT | Saros 146 02:55 GMT |

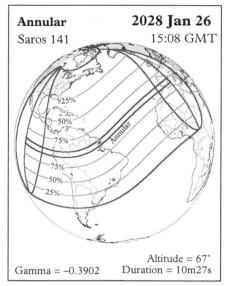

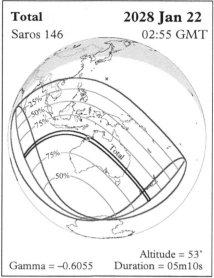

| | |
|---|---|
| Altitude = 67° | Altitude = 53° |
| Gamma = −0.3902 Duration = 10m27s | Gamma = −0.6055 Duration = 05m10s |

| | |
|---|---|
| **Partial** **2029 Jan 14** | **Partial** **2029 Jun 12** |
| Saros 151 17:13 GMT | Saros 118 04:05 GMT |

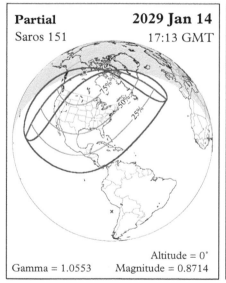

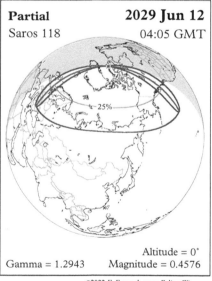

| | |
|---|---|
| Altitude = 0° | Altitude = 0° |
| Gamma = 1.0553 Magnitude = 0.8714 | Gamma = 1.2943 Magnitude = 0.4576 |

**Partial**                           **2029 Jul 11**

Saros 156                           15:36 GMT

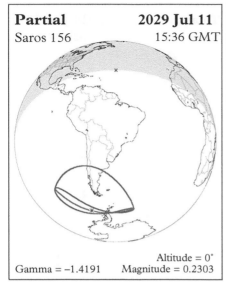

Altitude = 0°

Gamma = −1.4191     Magnitude = 0.2303

**Partial**                           **2029 Dec 05**

Saros 123                           15:03 GMT

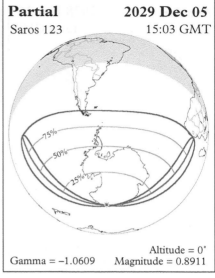

Altitude = 0°

Gamma = −1.0609     Magnitude = 0.8911

**Annular**                          **2030 Jun 01**

Saros 128                           06:28 GMT

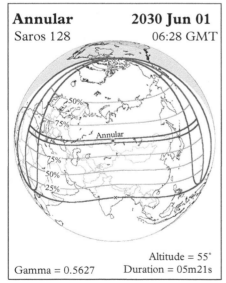

Altitude = 55°

Gamma = 0.5627      Duration = 05m21s

**Total**                             **2030 Nov 25**

Saros 133                           06:50 GMT

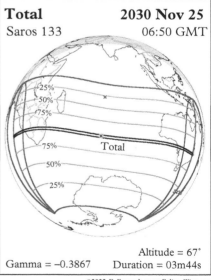

Altitude = 67°

Gamma = −0.3867     Duration = 03m44s

| Annular | 2031 May 21 |
|---|---|
| Saros 138 | 07:15 GMT |

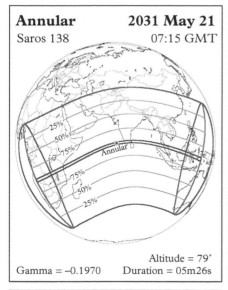

| | Altitude = 79° |
|---|---|
| Gamma = −0.1970 | Duration = 05m26s |

| Hybrid | 2031 Nov 14 |
|---|---|
| Saros 143 | 21:06 GMT |

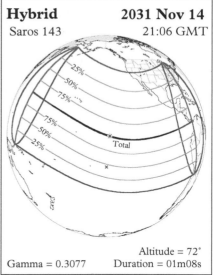

| | Altitude = 72° |
|---|---|
| Gamma = 0.3077 | Duration = 01m08s |

| Annular | 2032 May 09 |
|---|---|
| Saros 148 | 13:25 GMT |

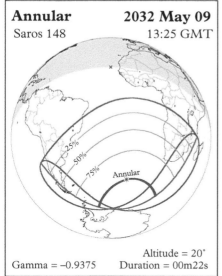

| | Altitude = 20° |
|---|---|
| Gamma = −0.9375 | Duration = 00m22s |

| Partial | 2032 Nov 03 |
|---|---|
| Saros 153 | 05:33 GMT |

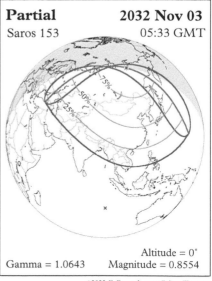

| | Altitude = 0° |
|---|---|
| Gamma = 1.0643 | Magnitude = 0.8554 |

©2022 F. Espenak, www.EclipseWise.com

| **Total** | **2033 May 30** |
|---|---|
| Saros 120 | 18:01 GMT |

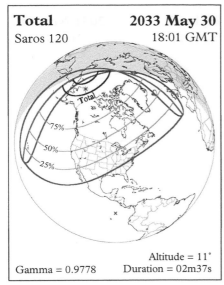

| | Altitude = 11° |
|---|---|
| Gamma = 0.9778 | Duration = 02m37s |

| **Partial** | **2033 Sep 23** |
|---|---|
| Saros 125 | 13:53 GMT |

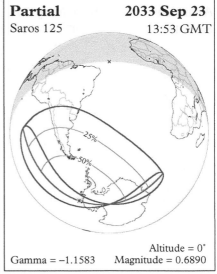

| | Altitude = 0° |
|---|---|
| Gamma = −1.1583 | Magnitude = 0.6890 |

| **Total** | **2034 Mar 20** |
|---|---|
| Saros 130 | 10:18 GMT |

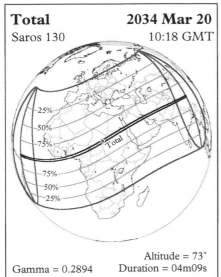

| | Altitude = 73° |
|---|---|
| Gamma = 0.2894 | Duration = 04m09s |

| **Annular** | **2034 Sep 12** |
|---|---|
| Saros 135 | 16:18 GMT |

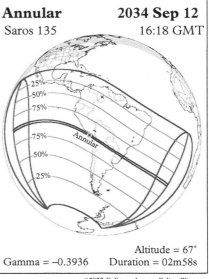

| | Altitude = 67° |
|---|---|
| Gamma = −0.3936 | Duration = 02m58s |

**Annular**                    **2035 May 09**
Saros 140                        23:05 GMT

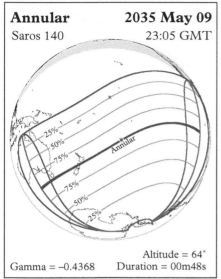

Altitude = 64°
Gamma = −0.4368     Duration = 00m48s

**Total**                      **2035 Sep 02**
Saros 145                        01:56 GMT

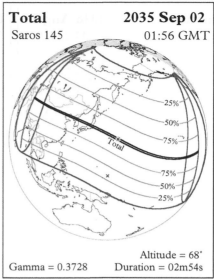

Altitude = 68°
Gamma = 0.3728     Duration = 02m54s

**Partial**                    **2036 Feb 27**
Saros 150                        04:46 GMT

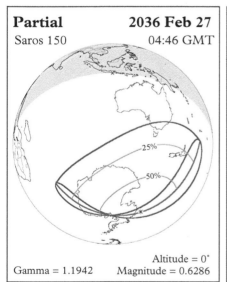

Altitude = 0°
Gamma = 1.1942     Magnitude = 0.6286

**Partial**                    **2036 Jul 23**
Saros 117                        10:31 GMT

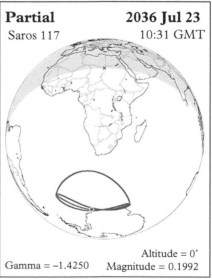

Altitude = 0°
Gamma = −1.4250     Magnitude = 0.1992

©2022 F. Espenak, www.EclipseWise.com

**Partial**       **2036 Aug 21**
Saros 155       17:24 GMT

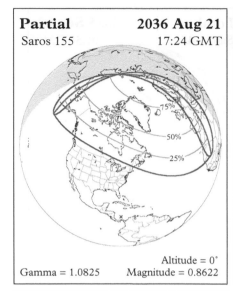

Gamma = 1.0825    Altitude = 0°
      Magnitude = 0.8622

**Partial**       **2037 Jan 16**
Saros 122       09:48 GMT

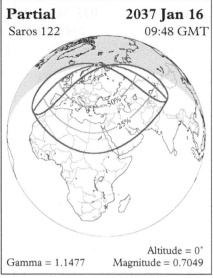

Gamma = 1.1477    Altitude = 0°
      Magnitude = 0.7049

**Total**       **2037 Jul 13**
Saros 127       02:39 GMT

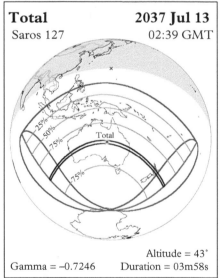

Gamma = –0.7246    Altitude = 43°
      Duration = 03m58s

**Annular**       **2038 Jan 05**
Saros 132       13:46 GMT

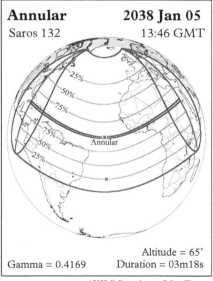

Gamma = 0.4169    Altitude = 65°
      Duration = 03m18s

**Annular**                    **2038 Jul 02**
Saros 137                       13:32 GMT

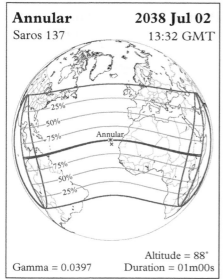

Gamma = 0.0397 | Altitude = 88°
Duration = 01m00s

**Total**                       **2038 Dec 26**
Saros 142                       00:59 GMT

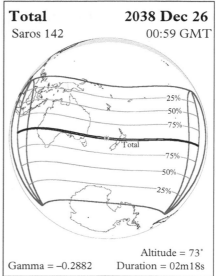

Gamma = −0.2882 | Altitude = 73°
Duration = 02m18s

**Annular**                    **2039 Jun 21**
Saros 147                       17:12 GMT

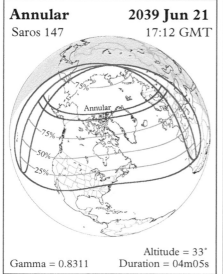

Gamma = 0.8311 | Altitude = 33°
Duration = 04m05s

**Total**                       **2039 Dec 15**
Saros 152                       16:22 GMT

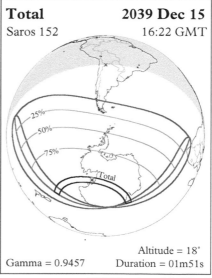

Gamma = 0.9457 | Altitude = 18°
Duration = 01m51s

**Partial**                        **2040 May 11**
Saros 119                          03:42 GMT

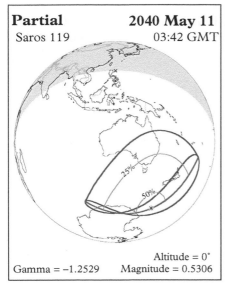

Gamma = −1.2529    Altitude = 0°
                   Magnitude = 0.5306

**Partial**                        **2040 Nov 04**
Saros 124                          19:08 GMT

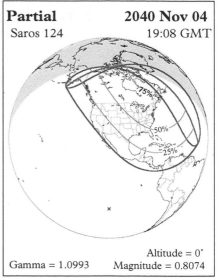

Gamma = 1.0993     Altitude = 0°
                   Magnitude = 0.8074

**Total**                          **2041 Apr 30**
Saros 129                          11:51 GMT

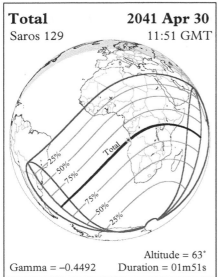

Gamma = −0.4492    Altitude = 63°
                   Duration = 01m51s

**Annular**                        **2041 Oct 25**
Saros 134                          01:35 GMT

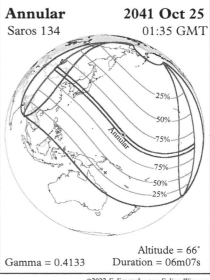

Gamma = 0.4133     Altitude = 66°
                   Duration = 06m07s

©2022 F. Espenak, www.EclipseWise.com

| **Total** | **2042 Apr 20** |
|---|---|
| Saros 139 | 02:16 GMT |

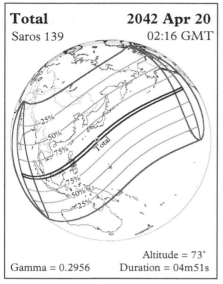

| | Altitude = 73° |
|---|---|
| Gamma = 0.2956 | Duration = 04m51s |

| **Annular** | **2042 Oct 14** |
|---|---|
| Saros 144 | 01:59 GMT |

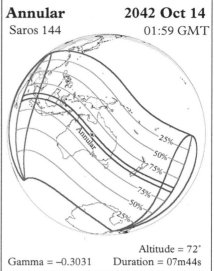

| | Altitude = 72° |
|---|---|
| Gamma = −0.3031 | Duration = 07m44s |

| **Total** | **2043 Apr 09** |
|---|---|
| Saros 149 | 18:56 GMT |

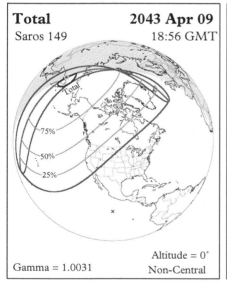

| | Altitude = 0° |
|---|---|
| Gamma = 1.0031 | Non-Central |

| **Annular** | **2043 Oct 03** |
|---|---|
| Saros 154 | 03:00 GMT |

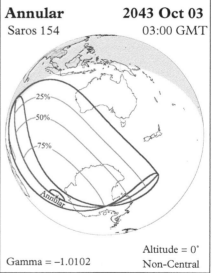

| | Altitude = 0° |
|---|---|
| Gamma = −1.0102 | Non-Central |

**Annular**            **2044 Feb 28**
Saros 121              20:23 GMT

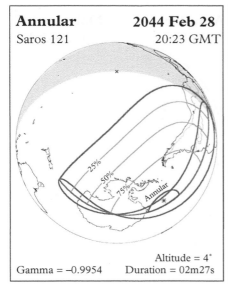

Gamma = −0.9954    Altitude = 4°
                   Duration = 02m27s

**Total**              **2044 Aug 23**
Saros 126              01:16 GMT

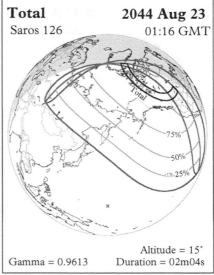

Gamma = 0.9613     Altitude = 15°
                   Duration = 02m04s

**Annular**            **2045 Feb 16**
Saros 131              23:55 GMT

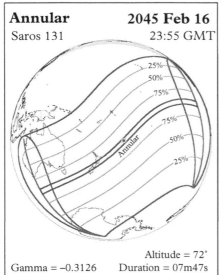

Gamma = −0.3126    Altitude = 72°
                   Duration = 07m47s

**Total**              **2045 Aug 12**
Saros 136              17:41 GMT

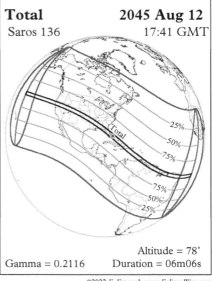

Gamma = 0.2116     Altitude = 78°
                   Duration = 06m06s

## Central Solar Eclipses for North & South America: 2017–2045

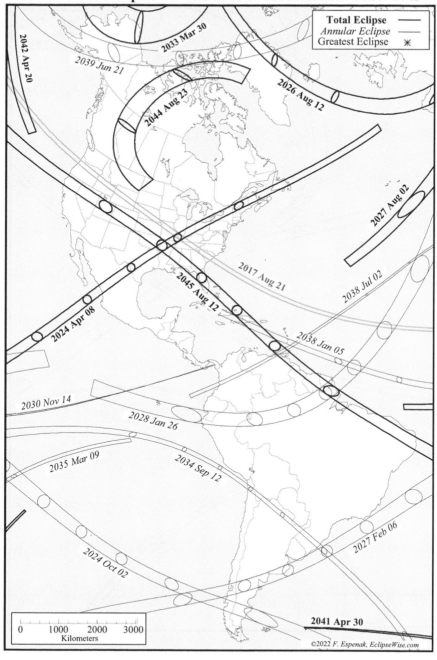

**Total Eclipse** ⎯⎯⎯
*Annular Eclipse* ⎯⎯⎯
Greatest Eclipse ✳

2042 Apr 20

2039 Jun 21

2033 Mar 30

2044 Aug 23

2026 Aug 12

2027 Aug 02

2017 Aug 21

2038 Jul 02

2045 Aug 12

2038 Jan 05

2024 Apr 08

2030 Nov 14

2028 Jan 26

2035 Mar 09

2034 Sep 12

2027 Feb 06

2024 Oct 02

0    1000    2000    3000
Kilometers

2041 Apr 30

©2022 F. Espenak, EclipseWise.com

## Central Solar Eclipses for Europe & Africa: 2024–2045

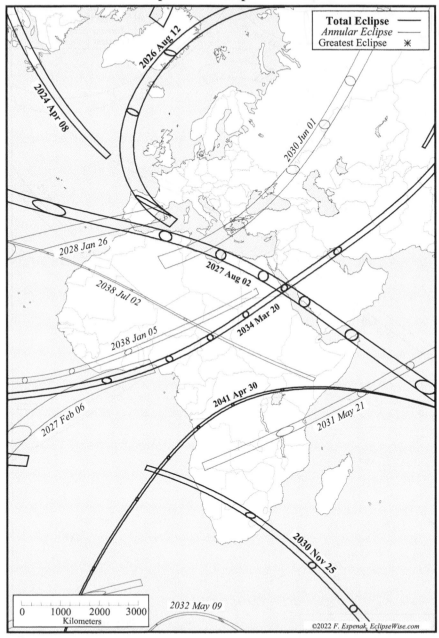

## Central Solar Eclipses for Asia & Australia: 2017–2045

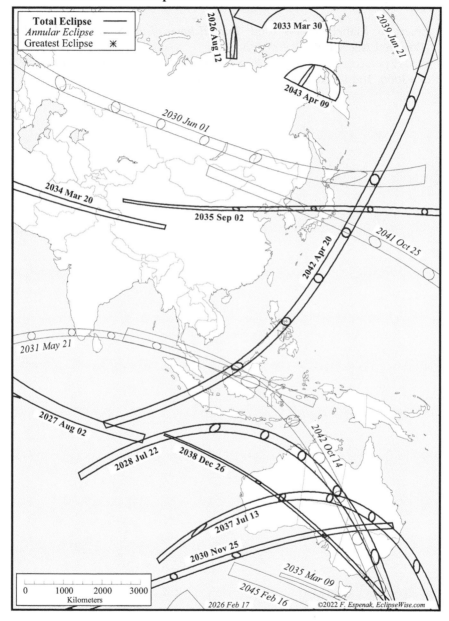

## NOTES AND REFERENCES

1    Epigraph: Mabel Loomis Todd: *Total Eclipses of the Sun*, revised edition (Boston: Little, Brown, 1900), page 25.
2    Fred Espenak and Jean Meeus: *Five Millennium Catalog of Solar Eclipses:–1999 to +3000*, 2nd Edition (Portal, AZ: Astropixels Publishing, 2021).

# Appendix B

# Total, Annular, and Hybrid Eclipses: 2024–2070

During the 47-year period 2024–2070, there are 106 eclipses of the Sun—partial, annular, hybrid (annular-total), and total.[1] Here's how they break down:

| Eclipse Type | Number of Eclipses |
| --- | --- |
| Partial | 38 = 35.8% |
| Annular | 33 = 30.6% |
| Total | 31 = 29.2% |
| Hybrid | 4 = 3.8% |

The table below gives the basic details for the 68 total, annular, and hybrid eclipses occurring during this period. When an eclipse path crosses the International Date Line, it can be seen on one of two dates depending on the observer's geographic position.

| Date | Maximum Duration[2] | Eclipse Type | Geographic Region of Visibility |
| --- | --- | --- | --- |
| 2024 Apr 8 | 4m28s | Total | Pacific Ocean, Mexico, USA (Texas, Oklahoma, Arkansas, Missouri, Kentucky, Illinois, Indiana, Ohio, Pennsylvania, New York, Vermont, New Hampshire, Maine), southeastern Canada, Atlantic Ocean |
| 2024 Oct 02 | 7m25s | Annular | South Pacific Ocean, Easter Island, Chile, Argentina, Atlantic Ocean |
| 2026 Feb 17 | 2m20s | Annular | Antarctica |

| Date | Maximum Duration[2] | Eclipse Type | Geographic Region of Visibility |
|---|---|---|---|
| 2026 Aug 12 | 2m18s | Total | Arctic Ocean, Greenland, Iceland, Atlantic Ocean, Spain |
| *2027 Feb 06* | *7m51s* | *Annular* | *Pacific Ocean, Chile, Argentina, Uruguay, Atlantic Ocean, Ivory Coast, Ghana, Togo, Benin, Nigeria* |
| 2027 Aug 02 | 6m23s | Total | Atlantic Ocean, Morocco, Spain, Algeria, Tunisia, Libya, Egypt, Saudi Arabia, Yemen, Somalia, Indian Ocean |
| *2028 Jan 26* | *10m27s* | *Annular* | *Pacific Ocean, Galapagos, Ecuador, Peru, Brazil, French Guiana, Portugal, Spain* |
| 2028 Jul 22 | 5m10s | Total | South Indian Ocean, Australia, New Zealand |
| *2030 Jun 01* | *5m21s* | *Annular* | *Algeria, Tunisia, Libya, Greece, Bulgaria, Turkey, Ukraine, Russia, Kazakhstan, northern China, Japan* |
| 2030 Nov 25 | 3m44s | Total | Namibia, Botswana, South Africa, Lesotho, Indian Ocean, Australia |
| *2031 May 21* | *5m26s* | *Annular* | *Angola, Zambia, DR Congo, Malawi, Tanzania, southern India, Sri Lanka, Thailand, Malaysia, Indonesia* |
| *2031 Nov 14/15* | *1m08s* | *Hybrid* | *Pacific Ocean (total), Panama (annular)* |
| *2032 May 09* | *0m22s* | *Annular* | *South Atlantic Ocean* |
| 2033 Mar 30 | 2m37s | Total | Russia (Siberia) USA (Alaska), Arctic Ocean |
| 2034 Mar 20 | 4m09s | Total | Atlantic Ocean, Nigeria, Cameroon, Chad, Sudan, Egypt, Saudi Arabia, Kuwait, Iran, Afghanistan, Pakistan, India, China |
| *2034 Sep 12* | *2m58s* | *Annular* | *Pacific, Chile, Bolivia, Argentina, Paraguay, Brazil, Atlantic* |
| *2035 Mar 09/10* | *0m48s* | *Annular* | *New Zealand, South Pacific Ocean* |
| 2035 Sep 02 | 2m54s | Total | China, North Korea, Japan, Pacific Ocean |
| 2037 Jul 13 | 3m58s | Total | Indian Ocean, Australia, New Zealand, South Pacific Ocean |
| *2038 Jan 05* | *3m19s* | *Annular* | *Cuba, Haiti, Dominican Republic, St. Lucia, St. Vincent, Barbados, Atlantic Ocean, Liberia, Ivory Coast, Ghana, Togo, Benin, Niger, Libya, Chad, Libya, Sudan, Egypt* |

| Date | Maximum Duration[2] | Eclipse Type | Geographic Region of Visibility |
|---|---|---|---|
| *2038 Jul 02* | *1m00s* | *Annular* | *Pacific, Colombia, Venezuela, Atlantic Ocean, Western Sahara, Mauritania, Mali, Algeria, Niger, Chad, Sudan, South Sudan, Ethiopia, Kenya* |
| 2038 Dec 25/26 | 2m18s | Total | Indian Ocean, Australia, New Zealand, Pacific Ocean |
| *2039 Jun 21* | *4m05s* | *Annular* | *Pacific, USA (Alaska), Canada, Greenland, Arctic Ocean, Greenland, Atlantic, Norway, Sweden, Finland, Estonia, Latvia, Lithuania, Belarus, Russia* |
| 2039 Dec 15 | 1m51s | Total | Antarctica |
| 2041 Apr 30 | 1m51s | Total | South Atlantic, Angola, DR Congo, Uganda, Kenya, Somalia, Indian Ocean |
| *2041 Oct 24/25* | *6m07s* | *Annular* | *Mongolia, China, North Korea, Japan, Pacific Ocean, Kiribati* |
| 2042 Apr 19/20 | 4m51s | Total | Indonesia, Malaysia, Philippines, Pacific Ocean |
| *2042 Oct 14* | *7m44s* | *Annular* | *Bay of Bengal, Thailand, Malaysia, Indonesia, Australia, New Zealand, Pacific Ocean* |
| 2043 Apr 09 | — | Total | Northeastern Russia/Siberia (shadow cone only grazes Earth) |
| *2043 Oct 03* | *—* | *Annular* | *South Indian Ocean (shadow cone only grazes Earth)* |
| *2044 Feb 28* | *2m27s* | *Annular* | *South Atlantic Ocean* |
| 2044 Aug 23 | 2m04s | Total | Greenland, Canada, USA (Montana, North Dakota) |
| *2045 Feb 16/17* | *7m47s* | *Annular* | *New Zealand, Cook Islands, Pacific Ocean* |
| 2045 Aug 12 | 6m06s | Total | Pacific Ocean, USA (California, Nevada, Utah, Colorado, Kansas, Oklahoma, Texas, Arkansas, Louisiana, Mississippi, Alabama, Georgia, Florida), Bahamas, Haiti, Dominican Republic, Trinidad & Tobago, Venezuela, Guyana, Suriname, French Guiana, Brazil, Atlantic Ocean |
| *2046 Feb 05/06* | *9m42s* | *Annular* | *Indonesia, Papua New Guinea, Pacific Ocean, Howland & Baker Islands, USA (Hawaii, California, Oregon, Nevada, Idaho)* |

| Date | Maximum Duration[2] | Eclipse Type | Geographic Region of Visibility |
|---|---|---|---|
| 2046 Aug 02 | 4m51s | Total | Brazil, Atlantic Ocean, Angola, Namibia, Botswana, South Africa, Swaziland, Indian Ocean, Kerguelen |
| *2048 Jun 11* | *4m58s* | *Annular* | *USA (Colorado, Nebraska, Kansas, Iowa, Missouri, Minnesota, Wisconsin, Illinois, Michigan), Canada, Greenland, Iceland, Atlantic Ocean, Norway, Sweden, Estonia, Latvia, Lithuania, Belarus, Ukraine, Russia, Turkmenistan, Uzbekistan, Afghanistan* |
| 2048 Dec 05 | 3m28s | Total | South Pacific Ocean, Chile, Argentina, Atlantic Ocean, Tristan da Cunha, Namibia, Botswana |
| *2049 May 31* | *4m45s* | *Annular* | *Pacific Ocean, Peru, Ecuador, Colombia, Brazil, Venezuela, Guyana, Suriname, Atlantic Ocean, Cape Verde, Senegal, The Gambia, Mali, Guinea, Burkina Faso, Ivory Coast, Ghana, Togo, Benin, Nigeria, Cameroon, Congo, DR Congo* |
| *2049 Nov 25* | *0m38s* | *Hybrid* | *Total for part of Indian Ocean and parts of Indonesia; annular for Saudi Arabia, Yemen, part of Indian Ocean, eastern Indonesia, Pacific Ocean* |
| *2050 May 20/21* | *0m21s* | *Hybrid* | *South Pacific Ocean* |
| 2052 Mar 30 | 04m08s | Total | Pacific Ocean, Kiribati, Mexico, USA (Texas, Louisiana, Alabama, Georgia, Florida, South Carolina), Atlantic Ocean |
| *2052 Sep 22/23* | *2m51s* | *Annular* | *Indonesia, East Timor, Australia, Pacific Ocean* |
| *2053 Mar 20* | *0m49s* | *Annular* | *Indian Ocean, Indonesia, Papua New Guinea* |
| 2053 Sep 12 | 3m04s | Total | Atlantic Ocean, Spain, Morocco, Algeria, Tunisia, Libya, Egypt, Saudi Arabia, Yemen, Indian Ocean, Indonesia (Sumatra) |
| 2055 Jul 24 | 3m17s | Total | South Atlantic Ocean, South Africa, south Indian Ocean |
| *2056 Jan 16/17* | *2m52s* | *Annular* | *Pacific Ocean, Mexico, USA (Texas)* |

| Date | Maximum Duration[2] | Eclipse Type | Geographic Region of Visibility |
|---|---|---|---|
| *2056 Jul 12* | *1m26s* | *Annular* | *Pacific Ocean, Colombia, Ecuador, Peru, Brazil* |
| 2057 Jan 05 | 2m29s | Total | South Atlantic Ocean, south Indian Ocean (no land) |
| *2057 Jul 01/02* | *4m22s* | *Annular* | *China, Mongolia, Russia (Siberia), Canada, USA (Alaska, Minnesota, Michigan)* |
| 2057 Dec 26 | 1m50s | Total | Antarctica |
| 2059 May 11 | 2m23s | Total | Pacific Ocean, Ecuador, Peru, Colombia, Brazil |
| *2059 Nov 05* | *7m00s* | *Annular* | *Atlantic Ocean, France, Spain, Mediterranean Sea, Sardinia, Sicily, Libya, Egypt, Sudan, Eritrea, Ethiopia, Somalia, Indian Ocean, Maldives, Indonesia (Sumatra)* |
| 2060 Apr 30 | 5m15s | Total | Brazil, Atlantic Ocean, Ivory Coast, Ghana, Togo, Benin, Burkina Faso, Niger, Nigeria, Chad, Libya, Egypt, Cyprus, Turkey, Syria, Armenia, Iran, Azerbaijan, Turkmenistan, Kazakhstan, Uzbekistan, Turkmenistan, Kyrgyzstan, China |
| *2060 Oct 24* | *8m06s* | *Annular* | *Atlantic Ocean, Guinea, Sierra Leone, Liberia, Ivory Coast, Angola, Namibia, Botswana, South Africa, Pacific Ocean* |
| 2061 Apr 20 | 02m37s | Total | Ukraine, Russia, Kazakhstan, Arctic Ocean, Svalbard (Norway) |
| *2061 Oct 13* | *03m41s* | *Annular* | *Chile, Argentina, Falklands, Antarctica* |
| *2063 Feb 28* | *07m41s* | *Annular* | *Indian Ocean, Indonesia (Sumatra), Malaysia, Singapore, Brunei, Philippines* |
| 2063 Aug 24 | 05m49s | Total | China, Mongolia, Russia (Siberia), Japan, Pacific Ocean |
| *2064 Feb 17* | *08m56s* | *Annular* | *Congo, Angola, DR Congo, Zambia, Tanzania, Seychelles, Indian Ocean, India, Nepal, Bangladesh, Bhutan, China* |

| Date | Maximum Duration[2] | Eclipse Type | Geographic Region of Visibility |
|---|---|---|---|
| 2064 Aug 12 | 04m28s | Total | Pacific Ocean, Chile, Argentina, Atlantic Ocean |
| *2066 Jun 22* | *04m40s* | *Annular* | *Russia (Siberia), USA (Alaska), Canada, Atlantic Ocean* |
| 2066 Dec 17 | 03m14s | Total | Indian Ocean, Australia, New Zealand, South Pacific |
| *2067 Jun 11* | *04m05s* | *Annular* | *Pacific Ocean, Kiribati, Ecuador, Peru* |
| *2067 Dec 06* | *00m08s* | *Hybrid* | *Guatemala. Belize, Honduras, Nicaragua, Colombia, Venezuela, Guyana, Brazil, Atlantic Ocean, Nigeria, Cameroon, Chad, Sudan* |
| 2068 May 31 | 01m06s | Total | Indian Ocean, Australia, New Zealand |
| 2070 Apr 11 | 04m04s | Total | Sri Lanka, Bay of Bengal, Myanmar, Thailand, Cambodia, Laos, Vietnam, Pacific Ocean |
| *2070 Oct 04* | *02m44s* | *Annular* | *Atlantic Ocean, Angola, Zambia, Zimbabwe, Mozambique, Madagascar, Indian Ocean* |

## NOTES AND REFERENCES

1   Based on Fred Espenak: *Thousand Year Canon of Solar Eclipses: 1501–2500* (Portal, Arizona, Astropixels Publishing, 2014; http://astropixels.com/pubs/).
2   Maximum duration of totality or annularity as seen from the central line in minutes and seconds.

# Appendix C

# Total Eclipses in the United States: 1900–2100

### Listing based on current boundaries of the lower 48 states.[1]

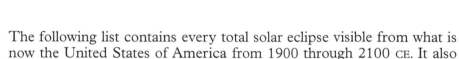

The following list contains every total solar eclipse visible from what is now the United States of America from 1900 through 2100 CE. It also includes hybrid eclipses if any section of the path crossing the USA is total. The path descriptions list every state inside the path of each eclipse along with other countries and geographic areas in the path.

| Date | Maximum Eclipse | Locations in the Path of Totality |
| --- | --- | --- |
| 1900, May 28 | 2m10s | Pacific Ocean, Mexico, Texas, Louisiana, Mississippi, Alabama, Georgia, South Carolina, North Carolina, Virginia, Atlantic Ocean, Portugal, Spain, Africa |
| 1918, June 8 | 2m23s | Pacific Ocean, Washington, Oregon, Idaho, Utah, Wyoming, Colorado, Kansas, Oklahoma, Arkansas, Louisiana, Mississippi, Alabama, Florida, Atlantic Ocean |
| 1923, September 10 | 3m37s | Pacific Ocean, California, Mexico, Belize, Caribbean Sea |
| 1925, January 24 | 2m32s | Minnesota, Wisconsin, Michigan, Canada, Pennsylvania, New York, Connecticut, Rhode Island, Atlantic Ocean |

| Date | Maximum Eclipse | Locations in the Path of Totality |
|------|-----------------|-----------------------------------|
| 1930, April 28 | 0m01s | Pacific Ocean, California, Oregon, Nevada, Idaho, Montana, Canada, Atlantic Ocean (hybrid eclipse: total in California, Oregon, Nevada, and Idaho) |
| 1932, August 31 | 1m45s | Arctic Ocean, Canada, Vermont, New Hampshire, Maine, Massachusetts, Atlantic Ocean |
| 1945 July 9 | 1m15s | Idaho, Montana, Canada, Greenland, Atlantic Ocean, Norway, Sweden, Finland, Russia, Kazakhstan |
| 1954, June 30 | 2m35s | Nebraska, South Dakota, Iowa, Minnesota, Canada, Atlantic Ocean, Greenland, Norway, Sweden, Lithuania, Belarus, Ukraine, Russia, Iran, Afghanistan, Pakistan, India |
| 1959, October 2 | 3m02s | New Hampshire, Massachusetts, Atlantic Ocean, Africa |
| 1963, July 20 | 1m40s | Pacific Ocean, Alaska, Canada, Maine, Atlantic Ocean |
| 1970, March 7 | 3m28s | Pacific Ocean, Mexico, Florida, Georgia, South Carolina, North Carolina, Virginia, Massachusetts, Canada, Atlantic Ocean |
| 1979, February 26 | 2m49s | Pacific Ocean, Washington, Oregon, Idaho, Montana, North Dakota, Canada, Greenland |
| 1991, July 11 | 6m53s | Pacific Ocean, Hawaii, Mexico, Central America, South America |
| 2017, August 21 | 2m40s | Pacific Ocean, Oregon, Idaho, Wyoming, Nebraska, Kansas, Missouri, Illinois, Kentucky, Tennessee, Georgia, North Carolina, South Carolina, Atlantic Ocean |
| 2024, April 8 | 4m28s | Pacific Ocean, Mexico, Texas, Oklahoma, Arkansas, Missouri, Illinois, Indiana, Kentucky, Ohio, Pennsylvania, New York, Vermont, New Hampshire, Maine, Canada, Atlantic Ocean |

| Date | Maximum Eclipse | Locations in the Path of Totality |
|------|-----------------|-----------------------------------|
| 2044, August 23 | 2m04s | Greenland, Canada, Montana, North Dakota |
| 2045, August 12 | 6m06s | Pacific Ocean, California, Nevada, Utah, Colorado, Kansas, Oklahoma, Texas, Arkansas, Mississippi, Louisiana, Alabama, Georgia, Florida, Caribbean, South America |
| 2052, March 30 | 4m08s | Pacific Ocean, Mexico, Texas, Louisiana, Alabama, Georgia, Florida, South Carolina, Atlantic Ocean |
| 2078, May 11 | 5m40s | Pacific Ocean, Mexico, Texas, Louisiana, Alabama, Georgia, Florida, South Carolina, North Carolina, Virginia, Atlantic Ocean |
| 2079, May 1 | 2m55s | Pennsylvania, New Jersey, New York, Connecticut, Rhode Island, Vermont, New Hampshire, Maine, Canada, Greenland, Arctic Ocean |
| 2099, September 14 | 5m18s | Pacific Ocean, Montana, North Dakota, Minnesota, Wisconsin, Michigan, Indiana, Ohio, West Virginia, Pennsylvania, Virginia, North Carolina, Atlantic Ocean |

## NOTES AND REFERENCES

1.  Based on Fred Espenak: "Total Eclipses in the USA: 1001–3000,"<http://www.eclipsewise.com/solar/SEcountry/SEinUSA-T.html>.

# Glossary

——◀o▶——

**annular eclipse**  A central eclipse of the Sun in which the angular diameter of the Moon is too small to cover the disk of the Sun completely and a thin ring (*annulus*) of the Sun's bright apparent surface surrounds the dark disk of the Moon. Thus an annular eclipse is actually a special kind of partial eclipse of the Sun. There are more annular eclipses than total eclipses.

**annular-total eclipse**  A solar eclipse that begins as an annular eclipse, changes to a total eclipse along its path, and then returns to annular before the end of the eclipse path. Also called a **hybrid eclipse**.

**anomalistic month**  The time it takes (27.55 days) for the Moon to orbit the Earth as measured from its closest point to Earth (perigee) to its farthest point (apogee) and back to its closest point again.

**anomalistic year**  The time it takes (365.26 days) for the Earth to orbit the Sun as measured from its closest point to the Sun (perihelion) to its farthest point (aphelion) and back to its closest point again.

**antumbra**  The extension of the Moon's shadow cone so that it forms a mirror image of itself. A region experiencing an annular eclipse lies in the antumbra of the Moon.

**aphelion**  The point for any object orbiting the Sun where it is farthest from the Sun.

**apogee**  The point for any object orbiting the Earth where it is farthest from the Earth.

**arc minute**  An angular measurement: 1 minute of arc is 60 seconds of arc and 1/60 of a degree of arc.

**arc second**  An angular measurement: 1 second of arc is 1/60 of a minute of arc and 1/3600 of a degree of arc.

**ascending node (of the Moon)**  The point on the Moon's orbit where it crosses the ecliptic (orbit of the Earth) going north.

**Baily's beads**  An effect seen just before and after the total phase of a solar eclipse in which the Moon hides all the light from the Sun's disk except for a few bright points of sunlight passing through valleys at the rim of the Moon.

**central eclipse (of the Sun)**  An eclipse in which the axis of the Moon's shadow touches the Earth. A central eclipse can be either total or annular.

**chromosphere**  The reddish lower atmosphere of the Sun just above the photosphere. The chromosphere is only about 600 miles (1,000 kilometers) thick and the temperature is about 7,100 °F (4,200 °C).

**contact (in a solar eclipse)**  Special numbered stages of a solar eclipse. For all eclipses (total, annular, and partial), **first contact** occurs when the leading edge of the Moon appears tangent to the western rim of the Sun, initiating the eclipse. In a *total eclipse*, **second contact** occurs when the Moon's leading edge appears tangent to the eastern rim of the Sun, initiating the total phase of the eclipse. In a *total eclipse*, **third contact** occurs when the Moon's trailing edge appears tangent to the western rim of the Sun, concluding the total phase of the eclipse. In an *annular eclipse*, **second contact** occurs when the Moon's trailing edge appears tangent to the Sun's western rim, initiating the ring of sunlight completely around the Moon. In an *annular eclipse*, **third contact** occurs when the Moon's leading edge appears tangent to the Sun's eastern rim, ending the ring of sunlight. **Fourth contact** (for all eclipses) occurs when the trailing edge of the Moon appears tangent to the eastern rim of the Sun, concluding the eclipse. Note that a partial eclipse has only first and fourth contacts (no second or third contacts). These same four contact points are used to describe lunar eclipses, planet transits across the face of the Sun, satellite transits across the face of a planet, and binary star transits across the face of one another.

**core (of the Sun)**    The central regions of the Sun where it produces its energy in a nuclear fusion reaction by converting hydrogen into helium at a temperature of 27 million °F (15 million °C).

**corona**    The rarefied upper atmosphere of the Sun that appears as a white halo around the totally eclipsed Sun. Speeds of atomic particles in the corona give it a temperature sometimes exceeding 2 million °F (1.1 million °C).

**coronagraph**    A special telescope that produces an artificial solar eclipse by masking the Sun's apparent surface with an opaque disk; invented by Bernard Lyot in 1930.

**coronal hole**    A region of the corona low in brightness and density. It is from coronal holes that solar particles escape most easily into space to become the solar wind.

**coronal mass ejections**    Vast bubbles of gas ejected from the corona into space. The bubbles can expand to a size larger than the Sun. Coronal mass ejections seem to be the principal cause of aurorae and a significant hazard to electric power grids and spacecraft electronics when these high-velocity particles and ropes of magnetic fields hit the Earth.

**degree of obscuration**    The fraction of the area of the Sun's disk obscured by the Moon at eclipse maximum, usually expressed as a percentage. (Degree of obscuration is not the same as the magnitude of an eclipse, which is the fraction of the Sun's diameter that is covered).

**descending node (of the Moon)**    The point on the Moon's orbit where it crosses the ecliptic (orbit of the Earth) going south.

**diamond ring effect**    The stage of a solar eclipse when only a tiny sliver of the Sun's photosphere shines along the edge of the Moon as the corona appears. The diamond ring effect occurs in the seconds before and after totality.

**draconic month**    The time it takes (27.21 days) for the Moon to orbit the Earth as measured from ascending node through descending node and back to ascending node again. Eclipses of the Sun and Moon can only take place near a node.

**eclipse limit (for the Sun)**    The maximum angular distance that the Sun can be from a node of the Moon and still be involved in an eclipse seen from Earth. For partial eclipses, the limit ranges from 15°21' to 18°31' according to the varying angular sizes of the Moon and Sun due to the elliptical orbits of the Moon and Earth. For a central eclipse, the maximum and minimum limits are 11°50' and 9°55'.

**eclipse magnitude**    The fraction of the Sun's *diameter* occulted by the Moon. The *eclipse magnitude* is less than 1.0 for *partial* and *annular* eclipses. The *eclipse magnitude* is equal to or greater than 1.0 for *total* eclipses.

**eclipse obscuration**    The fraction of the Sun's *area* occulted by the Moon. The *eclipse obscuration* is less than 1.0 for *partial* and *annular eclipses*. The *eclipse obscuration* is equal to or greater than 1.0 for *total* eclipses.

**eclipse season**    The period of time in which the apparent motion of the Sun places it close enough to a node of the Moon so that an eclipse is possible. The Sun crosses the ascending and descending nodes of the Moon in a period of 346.62 days, so eclipse seasons occur about 173.3 days apart. Depending on where the Sun is on the ecliptic (how fast the Earth is moving), a solar eclipse season may last from 31 to 37 days.

**eclipse year**    The time (346.62 days) it takes for the apparent motion of the Sun to carry it from ascending node to the descending node and back to the ascending node of the Moon.

**ecliptic**    The apparent annual path of the Sun around the star field as seen from the Earth as the Earth orbits the Sun in the course of a year. (Thus the ecliptic is the plane of the Earth's orbit around the Sun.) The Sun's apparent path is called the ecliptic because all eclipses of the Sun and Moon occur on or very close to this track in the sky.

**exeligmos**    An eclipse repetition cycle of 54 years 34 days, equal to three saros cycles and often called the triple saros. After one exeligmos cycle, a solar eclipse returns to almost the same longitude, but occurs about 600 miles (1,000 kilometers) north or south of its predecessor.

**filament**    A dark threadlike feature seen against the face of the Sun. A filament is a **prominence** seen from the top rather than at the edge of the Sun.

**flare**    Intense brightening in the upper atmosphere of the Sun which erupts vast amounts of charged particles into space. Flares can reach temperatures of 36 million °F (20 million °C).

**greatest eclipse**    The instant and location on the Earth's surface in a solar eclipse when the axis of the Moon's shadow cone passes closest to the Earth's center. This geometry is important in the computation of eclipses and is used as a standard in comparing eclipses. At this point along the path of a total eclipse, totality should last longest—and it does, to within a second or two. *Greatest eclipse* doesn't quite correspond to the *greatest duration* of totality because the calculation of *greatest eclipse* is made with a round Moon, ignoring mountains and valleys at the Moon's limb and because it ignores that the Earth isn't quite spherical.

**greatest duration**    The instant and location on the Earth's surface in a solar eclipse when the length of the total (or annular) phase of the eclipse is longest. *Greatest duration* does not occur at the point of *greatest eclipse* because of factors such as the relative motion of the Moon's shadow across the curvature of the Earth's surface. *Greatest duration* typically differs from *greatest eclipse* by a second or two in time and in geographic location by as much as 60 miles (100 kilometers) or so. *Greatest duration* can be further refined to include the effects of the Moon's irregular profile. This *corrected greatest duration* can differ from *greatest duration* by a couple of seconds.

**hybrid eclipse**    A solar eclipse that begins as an annular eclipse, changes to a total eclipse along its path, and then returns to annular before the end of the eclipse path. Also called an annular-total eclipse.

**inex**    A period of 10,571.95 days (29 years less 20.1 days) after which another eclipse of the Sun or Moon will occur (although not of the same type, such as total). This period equals 358 synodic months and 388.5 draconic months.

**lunation**    The time it takes (29.53 days) for the Moon to complete a phasing cycle (also called a synodic period).

**magnitude (of a solar eclipse)**    The fraction of the apparent diameter of the solar disk covered by the Moon at eclipse maximum. Eclipse magnitude is usually expressed as a decimal fraction: below 1.000 is a partial eclipse; 1.000 or above is a total eclipse. (The magnitude of a solar eclipse is not the same as the degree of obscuration, which is the percentage of the area of the Sun's disk that is covered.)

**mid-eclipse**    The instant in a central solar eclipse halfway between second and third contacts.

**New Moon**    The phase of the Moon when it is most nearly in conjunction with the Sun (also called dark-of-the-Moon). Solar eclipses can occur only at New Moon. (In ancient times, New Moon had a different meaning: the crescent Moon when it became visible after dark-of-the-Moon.)

**nodes**    The two points at which the orbit of a celestial body crosses a reference plane. The Moon crosses the orbital plane of the Earth (the ecliptic) going northward at the ascending node and going southward at the descending node.

**partial eclipse (of the Sun)**    An eclipse in which a portion of the Sun's disk is not covered by the Moon.

**penumbra**    The portion of a shadow from which only part of the light source is occulted by an opaque body. Seen from outside the shadow region, the penumbra is a fuzzy fringe to the dark umbra that declines in darkness outward from the umbra. A region experiencing a partial solar eclipse lies in the penumbra of the Moon.

**perigee**    The point for any object orbiting the Earth where it is closest to the Earth.

**perihelion**    The point for any object orbiting the Sun where it is closest to the Sun.

**photosphere**    The apparent "surface" of the Sun. It is actually a layer of hot gases only about 300 miles (500 kilometers) thick where the Sun's atmosphere changes from opaque to transparent, and visible light escapes from the Sun. The temperature of the photosphere is about 10,000 °F (5,500 °C).

**prominence**    An arch or filament of denser gas in the Sun's corona, shaped by the magnetic field of the Sun. Some prominences rise but most are descending, as if raining.

**regression of the nodes**    The westward shift of the Moon's nodes along the ecliptic due to tidal forces on the Moon's orbit exerted by the Sun and Earth. The regression of the nodes is responsible for the eclipse year being 18.62 days shorter than the seasonal (tropical) year. The nodes complete a westward regression entirely around the ecliptic in 18.6 years.

**saros**    An eclipse cycle of 6,585.32 days (18 years 11-1/3 days or 18 years 10-1/3 days if five leap years occur in the interval) in which an eclipse will occur that is very similar to the one that preceded it. The saros results from the near equivalence of 223 synodic months, 19 eclipse years, and 239 anomalistic months.

**shadow bands**    Faint flickers or ripples of light sometimes seen on the ground or buildings shortly before or after the total phase of a solar eclipse. Shadow bands are caused by light from the thin crescent of the Sun passing through parcels of rising and falling air that have different densities and hence act as lenses to bend the light continuously in varying amounts.

**solar constant**    The amount of power from the Sun falling on an average square meter of the Earth's surface (1.35 kilowatts).

**solar wind**    A stream of changed particles (mostly protons, electrons, and helium nuclei) ejected from the Sun which flows by the Earth at 720,000 to 1.8 million miles per hour (320 to 800 kilometers per second). When enhanced by flares, the particles collide with molecules in the Earth's upper atmosphere so intensely that they cause the upper atmosphere to glow by fluorescence in displays of the aurora (the northern and southern lights).

**spectrohelioscope**    A solar spectroscope that blocks unwanted colors so that an observer can view the Sun in the light of one spectral line at a time; invented by Jules Janssen in 1868.

**spectroscope**    A device (usually employing a prism or a diffraction grating) to spread out a beam of light into its component wavelengths for study. Spectroscopy can reveal the composition, temperature, radial velocity, rotation, magnetic fields, and other features of a light source.

**spicule**    A jetlike spike of upward-moving gas in the chromosphere of the Sun. Viewed near the edge of the Sun, spicules resemble a forest. Each spicule lasts 10 minutes or so and ejects material into the corona at speeds of 12–19 miles per second (20–30 kilometers per second).

**sunspot**    A darker area in the Sun's photosphere where magnetic fields are very strong. The temperatures of sunspots are 2,500–3,600 °F (1,400–2,400 °C) cooler than their surroundings, making them appear darker. Sunspots can last for a day up to several months.

**sunspot cycle**    A period averaging 11.1 years in which the number of sunspots increases, decreases, and then begins to increase again.

**synodic month**    The period of time (29.53 days) required for the Moon to orbit the Earth and catch up with the Sun again. Because the Moon's position with respect to the Sun determines the phase of the Moon, the synodic period is the time required for a complete set of phases by the Moon.

**transition region**    The thin, irregular layer that separates the chromosphere from the corona. In this layer the temperature rises suddenly from about 7,200 °F (4,000 °C) to about 1.8 million °F (1 million °C). The transition region is of variable thickness, sometimes no more than tens of miles.

**tritos**    A period of 3,986.6295 days (11 years less 31 days) after which another eclipse will occur (although not the same type, such as total). This period, less accurate than the saros or inex, equals 135 synodic months, 146.5 draconic months, and roughly 144.5 anomalistic months.

**total eclipse (of the Sun)**    An eclipse in which the angular size of the Moon is sufficient to totally cover the disk of the Sun. In a total eclipse, the umbral shadow of the Moon touches the surface of the Earth.

**umbra**    The central, completely dark portion of a shadow from which all of the light source is occulted by an opaque body. A region experiencing a total solar eclipse lies in the umbra of the Moon.

# Bibliography

------◀◉▶------

Alexander, Hartley Burr. *Latin-American Mythology*, volume 11 of *The Mythology of All Races*. Boston: Marshall Jones, 1920.

Allen, David and Allen, Carol. *Eclipse*. Sydney; Boston: Allen & Unwin, 1987.

Ananikian, Mardiros H. *Armenian Mythology*, volume 7 of *The Mythology of All Races*. Boston: Marshall Jones, 1925.

Andrews, Tamra. *Legends of the Earth, Sea, and Sky: An Encyclopedia of Nature Myths*. Santa Barbara, California: ABC-CLIO, 1998.

Arago, François. *Popular Astronomy*. 2 volumes. Translated by William Henry Smyth and Robert Grant. London: Longman, Brown, Green, Longmans, and Roberts, 1858.

Ashbrook, Joseph. *The Astronomical Scrapbook*. Edited by Leif J. Robinson. Cambridge: Cambridge University Press; Cambridge, Massachusetts: Sky Publishing, 1984.

Associated Press (No author given). "Antarctica Concerns Grow as Tourism Numbers Rise." March 16, 2013 <https://www.usatoday.com/story/travel/destinations/2013/03/16/antarctica-tourism-rise/1993181/>

Aveni, Anthony F. *Skywatchers of Ancient Mexico*. Austin: University of Texas Press, 1980.

Aveni, Anthony. *Stairways to the Stars: Skywatching in Three Great Ancient Cultures*. New York: John Wiley & Sons, 1997.

Baily, Francis. "On a Remarkable Phenomenon that Occurs in Total and Annular Eclipses of the Sun." *Memoirs of the Royal Astronomical Society*. Volume 10, 1838, pages 1–40.

Baily, Francis. "Some Remarks on the Total Eclipse of the Sun, on July 8th, 1842." *Memoirs of the Royal Astronomical Society*. Volume 15, 1846, pages 1–8.

Bakich, Michael. "25 Facts You Should Know about the August 21, 2017 Total Solar Eclipse." Blog, August 6, 2014. <http://cs.astronomy.com/asy/b/astronomy/archive/2014/08/05/25-facts-you-should-know-about-the-august-21-2017-total-solar-eclipse.aspx>.

Bakich, Michael. "Two Dozen Tips for the August 21, 2017 Total Solar Eclipse." Blog, August 8, 2014. <http://cs.astronomy.com/asy/b/astronomy/archive/2014/08/08/25-tips-for-the-august-21-2017-total-solar-eclipse.aspx>.

Bakich, Michael E. *Your Guide to the 2017 Total Solar Eclipse*. Berlin: Springer Nature (Patrick Moore Practical Astronomy Series), 2016.

Bakich, Michael E. *Your Guide to the 2024 Solar Eclipse*. Waukesha, Wisconsin: Kalmbach, 2022.

Baron, David. *American Eclipse: A Nation's Epic Race to Catch the Shadow of the Moon and Win the Glory of the World*. New York: Liveright Publishing (W. W. Norton). 2017.

Berman, Bob. *The Sun's Heartbeat and Other Stories from the Life of the Star that Powers Our Planet*. New York: Little, Brown, 2011.

Brewer, Bryan. *Eclipse*. Seattle: Earth View, 1978; 2nd edition 1991.

Bruce, Ian. *Eclipse: An Introduction to Total and Partial Eclipses of the Sun and Moon*. Harrogate, England: Take That, 1999.

Brunier, Serge and Luminet, Jean-Pierre. *Glorious Eclipses: Their Past, Present, and Future*. Translated by Storm Dunlop. Cambridge: Cambridge University Press, 2000.

Caesar, Ed. "Stonehenge: What Lies Beneath." *Smithsonian*. Volume 45, Number 5, September 2014, pages 30–41.

Calvin, William H. *How the Shaman Stole the Moon: In Search of Ancient Prophet-Scientists from Stonehenge to the Grand Canyon*. New York: Bantam Books, 1991.

Chambers, George F. *The Story of Eclipses*. Library of Valuable Knowledge. New York: D. Appleton, 1912.

Chang, Kenneth. "Life on Mars? Could Be, But How Will They Tell?" *New York Times*, March 29, 2005.

Clerke, Agnes M. *A Popular History of Astronomy During the Nineteenth Century*. 4th edition. London: A. and C. Black, 1902.

Codona, Johana L. "The Enigma of Shadow Bands," *Sky & Telescope*, volume 81 (May 1991), page 482.

Comte, Auguste. *The Essential Comte, Selected from Cours de philosophie positive*, translated by Margaret Clarke. London: Croom Helm, 1974.

Couderc, Paul. *Les éclipses*. (Que sais-je? series no. 940.) Paris: Presses Universitaires de France, 1961.

Covington, Michael. *Astrophotography for the Amateur*. Cambridge: Cambridge University Press, 1988.

Crommelin, Andrew C. D. "Results of the Total Solar Eclipse of May 29 and the Relativity Theory," *Nature*, volume 104, November 13, 1919, pages 280–281.

Crump, Thomas. *Solar Eclipse*. London: Constable, 1999.

De Sahagún, Bernadino. *Florentine Codex; General History of the Things of New Spain*. Book 7: *The Sun, Moon, and Stars, and the Binding of the Years*. Translated from the Aztec by Arthur J. O. Anderson and Charles E. Dibble. Santa Fe, New Mexico: School of American Research; and Salt Lake City: University of Utah, 1953, pages 36 and 38.

Dillard, Annie. "Total Eclipse," in *An Annie Dillard Reader*. New York: Harper Perennial, 1995.

Douglas, Allie Vibert. *The Life of Arthur Stanley Eddington*. London: T. Nelson, 1956.

Dunkin, Edwin. *Autobiography*, unpublished (compiled by Peter Hingley, Royal Astronomical Society) 1851.

Dyer, Alan. *How to Photograph the Solar Eclipse: A Guide to Capturing the 2017 Total Eclipse of the Sun*. E-book: <http://www.amazingsky.com/books.html>, 2017.

Dyson, Frank W., Crommelin, Andrew C. D., and Eddington, Arthur S. "Joint Eclipse Meeting of the Royal Society and the Royal Astronomical Society," *The Observatory*, volume 42, November 1919, pages 389–398.

Dyson, Frank W., Eddington, Arthur S., and Davidson, Charles R. "A Determination of the Deflection of Light by the Sun's Gravitational Field, From Observations Made at the Total Eclipse of May 29, 1919," *Philosophical Transactions of the Royal Society of London*, series A, volume 220, 1920, pages 291–333.

Dyson, Frank and Woolley, Richard v.d. R. *Eclipses of the Sun and Moon*. Oxford: Clarendon Press, 1937.

Espenak, Fred. *EclipseWise*. <http://eclipsewise.com/>.

Espenak, Fred. *Fifty Year Canon of Solar Eclipses: 1986–2035*. Washington, D.C.: NASA; Cambridge, Massachusetts: Sky Publishing, 1987. NASA Reference Publication 1178 Revised.

Espenak, Fred. *Road Atlas for the Total Solar Eclipse of 2017*. Portal, Arizona: Astropixels, 2015.

Espenak, Fred and Anderson, Jay. *Eclipse Bulletin: Total Solar Eclipse of 2017 August 21*. Portal, Arizona: Astropixels, 2015.

Espenak, Fred and Meeus, Jean. *Five Millennium Canon of Solar Eclipses: −1999 to +3000 (2000 BCE to 3000 CE)*. NASA Technical Publication 2006–214141. Greenbelt, Maryland: NASA Goddard Space Flight Center, 2006.

Farber, Jim. "Can't Get to Mars? Try Chile's Atacama Desert, The Driest Place on Earth and the Most Otherworldly." *New York Daily News*, January 16, 2012.

Fiala, Alan D., DeYoung, James A., and Lukac, Marie R. *Solar Eclipses, 1991–2000*. (US Naval Observatory circular 170.) Washington, D.C.: US Naval Observatory, 1986.

Flammarion, Camille. *The Flammarion Book of Astronomy*. Edited by Gabrielle Camille Flammarion and André Danjon. Translated by Annabel and Bernard Pagel. New York: Simon and Schuster, 1964.

Fomalont, Edward B. and Sramek, Richard A. "A Confirmation of Einstein's General Theory of Relativity by Measuring the Bending of Microwave Radiation in the Gravitational Field of the Sun," *Astrophysical Journal*, volume 199, August 1, 1975, pages 749–755.

Fowler, Alfred. "Sir Norman Lockyer, K. C. B., 1836–1920," *Proceedings of the Royal Society of London*, series A, volume 104, December 1, 1923, pages i–xiv.

Francillon, Gérard and Menget, Patrick, editors. *Soleil est mort: l'éclipse totale de soleil du 30 Juin 1973*. Nanterre: Laboratoire d'ethnologie et de sociologie comparative (Récherches thématiques, 1), 1979.

Frazer, James George. *Balder the Beautiful*, part 1, in *The Golden Bough*, volume 10. London: Macmillan, 1930.

Friedman, Herbert. *Sun and Earth*. New York: W. H. Freeman (Scientific American Library), 1986.

Gates, S. James Jr. and Pelletier, Cathie. *Proving Einstein Right: The Daring Expeditions that Changed How We Look at the Universe*. New York: Public Affairs, 2019.

Golub, Leon and Pasachoff, Jay M. *The Solar Corona*. Cambridge: Cambridge University Press, 1997.

Guillermier, Pierre and Koutchmy, Serge. *Total Eclipses: Science, Observations, Myths and Legends*. Translated by Bob Mizon. Berlin: Springer-Verlag and Chichester, United Kingdom: Praxis, 1999.

Hadingham, Evan. *Early Man and the Cosmos*. New York: Walker, 1984.

Harrington, Philip S. *Eclipse! The What, Where, When, Why, and How Guide to Watching Solar and Lunar Eclipses*. New York: John Wiley & Sons, 1997.

Harris, Joel and Talcott, Richard. *Chasing the Shadow*. Waukesha, Wisconsin: Kalmbach, 1994.

Held, Wolfgang. *Eclipses: 2005–2017*. Translated by Christian von Arnim. Edinburgh: Floris, 2005.

Herodotus. *The History*, volume 1, translated by George Rawlinson. Everyman's Library, volume 405. London: J. M. Dent, 1910.

Hetherington, Barry. *A Chronicle of Pre-Telescopic Astronomy*. Chichester, England: John Wiley & Sons, 1996.

Hoffleit, Dorrit. *Some Firsts in Astronomical Photography*. Cambridge, Massachusetts: Harvard College Observatory, 1950.

Hoffmann, Banesh, with the collaboration of Helen Dukas. *Albert Einstein, Creator and Rebel*. New York: Viking Press, 1972.

Holloway, Richard P. *Solar Eclipse 1999*. London: Calculus International, 1999.

Ion, Victoria. *Indian Mythology*. New York: Peter Bedrick Books, 1984.

Johnson, Samuel J. *Eclipses, Past and Future; with General Hints for Observing the Heavens*. Oxford: J. Parker, 1874.

Joslin, Rebecca R. *Chasing Eclipses: The Total Solar Eclipses of 1905, 1914, 1925*. Boston: Walton Advertising and Printing, 1929.

Keith, Arthur Berriedale. *Indian Mythology*, volume 6 of *The Mythology of All Races*. Boston: Marshall Jones, 1917.

Kippenhahn, Rudolph. *Discovering the Secrets of the Sun*. New York: John Wiley & Sons, 1994.

Knappert, Jan. *Indian Mythology: An Encyclopedia of Myth and Legend.* London: HarperCollins, 1991.

Koestler, Arthur. *The Sleepwalkers.* New York: Grosset & Dunlap, 1963.

Krupp, Edwin C. *Beyond the Blue Horizon: Myths and Legends of the Sun, Moon, Stars, and Planets.* New York: HarperCollins, 1991.

Kudlek, Manfred and Mickler, Erich H. *Solar and Lunar Eclipses of the Ancient Near East from 3000 B.C. to 0 with Maps.* Neukirchen-Vluyn, Germany: Butzon & Bercker Kevelaer, 1971.

Lang, Kenneth R. *Sun, Earth, and Sky.* New York: Springer, 1995.

Lang, Kenneth R. and Gingerich, Owen, editors. *A Source Book in Astronomy and Astrophysics, 1900–1975.* Cambridge, Massachusetts: Harvard University Press, 1979.

Le Bovier de Fontenelle, Bernard. *A Plurality of Worlds*, translated by John Glanvill. London: Nonesuch Press, 1929.

Legge, James (editor and translator). *The Chinese Classics*, volume 3, *The Shoo King [Shu Ching]*, Hong Kong: Hong Kong University Press, 1960.

Levy, Dawn. "An Eclipsed Vacation," *Los Altos* [California] *Town Crier*, August 7, 1991, pages 22–23.

Lewis, Isabel M. *A Handbook of Solar Eclipses.* New York: Duffield, 1924.

Lewis, Isabel M. "The Maximum Duration of a Total Solar Eclipse." *Publications of the American Astronomical Society.* Volume 6, 1931, pages 265–266.

Linsdau, Aaron. *Texas Total Eclipse Guide: Official Commemorative 2024 Keepsake Guidebook.* Self-published, 2017.

Little, Robert T. *Astrophotography: A Step-by-Step Approach.* New York: Macmillan, 1986.

Littmann, Mark, Espenak, Fred, and Willcox, Ken. *Totality: Eclipses of the Sun.* 3rd edition. New York: Oxford University Press, 2008.

Littmann, Mark, Espenak, Fred, and Willcox, Ken. *Totality: Eclipses of the Sun.* 3rd edition updated. New York: Oxford University Press, 2009.

Littmann, Mark and Willcox, Ken. *Totality: Eclipses of the Sun.* Honolulu: University of Hawaii Press, 1991.

Littmann, Mark, Willcox, Ken, and Espenak, Fred. *Totality: Eclipses of the Sun.* 2nd edition. New York: Oxford University Press, 1999.

Lockyer, William J. S. "The Total Eclipse of the Sun, April 1911, as Observed at Vavau, Tonga Islands," in Bernard Lovell, editor: *Astronomy*, volume 2, The Royal Institution Library of Science, Barking, Essex: Elsevier Publishing, 1970, pages 190–191.

Lovell, Bernard, editor. *Astronomy.* 2 volumes. The Royal Institution Library of Science. Barking, Essex; New York: Elsevier Publishing, 1970.

Lynn, William Thynne. *Remarkable Eclipses: A Sketch of the Most Interesting Circumstances Connected with the Observation of Solar and Lunar Eclipses, Both in Ancient and Modern Times.* London: Edward Stanford, 1896.

Marriott, Alice and Rachlin, Carol K. *Plains Indian Mythology*. New York: Thomas Y. Crowell, 1975.

Marschall, Laurence A. "A Tale of Two Eclipses." *Sky & Telescope*. Volume 57, February 1979, pages 116–118.

Maunder, Annie S. D. and Walter Maunder, E. *The Heavens and Their Story*. London: R. Culley, 1908.

Maunder, Michael. "Eclipse Chasing." in Patrick Moore, editor. *Yearbook of Astronomy*. New York: W. W. Norton, 1989, pages 139–157.

Maunder, Michael and Moore, Patrick. *The Sun in Eclipse*. London: Springer-Verlag London, 1997.

McEvoy, J. P. *Eclipse: The Science and History of Nature's Most Spectacular Phenomenon*. London: Fourth Estate, 1999.

McPherson, Florence Andsager, as told to Julie Andsager. *Florence's Memories—As Remembered at Age 89–90*. Self-published family history, 1998.

Meadows, A. J. *Early Solar Physics*. Oxford: Pergamon Press, 1970.

Meadows, A. J. *Science and Controversy: A Biography of Sir Norman Lockyer*. London: Macmillan, 1972.

Meeus, Jean. "The Frequency of Total and Annular Solar Eclipses at a Given Place," *Journal of the British Astronomical Association*, volume 92, April 1982, pages 124–126.

Meeus, Jean. *Elements of Solar Eclipses: 1951–2200*. Richmond, Virginia: Willmann-Bell, 1989.

Meeus, Jean. *Astronomical Algorithms*. Richmond, Virginia: Willmann-Bell, 1991.

Meeus, Jean. *Mathematical Astronomy Morsels*. Richmond, Virginia: Willmann-Bell, 1997.

Meeus, Jean. "The Maximum Possible Duration of a Total Solar Eclipse," *Journal of the British Astronomical Association*, volume 113, number 6, 2003, pages 343–348

Meeus, Jean, Grosjean, Carl C., and Vanderleen, Willy. *Canon of Solar Eclipses*. Oxford: Pergamon Press, 1966.

Menzel, Donald H. and Pasachoff, Jay M. *A Field Guide to the Stars and Planets*. 2nd edition. Boston: Houghton Mifflin, 1983.

Miller, Jon D. "Americans and the 2017 Eclipse: A Final Report on Public Viewing of the August Total Solar Eclipse," University of Michigan report for NASA, June 12, 2018.

Mitchell, Samuel A. *Eclipses of the Sun*. 5th edition. New York: Columbia University Press, 1951.

Mobberley, Martin. *Total Solar Eclipses and How to Observe Them*. New York: Springer-Verlag New York, 2007.

Montelle, Clemency. *Chasing Shadows: Mathematics, Astronomy, and the Early History of Eclipse Reckoning*. Baltimore: Johns Hopkins University Press, 2011.

Moskvitch, Katia. "Rover Explores Chile Desert to Aid Mars Life Hunt." Space.com, June 24, 2013.

Mucke, Hermann and Meeus, Jean. *Canon of Solar Eclipses: −2003 to +2526.* Vienna: Astronomisches Büro, 1983.

Nakayama, Shigeru. *A History of Japanese Astronomy—Chinese Background and Western Impact.* Cambridge, Massachusetts: Harvard University Press, 1969.

Needham, Joseph and Ling, Wang. *Science and Civilisation in China.* Volume 3: *Mathematics and the Sciences of the Heavens and the Earth.* Cambridge: Cambridge University Press, 1959.

Needham, Joseph and Ling, Wang. Colin A. Ronan, editor. *The Shorter Science and Civilisation in China.* Volume 2. Cambridge: Cambridge University Press, 1981.

Neugebauer, Otto. *The Exact Sciences in Antiquity.* 2nd edition. Providence: Brown University Press, 1957.

Newton, Robert R. *Ancient Astronomical Observations and the Accelerations of the Earth and Moon.* Baltimore: Johns Hopkins Press, 1970.

Nordgren, Tyler. *Sun Moon Earth: The History of Solar Eclipses from Omens of Doom to Einstein and Exoplanets.* New York: Basic Books, 2017.

O'Baugh, Stephen. "Eclipse Had Top Rating," *The Age* (Melbourne, Australia), October 24, 1976.

Olcott, William Tyler. *Sun Lore of All Ages.* New York: G. P. Putnam's Sons, 1914.

Oppolzer, Theodor von. *Canon of Eclipses.* Translated by Owen Gingerich. New York: Dover, 1962.

Osterbrock, Donald E., Gustafson, John R., and Shiloh Unruh, W. J. *Eye on the Sky: Lick Observatory's First Century.* Berkeley: University of California Press, 1988.

Ottewell, Guy. *The Under-Standing of Eclipses.* 3rd edition. Greenville, North Carolina: Universal Workshop, 2004.

Ottewell, Guy. *The Astronomical Companion.* 2nd edition. Greenville, North Carolina: Universal Workshops, 2010.

Ottewell, Guy. *Astronomical Calendar.* Greenville, North Carolina: Universal Workshop, published annually.

Pang, Alex SooJung-Kim. *Empire and the Sun: Victorian Solar Eclipse Expeditions.* Stanford, California: Stanford University Press, 2002.

Pankenier, David W. *Astrology and Cosmology in Early China: Conforming Earth to Heaven.* Cambridge: Cambridge University Press, 2013.

Pannekoek, Anton. *A History of Astronomy.* London: G. Allen & Unwin, 1961.

Parker Pearson, Mike and the Stonehenge Riverside Project. *Stonehenge—A New Understanding: Solving the Mysteries of the Greatest Stone Age Monument.* New York: The Experiment, 2014.

Parrinder, Geoffrey. *Africa Mythology*. New York: Peter Bedrick Books, 1986.

Pasachoff, Jay M. "Halley as an Eclipse Pioneer: His Maps and Observations of the Total Solar Eclipses of 1715 and 1724," *Journal of Astronomical History and Heritage*, volume 2, no. 1, 1999, pages 39–54.

Pasachoff, Jay M. and Covington, Michael A. *The Cambridge Eclipse Photography Guide*. Cambridge: Cambridge University Press, 1993.

Pepin, Robert O., Eddy, John A., and Merrill Russ B., editors. *The Ancient Sun: Fossil Record in the Earth, Moon and Meteorites*. Proceedings of the Conference on the Ancient Sun; Boulder, Colorado; October 16–19, 1979. New York: Pergamon Press, 1980.

Percival, Chap. *Go See the Eclipse and Take a Child with You*. Sarasota, Florida: Bee Ridge Press, 2015.

Pitts, Michael W. "Stones, Pits and Stonehenge," *Nature*, volume 290, March 5, 1981, pages 46–47.

Rao, Joe. *Your Guide to the Great Solar Eclipse of 1991*. Cambridge, Massachusetts: Sky Publishing, 1989.

Reynolds, Michael D. and Sweetsit, Richard A. *Observe: Eclipses*. Washington, D.C.: Astronomical League, 1995.

Robinson, Leif. "Foreword" in Mark Littmann, Fred Espenak, and Ken Willcox, *Totality: Eclipses of the Sun*, 2nd edition. New York: Oxford University Press, 1999.

Rogers, Michael. "Totality—A Report," *Rolling Stone*, October 11, 1973. Reprinted in Gannon, Robert, editor: *Best Science Writing: Reading and Insights*. Phoenix, Arizona: Oryx Press, 1991, pages 168–185.

Ruggles, Clive. "Stonehenge for the 1990s," *Nature*, volume 381, May 23, 1996, pages 278–279.

Ruggles, Clive. *Astronomy in Prehistoric Britain and Ireland*. New Haven, Connecticut: Yale University Press, 1999.

Ruggles, Clive. *Ancient Astronomy: An Encyclopedia of Cosmologies and Myth*. Santa Barbara, California: ABC Clio, 2005.

Russo, Kate. *Total Addiction: The Life of an Eclipse Chaser*. Heidelberg, Germany, Springer, 2012.

Sands, Charles P. *Chasing the Shadow: The Dynamics of Eclipses*. Frederick, Maryland: PublishAmerica, 2005.

Schatz, Dennis and Franknoi, Andrew. *Solar Science: Exploring Sunspots, Seasons, Eclipses, and More*. Arlinton, Virginia: NSTA Press, 2016.

Scheub, Harold. *A Dictionary of Africa Mythology: The Mythmaker as Storyteller*. New York: Oxford University Press, 2000.

Schwegman, John E. "A Prehistoric Cultural and Religious Center in Southern Illinois." <http://www.southernmostillinoishistory.net/kincaid-mounds.html>.

Sébillot, Paul Y. *Le folk-lore de France*. Volume 1: *Le ciel et la terre*. Paris: Librairie orientale & américaine, 1904.

Shapiro, Irwin I. "New Method for the Detection of Light Deflection by Solar Gravity," *Science*, volume 157, August 18, 1967, pages 806–807.

Silverman, Sam and Mullen, Gary. "Eclipses: A Literature of Misadventures." *Natural History*, volume 81, June–July 1972, pages 48–51, 82.

Steel, Duncan. *Eclipse: The Celestial Phenomenon Which Has Changed the Course of History*. London: Headline, 1999, and Washington, D.C.: National Academies Press, 2001.

Stegemann, Viktor. "Finsternisse," in Hanns Bächtold-Stäubli, editor. *Handwörterbuch des Deutschen Aberglaubens*. Volume 2. Berlin: W. de Gruyter, 1930. Columns 1509–1526.

Stephenson, F. Richard and Clark, David H. *Applications of Early Astronomical Records*. Monographs on Astronomical Subjects, 4. New York: Oxford University Press, 1978.

Stephenson, F. Richard. *Historical Eclipses and Earth's Rotation*. Cambridge: Cambridge University Press, 1997.

Thompson, J. Eric S. *A Commentary on the Dresden Codex; A Maya Hieroglyphic Book*. Philadelphia: American Philosophical Society, 1972.

Todd, Mabel Loomis. *Total Eclipses of the Sun*. Revised. Boston: Little, Brown, 1900.

Van den Bergh, George. *Periodicity and Variation of Solar (and Lunar) Eclipses*. Haarlem: H. D. Tjeenk Willink, 1955.

Vaquero, José M. and Vázquez, Manuel. *The Sun Recorded Through History*. Heidelberg, Germany: Springer, 2009.

Wentzel, Donat G. *The Restless Sun*. Washington, D.C.: Smithsonian Institution Press, 1989.

Werner, Edward T. C. *A Dictionary of Chinese Mythology*. New York: Julian Press, 1961.

Williams, Sheridan. *2012 & 2013 Solar Eclipses with the Transit of Venus*. London: Bradt Travel Guides, 2012.

Williams, Sheridan. *Solar Eclipses 2024–2027: Where and When to Experience Totality*. Chalfont Saint Peter, United Kingdom: Bradt Travel Guides, 2023.

Zahn, Jean-Paul and Stavinschi, Magda, editors. *Advances in Solar Research at Eclipses from Ground and from Space*. Dordrecht, Netherlands: Kluwer, 1999.

Zeiler, Michael and Bakich, Michael. *Atlas of Solar Eclipses—2020 to 2045*. Santa Fe, New Mexico: Great American Eclipse, 2020.

Zirker, Jack B. *Total Eclipses of the Sun*. 2nd edition. Princeton: Princeton University Press, 1995.

# Index

Abu Simbel, 272
Academy of Sciences, 96–8
accelerated systems, 109, 111
AccuWeather, 249
Adirondack Mountains, 235
Agate Fossil Beds National Monument, 212–5
*Age, The*, 167–8
Airy, George B., 87, 91, 127, 134–5
Alaska, 235, 263, 269, 272, 287, 289–94
    Anchorage, 290
    Barrow, 293
    Fairbanks, 290
    Kohklux, 51–2
    Kotzebue, 291, 293
    mythology, 51–2
    Nome, 290–1
    Seward Peninsula, 290
    Utqiaġvik, 293–4
Alfonso X, 9
Algeria, 95, 267*f*–8*f*, 269, 272
American Astronomical Society, 211
American Paper Optics, 152, 156
Anderson, Jay, 125–6, 140, 144, 147, 169, 208,
        247, 251, 253–5, 257–9, 260 n.4,
        260 n.6
Andes Mountains, 57
Andrew, Julie, 126
Andrews, Paul, 126
Andsager, Julie, 161
Angel Mounds, 231
Angola, 281
*Annalen der Physik*, 109
anomalistic,
    month, 22–3, 25–7
    year, 27
Antarctica, 43, 218, 247
Antigua, 107, 164

Apep, 47–8
Apollo, 295
Appalachian Mountains, 260 n.6
Appalachian Trail, 239
Arago, Francois, 69, 127, 140
Arctic National Park and Preserve, 292
Argentina, 203
Argentine National Observatory, 121 n.5
Aristotle, 104
Arkansas, 230*f*
    Fort Smith, 228
    Hot Springs, 234, 259
    Jonesboro, 234
    Little Rock, 228, 234, 259
Armenia, 47, 49
Armstrong, Neil, 212
Aroostock River, 239
Aruba, 165–6, 170
Associated Press, 166
Astronomical Society of London, 86
*Astronomy*, 126, 190
*Astronomy Now*, 190
astrophotography, 95
Astrophysical Observatory, Potsdam, 121 n.7
AstroPhysics, 198 n.4
Aswan High Dam, 272
Atlantic walrus, 241
atoms, 98, 109
aurora, 74*f*, 77
Australia, 49*f*, 130, 164, 166–7, 218, 263, 273,
        277–80, 284–7
    Buckleboo, 285
    Chinchilla, 287
    Christmas Island, 277–8
    Cocos (Keeling) Islands), Territory of the,
        277–8
    Come By Chance, 278

Australia (*Continued*)
  Dubbo, 279
  Elong Elong, 279
  Hungerford, 285–6
  Melbourne, 164, 167–8
  Mudgee, 279
  New South Wales, 285
  Packsaddle, 285
  Perth, 278
  Pinkawillinie, 285
  Streaky Bay, 285
  Sydney, 166, 245, 279
  Wingadee, 278
  Wyndham, 278
Australian Optometrical Association, 167–8
Austria, 91
Aveni, Anthony, 40
Awawak, 50
Aymara Indians, 58
Aztecs, 41, 53–4, 170

Baader Planetarium AstroSolar Safety Film,
    154, 190
BabyCenter blog, 169
Babylonians, 22, 28, 29 n.6, 29 n.10, 39
  definition of circle, 28 n.4
    eclipse period, 42 n.12
    eclipse predictions, 29 n.7, 33, 38, 60
Baily, Francis, 45, 85–6, 88–9, 94, 103, 105 n.2
Baily's beads, 2, 3*f*, 86–9, 94, 105 n.2, 132–7,
    140–3, 145, 178, 187*f*, 212
Baluarte Bridge, 226
Baluarte River gorge, 226
Barbary apes, 268
Barbary pirates, 269
Barbovschi, Alex, 216–8
Barker, David, 206–8
Baruch College, City University of
    New York, 262
Bay of Fundy, 239
BBC Television, 212
Beattie, John, 126, 136
Benin, 50
Berber, 48
Bering Land Bridge, 290
Berkowski, 90
Berry, Richard, 126, 130

bilbies, 286
binoculars, 157–8
Black Hills, 215
Bloom, Marlene, 218
Bogen/Manfrotto, 182, 198 n.5
Bohemia, 54
Bolivia, 103*f*, 137
  Altiplano, 57–8
  La Paz, 57
  Sevaruyo, 57
Botswana, 282–4
Brazil, 121 n.5, 146, 203
  Rio de Janeiro, 149
  Sobral, 115, 117, 118*f*
British Empire, 277–8
Brooks Range, 292
Brown, Eric, 219
Brown, Ilana, 219
Brown, Janice, 219
Bruns, Carol, 211*f*
Bruns, Donald, 209–2, 220 n.8
Buchman, Joe, 126, 164, 170
Buchman, Kristian, 126, 163
Buddhism, 49–50
Bunsen, Robert, 97
Buryat, 46
Busch, August Ludwig, 90
Byrd, Mary Emma, 141

calendar year, 20, 22
California, 169, 276, 290
  Los Angeles, 234, 281
  San Diego, 283
  San Francisco, 234
Cambodia, 165
  Angkor Wat, 294 n.6
Camino de Santiago, 265
Campbell, Wallace, 115
Canada, 223, 249
  Aux Basques, Newfoundland, 238
  Belleville, Ontario, 238
  Bonavista, Newfoundland, 247
  Breton Island, Nova Scotia, 241
  Brockville, Ontario, 238
  Channel Port, Newfoundland, 238
  Come By Chance, Newfoundland, 243
  Cornwall, Ontario, 238

Drumondville, Quebec, 238
Fredericton, New Brunswick, 238–9
Gander, Newfoundland, 238, 241, 243
Granby, Quebec, 238
Hamilton, 223, 233, 238
Jasper, Alberta, 51*f*
Kingston, Ontario, 235, 238
Leamington, 233
Magdalen Islands, Quebec, 241
Mansonville, Quebec, 247
Moncton, New Brunswick, 239
Montreal, Quebec, 223, 235, 238, 249, 257
mythology, 50–2
New Brunswick, 240*f*, 258
Newfoundland, 223, 242*f*, 255
North Cape, Prince Edward Island, 259
Ontario, 232*f*, 233, 235
Point Escuminac, New Brunswick, 258–9
Prince Edward Island, 239, 241, 242*f*, 258
Quebec, 235, 240*f*
Quebec City, 164
Richibucto Head, New Brunswick, 258–9
Seacow Pond, Prince Edward Island, 241
Sherbrooke, Quebec, 238
Sunnyside, Newfoundland, 243
Tignish, Prince Edward Island, 258–9
Toronto, Ontario, 235, 238
Winnipeg, 169
Canon, 185, 188, 198 n.3
*Canon of Eclipses* (von Oppolzer), 18
Carhenge, 213
Cassini, Gian Domenico, 104
Cassini, Jacques, 69
Cedar Lake, 231
Chambliss Amateur Achievement
    Award, 211
Chesapeake Bay, 276
Chile, 139*f*
China, 37–9, 89, 170, 276, 299*f*
    definition of circle, 28 n.4
    eclipse period, 42 n.12
    eclipse predictions, 37–9, 165
    Jinta, 137*f*, 187*f*
    mythology, 46, 49
Chippewa Indians, 53
Chou, Ralph, 203
Christmas Island, 277–8

chromosphere (of Sun), 72*f*, 74, 78–9, 80 n.4,
    91, 98, 133, 140–3, 178
    color, 75
    lower levels, 74
    reversing layer, 101–2
    spicules, 75, 91
    temperature, 74–6
    upper levels, 74–5
Civil War, 101
Clavius, Christoph, 63
Clemson University, 219
Clerke, Agnes M., 88
Cleveland Browns, 233
*Close Encounters of the Third Kind*, 215
cloudiness index, 257
Cocos (Keeling) Islands), Territory of the,
    277–8
College of New Jersey, 102
Collegio Romano (Gregorian University), 93
Colorado, 234
Columbia, 237
Columbia University, 218
Come By Chance Cemetery, 278
Community Services Centre, 167
Comte, Auguste, 97
Confucius, 41 n.8
constellations, 28
convergence, 298
*Conversations on the Plurality of Worlds*
    (de Fontenelle), 52
corona (of Sun), 2, 31, 78–9, 80 n.4, 87–9,
    116, 133–4, 136–9, 142, 199, 203, 205,
    208–9, 213, 222, 262
    color of, 137
    drawings of, 48*f*, 72*f*, 88*f*
    loops, 78*f*
    magnetic fields, 75, 77, 78*f*, 81 n.10
    photographing, 93–4, 176, 178, 181–2, 186,
        191, 192*f*, 194*f*
    photographs of, 4*f*, 10*f*, 49*f*, 191*f*
    shape, 102–3, 143
    temperature, 76, 78*f*, 80 n.7, 101
    visible light, 77
    x-ray pictures of, 77
coronal holes, 77
coronal mass ejection (CME), 74*f*, 77, 79–81
    nn.8–9, 129

coronium, 101
Costa Rica, 130
Cottingham, Edwin T., 115–7
Covid-19, 218, 286
Croatia, 62
Crocker, Tony, 208–9
Crommelin, Andrew C. D., 115, 117, 118*f*
Cuba, 89
Currawinya National Park, 286
Curtis, Heber D., 115

daguerreotypes, 90–2
*Daily Telegraph* (Syndey), 166–7
D'Allonville, Jacques Eugène, 63
d'Arezzo, Ristoro, 63
Dartmouth College, 101–2
Davidson, Charles R., 115, 117, 118*f*
Davidson, George, 51–2
de Ferrer, José Jacques, 89, 105 n.2
de Fontenelle, Bernard Le Bovier, 52
De La Rue, Warren, 92–3, 98, 106 n.10
De Sitter, Willem, 114
Delaware, 239
Del Priore, Lucian V., 154–6
Denmark, 103–5
    Copenhagen, 105
Descartes, René, 104
Despommier, Dickson, 218
Devil's Tower National Monument, 213–5
di Cicco, Dennis, 126, 130, 170
diamond ring effect, 2, 134–6, 140, 142–3,
        158, 164, 170, 204–5, 217*f*, 218,
        222, 252*f*
    photographing, 176, 178, 182, 184*f*, 189*f*,
        194
Dillard, Annie, 127
Dittrich, Tony, 220 n.8
divergence, 298
Dom tribe, 48
Dresden Codex, 39–40
Dunkin, Edwin, 223
DuPont, 198 n.6
Dyson, Frank, 115–6

Earhart, Amelia, 243
Earth
    at aphelion, 24
    axis tilt, 28, 34–5

diameter, 28
distance from Sun, 11–2, 15, 24, 28
gravitational field, 114
magnetosphere, 74, 76*f*
orbit, 11, 15
orbit, plane of, 18*f*
rotation, 27, 295
shadow, 36–7
shape, 236
size, 72*f*, 74
umbra, 10*f*
East Carolina University, 31
East Richford Cemetery, 238
East Richford Slide Road, 237–8
Easter Island, 136, 247
eclipse(s),
    animal behavior during, 63–4, 130, 142–3,
        161, 163–4, 208, 261
    chasers, 85–105, 164, 170, 252*f*
    cure for, 64–6
    cycles, 22, 29 n.8
    creating the ultimate, 27
    equipment checklist for viewing, 146
    fear of, 164–6, 169–70
    human behavior during, 59–63, 142–4,
        146–7, 164–9, 199, 219, 262
    insect behavior during, 164
    junkies, confessions of, 144–7
    misinformation about, 166–70
    in mythology, 45–54
    observing a total, 125–47
    omens, 38
    personalities of, 133
    photographing, 175–98
    plant behavior, 142–3
    predicting, 22–5, 29 n.7, 33, 37–41, 60
    rhythms, 22–5, 60
    trip, planning, 145–6, 159, 245, 259
    weather and, 247–60
eclipse alert, 17
eclipse half-year, 39
eclipse limits "danger zone," 16–20, 26, 39
eclipse mythology themes
    corruption and death, 50–2
    danger to humanity, 50–1
    monster attempting to eat/destroy the Sun,
        45–50
    participatory events and, 53–4

sacrifice and, 53–4
 Sun and Moon conflicts, 45, 50–2
 Sun and Moon as lovers, 45, 50
 Sun grows ill/sad/neglectful, 45, 51–2
eclipse season, 17, 20, 22
eclipse stamp, 197
eclipse year, 20, 25, 39
ecliptic, 15–6, 18*f*, 41 n.5
Edberg, Steve, 125–6, 133, 137, 139, 140,
    144, 169
Eddington, Arthur S., 109, 114–17, 119*f*,
    210–1
Edlén, Bengt, 101
Egypt, 33, 47*f*–48*f*, 267–72, 276
 Abydos, 272
 Cairo, 270
 Dendera, 272
 Edfu, 272
 Giza, 270
 Karnak, 47*f*, 270
 Kom Ombo, 272
 Luxor, 215, 270
 mythology, 47–8
 Qara, 270
 Sallum, 10*f*
 Siwa, 270
 temple alignment, 98
 Thebes, 270
Einstein, Albert, 40, 109–22, 210, 212
Einstein Institute, 121 n.7
El Faro, 226
electrons, 79, 118
Empire State Building, 231, 281, 293
Empty Quarter, 272–3
End of the Mountains, 64
England, 127
 London, 63, 98
*Entriens sur la pluralité des mondes* (de
    Fontenelle,) 52
Erie Canal, 235
Eskimos, 51
Espenak, Fred, 163, 197, 259
Espenak, Pat, 163, 182
Etosha, 283
European Space Agency (ESA), 80 n.6,
    81 n.8
 Comet Interceptor, 212, 220 n.10

*Gaia* space observatory, 210
 Plato Space Telescope, 212, 220 n.9
*exeligmos*, 29 n.10
eye, cross section of, 155*f*
eye damage caused by solar eclipses,
    154–7, 167
eye safety and protection, 94, 145–6, 151–60,
    173, 203
 binoculars, 157–8
 pinhole protection method, 153
 solar eclipse glasses, 151–3, 159*f*,
    196, 262
 solar eyepieces, 156
 solar filters, 107, 129, 145–6, 154, 159, 178,
    189–90, 196
 telescopes, 156, 158
 welder's goggles, 153
 when to wear, 151, 158

Felts, Susannah, 201
Fiala, Alan, 126, 131, 146–7
Finger Lakes, 235
Flammarion, Camille, 69
Fleenor, George, 126, 144, 163–6, 173
Fleenor, Stephanie, 165–6
Florentine Codex, 41
Florida, 276
 Bradenton, 173
 Key West, 239
 Miami, 234, 281
 Sarasota, 173
Fomalont, Edward B., 120
Fon, 50
forbidden lines, 101
Fort Sill, 63
France, 52, 114, 140, 168*f*
 Montpellier, 63
 Paris, 63, 69, 94–6, 165*f*
 Perpignan, 64, 127
Franco-Prussian War, 95
Freitag, Ruth S., 126, 137, 158–9
Freundlich, Erwin, 112–3, 120 n.3, 121 n.7
Freundlich, Käte, 112–3
Fujifilm, 188

Ganges River, 52
Gateway of Ptolemy, 47*f*

Gê, 50
Georgetown University, 93
Georgia
    Atlanta, 234
Germany, 101, 113–4
    Berlin, 112
    mythology, 50, 53
Giant City State Park, 231, 243 n.6
Gibraltar, 268
Gitzo, 198 n.5
Google, 43, 126, 190, 247
Grand Teton National Park, 208
Grant, Robert, 90–1
gravitational redshift, 114, 121 n.10
Great Britain, 114–5
Great Kobuk Sand Dunes, 291
Great Lakes Science Center, 233
Great Northern Highway, 278
Greece,
    ancient, 29 n.10, 49, 60–2
    Athens, 62
    eclipse prediction, 60
Green, William Ellis, 168
Greenland, 264
Grotrian, Walter, 101
Gulf of Aden, 273
Gulf of Guinea, 115
Gulf of Mexico, 255
Gulf of St. Lawrence, 239

Haiti, 46, 166
Halley, Edmond, 29 n.8, 90, 105 n.2, 295
Hamilton College, 140
Hampton (Lord), 251f
Harkness, William, 101
harp seals, 241
Harrison, William Henry, 65–6
Hawaii, 154
helium, 72f, 80 n.3, 80 n.4
    atom, 71
    discovery of, 98–101
Herald (Melbourne), 168
Herodotus, 59–60
Herschel, John, 93
Herschel, John F.W., 93, 106 n.11
Herschel, William, 93, 106 n.11
High Dynamic Range (HDR) composite, 4f

Hinduism, 48–9, 52
Hinode, 81 n.8
Hitler, Adolph, 121 n.7
Holiday, Henry, 163
Hollweg, Joseph V., 80 n.7, 126, 144–5
Horn of Africa, 273
Hsi and Ho, 37, 38f, 41–2 nn.8–9, 165
Hubble Space Telescope, 31
Hudson River, 235
Hyades, 115
hydrogen, 71, 72f, 75, 80 nn.3–4

Iceland, 264–5
    Reykjavik, 265
Idaho
    Redfish Lake, 141f
    Stanley, 141f
Iditarod, 290
Iguazú Falls, 203
Illinois, 230f
    Carbondale, 229, 231, 234, 259
    Chester, 229
    Chicago, 216–7, 234
    Makanda, 231
    Marion, 216–8, 229, 231
    Metropolis, 229
India, 93, 103f, 146, 163, 165, 273, 276
    Guntur, 95, 97
    Hyderabad, 145f
    Mangalore, 165
    mythology, 46, 49, 52
Indiana, 230f, 232f
    Bloomington, 234
    Evansville, 231, 234
    Indianapolis, 223, 231, 233–4, 249
    limestone, 231
    Marion, 234
    Muncie, 234
    Terre Haute, 234
    Vincennes, 231
Indiana Territory, 65
Indiana University, 231
Indianapolis Motor Speedway, 231, 233
Indonesia, 147, 236
International Bridge, 226
Interstate Highway 10, 226
Interstate Highway 30, 228

Interstate Highway 35, 228
Interstate Highway 40, 228
Interstate Highway 95, 239
Iñupiat, 290, 293
iPhone 5s, 176*f*
Iran, 163
Italy, 93
    Pavia, 87
    Turin, 87

Jackson Hole Mountain Resort, 208
James (Saint), 266
Janssen, Pierre Jules César, 93–8,
    102–3
Java, 103*f*, 147
Jesuits, 92–3
Joslin, Rebecca, 141, 263
JPEG, 180–2
    compressed, 180
    quality, 180–1
    RAW vs, 180–1
    storage chart, 181*f*
Jubier, Xavier, 43, 126, 170, 247
Jupiter, 49, 73
    eclipses, 103–5
    size, 74

Kaiser Wilhelm Institute, 112, 121 n.7
Kalahari Desert, 283
Kalina tribe, 53
Kansas, 161
Karnak Temple, 270–1
Keeling, William, 277
Kentucky, 230*f*
    Paducah, 229, 231, 234, 243 n.3, 259
Kenya, 152*f*
Kepler, Johannes, 89, 104
Ketu, 46, 49
    Chi-Tu, 49–50
Kidger, Mark, 212–5
Kidger, Lourdes, 212–5
Kidger, Paula, 212–5
Kincaid Mounds State Historic Park,
    229, 231
Kiowa tribe, 64
Kirchhoff, Gustav, 97
Kobuk River, 291

Kobuk Valley National Park, 291
Kukulcán, 53

Ladenburg, Rudolf, 112
Lake Argyle, 278
Lake Champlain, 235, 237
lake effect, 235, 257–8
Lake Erie, 233, 235, 237, 257–9
Lake Gairdner National Park, 285
Lake Huron, 235
Lake Michigan, 235
Lake Ontario, 233, 235, 237, 258–9
Lake Superior, 235
Langley, Samuel P., 48*f*
Lapland, 251*f*
Larivière, Jean Marc, 123, 126
Le Verrier, 120 n.3
Leo, 89
Leonid meteor storm, 68 n.17
Lesotho, 282*f*, 284
Levy, Dawn, 126, 221–2
Lewis, Isabel Martin, 159
Library of Congress, 126
Libya, 183*f*, 192*f*, 267*f*, 269–71
    Benghazi, 270
    Jalu, 3*f*, 194*f*
    Tripoli, 269–70
Lick Observatory, 115, 121 n.7, 121 n.11
light
    displacement/deflection/bending, 111*f*,
        112–20, 121 n.5, 122 n.17, 133,
        210*f*, 211*f*, 220 n.9
    speed of, 103–4
Lindbergh, Charles, 243
Lindsey, Charles, 126
Littmann, Carl, 103–5
Lockyer, William J. S., 64, 95, 97–8, 101–2
Loki, 45
Lorentz, Hendrik Antoon, 118
Loring, Jeanne, 206–8
Louisiana, 234
Lovi, George, 126, 137, 147, 157,
    159, 169
lunar eclipse, 22, 25, 33, 35, 37–9, 41 n.7
    2022 November, 218
    cause of, 10*f*, 39
    ratio to solar eclipse, 18

Luxor Temple, 271
Lydians, 59, 61*f*, 273

Maine, 144, 238–9, 240*f*
    Greenville, 259
    Houlton, 239, 247
    Millinocket, 259
Majorca (Mallorca), 266*f*, 267
    Palma, 267
Makepeace, David, 126, 151, 163
Malaysia, 93
Maldives, 179
Malta, 129*f*
Maraldi, Giacomo Filippo, 89
Mars, 49, 95
Marschall, Larry, 126, 132
Martin, Katherine, 169–70
Martin-Leake, Lilian, 91*f*
Mason, Helen, 80 n.6
Massachusetts, 228, 234, 281
Mataguaya, 46–7
Maturnus, Julius Firmicus, 90
Mauritania, 199
Maya, 39–41, 53
McPherson, Florence Andsager, 161
Meade Instruments, 156, 198 n.4
Medes, 59, 61*f*, 273
medical x-ray film, 157
Meeus, Jean, 27
Melbourne Zoo, 164
memory cards, 175–6, 180, 191, 196
    compact flash (CF), 180
    Memory Stick, 180
    Micro SD, 180
    secure digital (SD), 180
    storage chart, 181*f*
    xD, 180
Mercury, 3, 49, 143
    orbit, 113–4, 120 n.3, 211–2
    perihelion, 113
    Vulcan, 114, 211–2
*Metropolis News*, 243 n.2
*Metropolis Planet*, 229, 243 n.2
Mexico, 103*f*, 208, 223, 225*f*, 237, 243 n.6,
        249, 254
    Baja California, 163, 166
    Beach of Goats, 164

Central Plaza, 255
Durango, 223, 226, 238, 255, 259
Mazatlán, 164, 223, 226, 238, 247, 254–5, 259
mythology, 41, 53–4
Piedras Negras, 247
San Blas, 144, 166
Socorro, 223
Torreón, 223, 226, 238, 249, 255, 259
Mexican Highway 40D, 226
Miami, 268
Michigan, 233–4
Mid-Atlantic Ridge, 265
Miller, Jon, 201
Milton, John, 59
Minamato, 66 n.3
Minolta, 198 n.3
Missisquoi River, 237
Mississippi River, 21*f*, 228–9
Mississippian Culture, 229, 231
Missouri, 230*f*
    Cape Girardeau, 229, 234
    Kansas City, 249
    New Madrid, 228
    St. Louis, 228, 249
    Ste. Genevieve, 229
Mitre Peak, 280
Molopo River, 283
Mongolia, 46, 163
Montana, 136, 290
month,
    anomalistic, 22–23, 25–7, 29 n.9
    definition, 13
    synodic, 13–4, 16
Moon, 9, 36
    angular diameter, 13, 138
    angular momentum, 295
    angular size, 9, 11–3, 27, 134*f*, 143
    apogee, 23
    apparent size, 12
    cycle, 39
    diameter, 28
    distance from Earth, 12–3, 15, 20, 28,
        236, 296
    Earth-facing side, 14
    formation of, 300 n.2
    full, 14, 36
    movement across Sun, 91

new, 14, 16, 36
nodes, 15–20, 23, 28, 29 n.11, 39, 49–50
node, motion of, 20
node, regression, 18*f*, 20, 28
orbit, 10*f*, 12–5, 18*f*, 19, 22–3, 27, 34–5, 41
orbit, plane of, 18*f*
orbital spin, 28
orbital velocity, 14*f*
perigee, 23, 25–7
shadow, 2–4, 13, 14*f*, 23, 27–8, 131, 133,
    136, 138, 141–5, 147, 163–4, 178, 195,
    204–5, 208, 233, 254, 264–5, 284–5,
    288, 297–9
sierra, 91, 98
size, 11
synodic period, 16, 22, 25–6
umbra, 10*f*–11*f*, 13, 18, 23
Moon-got-bit-by-bear, 48
Morocco, 267–9
Mössbauer effect, 114
Mount Hood, 205*f*
Mount Jefferson, 204–5
Mount Katahdin, 238–9
Mount Newberry, 205*f*
Mount Royal, 235
Mount Rushmore, 213, 215
*MS Paul Gauguin*, 252*f*
Murphy, Tony, 167

Namib Desert, 281–3
Namib-Naukluft National Park, 281
Namibia, 281–4
    Windhoek, 282
NASA, 80 n.6, 81 n.8
    GOES-7 weather satellite, 298*f*
    Orbiting Solar Observatory, 80 n.8
    Solar Dynamics Observatory (SDO), 76*f*
    Solar Maximum Mission, 80–81 n.8
    Solar–Terrestrial Relations Observatory
        (STEREO), 81 n.8
    Space Shuttle, 81 n.8
    Transition Region and Coronal Explorer
        (TRACE), 81 n.8
National Historical Landmarks, 231
National Quilt Museum, 229
National Radio Astronomy Observatory,
    122 n.17

National Weather Service, 249
*Nature*, 98
Navagrahas, the Nine Seizers, 49
Nebraska, 213
Nefertari, 271
Netherlands, 114, 118
New Hampshire, 240*f*, 238, 284
New Jersey, 218
New Madrid (Missouri) earthquakes, 228
New York, 232*f*, 240*f*, 258
    Albany, 235
    Buffalo, 223, 233–5, 249
    Dunkirk, 257, 259
    Jamestown, 234
    Kinderhook, 89
    New York, 228, 234–5
    Oswego, 258–9
    Plattsburgh, 234–5
    Rochester, 223, 234–5, 249, 258–9
    Syracuse, 223, 234–5
    Watertown, 235
New Zealand, 277, 280–1, 285
    Dunedin, 280–1
    Milford Sound, 280
    North Island, 245
    Queenstown, 280
    South Island, 280
Newkirk, Gordon, 103*f*
Newton, Isaac, 40, 113, 118, 120 n.3, 295
Newton, Robert R., 66 n.4
Niagara Falls, 233, 235, 258–9
Niagara River, 233
Nikon, 198 n.3
Nile River, 271–2
Nome Gold Rush, 290
Norse, 45–6
North Africa, 269–70
North Carolina, 170
North Pole, 290*f*
Northeast Greenland National
    Park, 264
Northumberland Strait, 258
Nuxalk (Bella Coola), 52

O'Baugh, Stephen, 164
O'Mara, Liz, 208–9
O'Neil, Julie, 201

Ohio, 232–3
  Akron, 234
  Cincinnati, 223, 233, 249
  Cleveland, 223, 233–4, 237, 249, 257–9
  Columbus, 223, 233, 249
  Dayton, 233–4
  Lima, 234
  Mansfield, 234
  Sandusky, 257, 259
  Toledo, 234
Ohio River, 21f, 229, 231
Oklahoma, 64, 228
Oman, 272
Ontario Science Center, 203
Oregon, 177f, 220 n.8
  artists, 207f, 208
  Madras, 204–6
  Portland, 205
  Prairie City, 206–7
  Salem, 4f, 203
Orrall, Frank, 126, 137
Ouachita National Forest, 228

Panama, 289
Panasonic, 188
Paris Observatory, 69, 95, 104
Paroo River, 286
Pasachoff, Jay M., 126–7, 146, 169
Pennsylvania, 232f
  Erie, 233–4, 257, 259
  Philadelphia, 234
Pentagon, the, 231
Pentax, 188, 198 n.3
penumbra, 11f, 18
Pericles, 62
Period Table, 101
Perrine, Charles Dillon, 121 n.5
Persia, 49, 60
Peru, 94
Peters (Dr.), 140
Petitcodiac River, 239
Philippines, 103f
Phoenix, 198 n.3
photographic neutral-density filters, 157
photography/photographing eclipses, 175–98
  auto-exposure feature, 178, 187, 194
  automatic flash feature, 178–9

autofocus feature, 187
bridge camera, 187–8
burst mode, 185
cable release, 188–9
camera movement/vibration, 184–5, 188, 194–5
CCD camera, 210, 211f
checklist for, 196
close-up photos, 178–80
compact cameras, 176–7
crop cameras, 185, 186f
digital, 175–7, 180–5
digital "film," 180–2
digital, quick and easy, 182
digital magnification factor, 185
digital single-lens reflex (DSLR) camera, 185–8, 192f, 195
digital single-lens reflex (DSLR) camera, crop, 186, 188
digital single-lens reflex (DSLR) camera, full-frame, 186
dynamic range: RAW vs. JPEG, 180–1
electronic image sensors, 175
exposure tables, 193f
exposure test, 190–1
HD video, 176f
ISO value, 183–5, 191–3, 195
landscape, 178
lens aperture, 183
lens f-numbers, 191–3
lens, single-focal length, 176
lens, superzoom, 187–8, 195
lens, telephoto, 180, 185–7, 190–1, 195
lens, zoom, 176, 178, 195
manual exposure mode, 191
memory cards, 175–6, 180, 191, 196
mirrorless interchangeable-lens cameras (MILC), 185–7, 198 n.3
pinhole crescents, 179
planning, 175, 182, 196
point-and-shoot cameras, 176–7
at sea, 194–5
sensor sensitivity, 183–5
shutter speed, 183, 185, 191–3, 195
simple cameras, 176–7
smart phones, 176f, 177
solar filters, 180, 182, 188–90, 195

spot meter, 190
suggested targets, 178
super telephotos, 185–7
telescopes, 180, 185–7
tripods, 175, 177, 180, 182, 184, 186, 188–9,
    198 n.5
wide-angle photos, 178–9
photons, 72–3
Photoshop, 189*f*, 191*f*, 194*f*
photosphere (of Sun), 72–4, 78, 79, 92, 133,
    142–3
blocked, 101
granulations, 74
magnetic fields, 74–5
temperature, 76
Piedras Negras, 226
Pike, David, 80 n.6
Pike's Peak, 48*f*
pinhole
    camera, 128–9, 146, 153, 179
    projector, 153–4
    projections, 179
Pliny, 29 n.8
Pogson, Normon, 93
Poitevin, Joanne, 164, 170
Poitevin, Patrick, 126, 164, 166, 170
polarizing filters, 157
Polynesia, 46
Pomo Indians, 48
Pontifical Observatory, 93
Porland Community College, 220 n.8
Portugal, 115, 116*f*
Pound, Robert V., 114
Presque Isle, 234, 239
Presque Isle Stream, 239
Princeton University, 102
principal of equivalence, 109–12
prominence(s) (of Sun), 3*f*, 72*f*, 74*f*, 75,
    77–8, 89–92, 97*f*, 116*f*, 129, 133–4,
    140–3
    drawings of, 88*f*
    mapping, 96
    photographing, 93–4, 178, 186
    spectrum, 94–6
    temperature, 75
protons, 71, 79*f*
Prussia, 90

Qattara Depression, 270
Quaglia, Luca, 126, 261
quasars, 120
Queensland Police, 286
Quetzalcóatl, 53

Ra, 47–8
radial density filter, 103*f*
radio telescopes, 120, 122 n.17
radio waves, 120, 122 n.17
Rahu, 46, 49
    Arakho, 46
    Kala Rau, 46
    Lo-Hou, 49–50
Ramsay, William, 99
Rao, Joe, 126
Rayet, Georges, 93
RAW format, 182
    brightness values, 181
    vs. JPEG, 180–1
Rebka, Glen A., Jr., 114
Red River, 228
Red Sea, 60, 93, 272
Red Skelton Museum of American
    Comedy, 231
Remarkables, The, 280
Richardson, Tony, 209*f*
Rio Grande, 226, 255
Robinson, Leif J., 126, 139, 297
Rock and Roll Hall of Fame, 233
Rock of Gibraltar, 268
*Rocky*, 236
Rocky Mountains, 226, 231, 235,
    254–5
Rogers, Michael, 126, 165, 199
*Rolling Stone*, 199
Römer, Ole, 103–5
Ropski, Barbara, 126
Ropski, Gary, 83, 126
Rössing Mine, 282
Roth, Walter, 126, 130
Rotstein, Dimitry, 204–6
Royal Astronomical Society, 86, 118–9
Royal Observatory, 112, 121 n.7
Royal Observatory of Belgium, 166
Royal Society, 118–9
Royal Victoria Eye and Ear Hospital, 167

Russia, 112–3, 121 n.11, 264
  Anadyr, 290
  Crimea, 113, 121 n.7
  mythology, 46
  Siberia, 103*f*, 263, 289–90
  Uelen, 290

Samburu, 152*f*
Sahara Desert, 269–72, 282
Samsung, 188
Santiago de Compostela, 265–6
saros, 26, 29 n.8, 29 n.11, 42 n.12, 60
  cycle, 22, 25, 29 n.10
  duration, 26
  family, 26, 236–7, 277
Saros 120, 277, 294
Saros 126, 277
Saros 133, 277
Saros 136, 273–7
Saros 139, 236–7
Saros 143, 277
Saros 146, 277
Saturn, 49
  Dione, 104*f*
  eclipse, 104*f*
  shadow, 105*f*
  Tethys, 104*f*
Saudi Arabia, 267–70, 272–3
  Jeddah, 272
  Mecca, 272
Science Museum, 98
Schleck, Barbara, 83
Schmidt-Cassegrain telescope, 121 n.7
Schneider, Glenn, 31, 126, 136, 149, 164
Secchi, Angelo, 92–3, 106 n.11
Sencis, 53
shadow bands, 89, 132–5, 142, 144,
    209, 262
  filming, 195
*Shadow Chasers*, 123
Shawnee, 65–6
*shih*, 38
*Shu Ching*, 37
Sicily, 90
Sierra Madre Occidental, 226, 254–5
Sierra Madre Orientale, 226, 255
Sigman, 198 n.3

Simmons, Mike, 125–6, 144, 163
Skelton, Red, 231
Sköll, 46
*Sky & Telescope*, 126, 156, 190
*Sky at Night*, 190
*SkyNews*, 190
Slik, 198 n.5
Smith College, 141
Smithsonian Institution, 57
Smyth, Charles Piazzi, 144
Snake River, 208
Solar and Heliospheric Observatory (SOHO),
    80 n.6, 81 n.8
solar eclipse, 22, 33, 35–7, 39
  frequency of, 20, 22–3
  ratio to lunar eclipse, 18–9
solar eclipse, 2024, 21*f*, 26, 170, 197, 208, 211,
    218, 220 n.8, 222, 247
  eclipse times for, 234*f*, 238*f*
  path of, 21*f*, 224–3, 247–50
  sunniest spots to watch, 259
  weather and, 254–5
solar eclipse, annular, 16, 17*f*, 20, 26, 66 n.4,
    87, 123, 128*f*, 130, 152*f*, 189
  1501, 236
  1504, 274
  1594, 274
  1820, 86
  1836 May 15, 86, 103
  1930 April 28, 87*f*
  2005 October 3, 12, 129*f*, 212
  2010 January 15, 165, 179*f*
  2012 May 20, 169
  2026, 277
  2027, 277
  2028, 277
  2030, 277
  2031, 277, 288–9
  2033, 277
  beginning, 12*f*
  cause of, 11*f*
  ending, 12*f*
  frequency of, 23
  middle, 12*f*
  occurrence, 12–3
  points of contact in, 128*f*
solar eclipse, central, 16, 20

Solar Eclipse Committee, 167
solar eclipse glasses, 151–3, 159*f*, 262
    filters in, 151–2
solar eclipse, hybrid, 17*f*, 275
    1627, 236
    2005, 195
    2026, 277
    2027, 277
    2028, 277
    2030, 277
    2031, 277, 288
    2033, 277
    frequency of, 23
solar eclipse, partial, 17*f*, 18, 26, 123, 127,
        128*f*, 151
    1185 May 1, 127
    1360 June 14, 273
    1378, 274
    1486, 274
    1962, 154, 178
    2001 December 14, 163, 173, 261
    2014 October 23, 51*f*, 73*f*
    2026, 277
    2027, 277
    2028, 277
    2030, 277
    2031, 277
    2033, 277
    2622 July 30, 276
    cause of, 11*f*, 19
    eye damage caused by, 154–5
    frequency of, 23
    photographing, 173, 190–1
    points of contact in, 128*f*
    viewing, 151–60
solar eclipses, total, 14*f*, 16, 17*f*, 20, 22, 26, 38,
        123, 128*f*, 151
    603 BCE May 18, 60
    585 BCE May 28, 59–60, 61*f*, 273
    478 BCE February 17, 62
    430 BCE August 3, 62
    334 July 17, 90
    868 December 22, 89
    1080 June 20, 275
    1183 November 11, 66 n.3
    1239 June 3, 62
    1241 October 6, 62

1406 June 16, 52
1560 August 21, 63
1612 November 22, 274–5
1654 August 12, 63
1706, 63
1703 January 17, 275
1715 May 3, 63, 90, 100*f*
1724 May 22, 69
1803, 89
1806 June 16, 65–6, 100*f*
1834 November 30, 64, 68 n.17
1836 May 15, 100*f*
1842 July 8, 64, 87, 89, 100*f*, 105 n.2,
        127, 140
1843, 236
1851 July 28, 90–1, 100*f*, 144, 243 n.1
1860 July 18, 92–3, 100*f*
1865 April 25, 275
1868 August 18, 93, 95–7, 99, 100*f*
1869 August 7, 51–2, 99, 100*f*, 101
1870 December 22, 95, 100*f*, 101
1871, 100*f*, 102, 163
1878, 100*f*, 102, 211
1896 August 9, 251*f*
1901 May 18, 275
1905, 141, 265
1911, 64
1912 October 10, 121 n.5, 165*f*
1914 August 21, 112, 121 n.11
1918 June 8, 115, 121 n.11, 161
1919 May 29, 100*f*, 109, 116*f*, 118*f*, 120,
        210, 275
1929, 121 n.7
1937, 275
1952, 236*f*
1955 June 20, 276
1963, 144–5, 203
1966 November 12, 103*f*, 137
1970 March 7, 103*f*, 154, 169–70, 219, 236
1972, 164
1973 June 30, 53, 130, 199, 276
1976 October 23, 130, 164, 166–8
1979, 127, 136, 169
1980 February 16, 103, 145*f*, 146
1981 July 13, 103*f*
1983 June 11, 103*f*, 147
1988 March 18, 103*f*, 236

solar eclipses, total (*Continued*)
  1991 July 11, 133, 163–4, 166, 221, 223,
     276, 298*f*
  1992, 146, 149
  1994 November 3, 57–8, 203
  1995, 163
  1998, 107, 163–4, 166, 170
  1999 August 11, 163, 168*f*, 261, 265
  2001, 163
  2002, 169
  2005 April 9, 252*f*
  2006 March 29, 3*f*, 10*f*, 144, 183*f*, 192*f*,
     194*f*, 236
  2008 August 1, 137, 163, 170, 187*f*, 299*f*
  2009 July 22, 276
  2010 July 11, 136*f*, 247
  2012 November 12, 49*f*, 169
  2013, 170
  2015 March 20, 176, 247
  2017 August 21, 2*f*–4*f*, 21*f*, 141*f*, 177*f*, 184*f*,
     189*f*, 191*f*, 201–20, 222, 229, 231,
     249, 292
  2019 July 3, 139
  2020, 218
  2021, 218–9, 247
  2023, 218
  2024 April 8, 21*f*, 26, 170, 197, 208, 211,
     218, 220 n.8, 222–43, 292
  2026 August 12, 263–67, 277
  2027 August 2, 21*f*, 26, 215, 263, 267–73,
     276–77
  2028 July 22, 263, 277–82
  2029, 281
  2030 November 25, 263, 277, 281–89
  2031 November 14, 263, 277, 288
  2033 March 30, 263, 277, 289–94
  2037, 245, 287
  2038, 245, 287
  2042, 236*f*, 237
  2044 August 23, 241, 292
  2045 August 12, 241, 276, 292
  2052 March 30, 241, 292
  2060, 237
  2078 May 11, 237, 241, 292
  2079 May 1, 241, 292
  2099 September 14, 241, 276, 292
  2150 June 25, 275

  2186 July 16, 237
  2496 May 13, 276
  2514 May 25, 276
  2601, 237
  causes of, 9–13, 16, 22–3, 39
  first photograph of, 90*f*
  partial phases of, 3*f*
  path of 2017, 21*f*, 202*f*, 229
  points of contact in, 128*f*
  stages of, 142–3
  viewing, 151–60
solar eyepieces, 156
solar filters, 107, 129, 145–6, 159, 178, 196,
    198 n.6
  aluminized polyester (Mylar), 152, 190,
    198 n.6
  black polymer, 151–2, 154, 190
  for cameras, binoculars, and telescopes, 154,
    180, 182, 185, 189–90
  companies that carry, 152, 156–7
  metal-coated glass, 189–90
solar flares, 73*f*, 75–7
solar tornados, 75, 80 n.6
solar wind, 77–8
Somalia, 267–9, 273
South Africa, 282*f*, 283
  Big Hole, The, 284
  Cape Town, 43, 281
  Durban, 284
  Johannesburg, 283
  Kimberly, 284
  Welkom, 283
South Carolina, 219
South Dakota, 213, 215
space blankets, 198 n.6
space weather, 78
Spain, 12*f*, 141, 263, 265–70
  Barcelona, 266
  Burgos, 265
  Cádiz, 269
  conquistadors, 39
  Gijon, 266
  Jerez, 101
  León, 266
  Madrid, 212, 266
  Málaga, 269
  Oviedo, 26

Pamplona, 266
Valencia, 267
spectral analysis, 95
spectral lines, 94–5
spectroscopes/spectroscopy, 90, 93–7, 101
Spielberg, Steven, 125
*Spring and Autumn Annals* (Ch'un-ch'iu), 38
Sramek, Richard A., 120
Station Nord, 265
Stonehenge, 33–7, 213
    aerial view of, 34*f*
    center, 33–5
    construction of, 36*f*
    Heel Stones, 33–5, 41 n.6
    positions of Moon in, 34–6
    positions of Sun in, 34–5
    Sarsen Circle, 35, 41 n.6
    Station Stones, 34–5
    Trilithons, 35
Strait of Gibraltar, 269
Sudan, 294 n.4
Sumatra, 121 n.7
Sun
    angular diameter, 13, 138
    angular distance, 16
    angular size, 9–13, 27, 139
    aphelion, 27
    apparent size, 12
    atmosphere, 72*f*, 78, 87–88, 96, 98
    center, 72*f*
    convection zone, 72*f*
    core, 71–73, 76
    cross-section, 72*f*
    diameter, 28
    distance from Earth, 11–12, 15, 24, 28
    gravitational field, 77, 111*f*, 118*f*, 120
    layers, 73–5
    lifespan, 80 n.3
    magnetic fields, 73–5, 77, 80 n.7
    magnetic regions, 76
    mass, 71, 77
    plasma, 77
    size, 11–12
Sun Dance ceremony, 64, 67–8 n.17
sunglasses, standard, 157
Sun-got-bit-by-bear, 48
sunniness index, 257, 259

sunspot(s), 51*f*, 73*f*, 74, 79, 92,
        106 n.10, 129
    cycle, 134
    maximum, 81 n.9, 103*f*, 134
    minimum, 81 n.9, 103*f*, 134
Superman Square, 229
Suriname, 53
Svalbard, 176*f*
Swan, William, 90–1
Sweden, 101
Sydney Harbor Bridge, 245, 279
Sydney Opera House, 279

Tahiti, 50
Taira, 66 n.3
Taj Mahal, 163
Takahashi, 198 n.4
Tamron, 198 n.3
Tau Herculid meteors, 218
Tele-Vue, 198 n.4
telescopes, 159
    apochromatic lenses, 186–7
    extra-low-dispersion lenses, 186
    finder scopes, 158, 190, 196
    modern refracting, 186, 198 n.4, 210, 211*f*
    photographing with, 180, 185–7
    solar filters for, 159, 190, 196
Temple of Amun, 270
Temple of Hathor, 272
Temple of Horus, 272
Temple of Kom Ombo, 272
Temple of Seti I, 272
Tenant, James Francis, 93
Tennessee, 228
    Madison, 218
    Nashville, 249
Tenskwatawa, 65–6
Texas, 255, 269, 272, 290
    Austin, 223, 228, 234, 249
    Brady, 259
    Dallas, 223, 228, 234, 249, 257
    Del Rio, 259
    Eagle Pass, 226, 247, 259
    Fort Worth, 223, 228, 234, 249
    Houston, 234
    Kerrville, 226
    Killeen, 28

Texas (*Continued*)
  San Antonio, 223, 226, 234, 249
  Texarkana, 234
  Waco, 228, 234
Texas State Capitol Building, 228
Thales, 60, 86
theory of relativity, 71, 109–22
  general theory of relativity, 113–22,
    210, 212
  proving, 115–20, 121 n.7
  special theory of relativity, 109–13
theory of universal gravitation, 113–5,
    120 n.3, 295
Thomas (Archdeacon), 62
Thomson, Joseph John, 118–9
Thousand Oaks Optical, 152, 156–7
Thucydides, 62
tides, 20, 295, 299
*Times of India*, 165
Tlingit Indians, 50
Todd, Mabel Loomis, 131, 140–1
Tokina, 198 n.3
Tonga, 64
Top of the World Hardware Store, 294
Top of the World Hotel, 294
totality, 49, 65*f*, 78–9, 189*f*
  beginning, 135–7, 176, 182, 184*f*,
    218, 222
  central line of, 132
  changing environment during, 130
  clouds during, 131, 253–4
  color change during, 131, 142–3
  crescent sun, 127–30, 134*f*, 142, 144, 153,
    185, 262
  during, 138–40
  end of, 7, 140–41, 143–4, 158, 176, 178,
    187*f*, 204–5, 262
  experience of, 1–5
  first contact, 128*f*, 127, 142
  fourth contact, 128*f*, 143
  light level, 138, 142–3, 145*f*
  midpoint, 4
  photographing, 191–4
  reactions to, 7, 31, 83, 127
  second contact, 128*f*, 142
  shadows during, 130, 136
  sounds of, 199
  stages of, 142–3
  surprise of, 87–88
  temperature during, 131, 142–3
  third contact, 128*f*, 141*f*, 143–4, 182
  unexpected, 161
  value of, 43
  viewing, 151–60
transition region (of the Sun),
    temperature, 75–6
Transylvania, 50–1
Tropic of Cancer, 134
Tropic of Capricorn, 134
Tunisia, 267*f*
  Sfax, 269
  Tunis, 269
Turkey, 144
Tutankhamun, 271
Tuthill, Roger, 125–6, 146–7

Uganda, 170
United Airlines flight #632, 218
United Arab Emirates, 272
University of Arizona, 31, 126
University of Oregon, 206
University of Toronto, 203
University of Waterloo, 203
US Coast Survey, 51–2
US Department of State Travel
    Advisory, 269
US Geological Survey, 228
US Postal Service, 197
US Route 1, 239
Utah, 234

Valley of the Kings, 271
Valley of the Queens, 271
Vatican Observatory, 93
Venezuela, 237
Venus, 3, 49, 143
Vermont, 235, 240*f*, 284
  Burlington, 234, 237
  East Richford, 237–8
  Montpelier, 234, 237
video, shooting, 195
Virginia, 169
Vishnu, 46
Vivitar, 198 n.3
Vixen, 198 n.4
von Littrow, Karl Ludwig, 91

von Oppolzer, Theodor, 18
Vulcan, 114, 211–2

Wabash River, 231
Waldron, C., 167
Washington
    Goldendale, 115
    Seattle, 234
Washington, D.C., 234
Waters, Terry, 209f
weather, 247–60
    2024 eclipse and, 254–6
    clouds, 251–8
    forecasts, 249–51, 256f, 258
    lake effect cloud clearing, 257–8
    mountains and, 254, 260 n.6
    "rain shadow effect," 254
    thunderstorms, 252–3
Weather Channel, 249
Weather Spark, 272
Weather Underground, 249
welder's goggles, 153
West Virginia, 122 n.17
Western Reserve College, 101
wet-plate (collodion) photographic
        process, 92–3
white dwarf stars, 121 n.10
Wichita Mountains, 64
Willcox, Ken, 57–8, 126
William Optics, 198 n.4
Williams, Sheridan, 7, 107, 126, 245

Wilson, Alexander, 106 n.10
World War I, 113–5, 121 n.11
World War II, 233, 243
Wright, Orville, 233
Wright, Wilbur, 233
Wu Ching (Five Classics), 41 n.8
Wyoming, 213–5
    Casper, 2f, 209–12
    Chugwater, 215
    Fort Laramie, 215
    Glendo, 3f
    Jackson Hole, 208–9
    Riverton, 184f

Xerxes, 60–2
Xolotl, 54

Yahoo, 190
Yao (Emperor), 37, 38f
Yeager, Chuck, 243
Yemen, 267–9, 272–3
    Sana'a, 273
Young, Charles Augustus,
        101–2

Zambia, 163, 262
    Lusaka, 261
Zeiler, Michael, 126, 136–7
Zimbleman, Jim, 57
Zirker, Jack B., 33, 126, 137
zodiac, 22, 28